Laws of Nature

Philosophie und Wissenschaft
Transdisziplinäre Studien

Herausgegeben von
Carl Friedrich Gethmann
Jürgen Mittelstraß

in Verbindung mit

Dietrich Dörner, Wolfgang Frühwald, Hermann Haken,
Jürgen Kocka, Wolf Lepenies, Hubert Markl, Dieter Simon

Band 8

Walter de Gruyter · Berlin · New York
1995

Laws of Nature
Essays on the Philosophical, Scientific and Historical Dimensions

Edited by
Friedel Weinert

Walter de Gruyter · Berlin · New York
1995

⊗ Printed on acid-free paper which falls within the giudelines
of the ANSI to ensure permanence and durability

Library of Congress Cataloging-in-Puplication Data

Laws of nature : Essays on the philosophical, scientific and historical dimensions /
edited by Friedel Weinert.
(Philosophie und Wissenschaft, Transdisziplinäre Studien ; Bd. 8)
Includes bibliographical references and index.
ISBN 3-11-013918-9
1. Science–Philosophy. 2. Philosophy and science. 3. Nature.
I. Weinert, Friedel 1950– . II. Series.
Q175.L328 1995
501–dc20 95-13649
CIP

Die Deutsche Bibliothek – Cataloging-in-Puplication Data

Laws of nature : essays on the philosophical, scientific and historical dimensions /
ed. by Friedel Weinert. – Berlin ; New York : de Gruyter, 1995
(Philosophie und Wissenschaft ; Bd. 8)
ISBN 3-11-013918-9
NE: Weinert, Friedel [Hrsg.], GT

Printed in Germany
Typesetting: Converted by Knipp Satz und Bild digital, Dortmund
Printing: Ratzlow Druck, Berlin
Binding: Lüderitz & Bauer, Berlin
Cover design: Rudolf Hübler, Berlin

Preface

A number of recent publications both from philosophers of science and working scientists manifests a rising interest in the nature of laws. After the numerous investigations into the nature of scientific theories and the growth of scientific knowledge, it is not surprising that more and more writers should turn towards an exploration of laws of nature.

In fact, with the continuing debate between empiricist and realist philosophies of science and the very important rise of studies into the experimental nature of science, it is to be expected that both scientists and philosophers will pay increasing attention to such questions as the ones posed in this volume: "What is a law of nature?" and "What is the relationship between laws of nature and laws of science?"

From the viewpoint of working scientists laws of nature – or rather the known statements of these laws in the form of laws of science – obviously play a crucial role both in their research and their teaching, since the function of scientific laws can be seen as a hinge between the experimental basis of science and the theoretical understanding of the experimental findings.

From the point of view of philosophers of science, a proper appreciation of laws of nature and laws of science, will, given their growing awareness of the importance of laws in the working of science, constitute a significant step towards a more comprehensive understanding of the essence of the scientific enterprise.

Given the nature of these questions, it seems imperative that any such undertaking should opt for a transdisciplinary approach, in which both philosophers, scientists and historians present their views from their respective angles. This is the aim of this volume. The French physicist Bernard d'Espagnat expressed this spirit in a succinct formula when he said that science was not only there to be learned, but also to be contemplated.

It is hoped that this collection of essays will provide a good survey of the present level of discussions on the nature of laws. Hopefully, it will also act as a valuable teaching aid in philosophy of science courses and stimulate further reflections on this fascinating topic.

It has taken some time for this volume to emerge in its present form. The editor would like to thank all the contributors for their interest and support. A special thank you is also due to Professor Jürgen Mittelstraß for his encouragement. The editor has also received support from a number of colleagues over the years – Robert Nola, Chris Parkin and Kim Sterelny – whom he wishes to thank.

Friedel Weinert

Acknowledgements

The editor would like to thank: The Johns Hopkins University Press for its permission to reprint Jane Ruby's article 'The Origins of Scientific "Law"' which was first published in the *Journal of the History of Ideas,* Vol. XLVII (1986), pp. 341-59; Brockman, Inc. for permission to reprint excerpts from Paul Davies, *The Mind of God,* first published by Simon & Schuster (1992) and "What are the Laws of Nature" first published in *The Reality Club No. 2,* ed. J. Brockman, Lynx Publications (1989); the *New Scientist* for its permission to reprint excerpts from P. Davies, 'Law and Order in the Universe' first published in the October issue 1988; and Addison-Wesley Publishing Company for permission to reprint excerpts from Wojciech H. Zurek, *Complexity, Entropy and the Physics of Information,* 1990.

Table of Contents

I. Introduction

Friedel Weinert

Laws of Nature – Laws of Science

This volume brings together a number of essays dealing with various aspects of natural laws. In particular, the essays deal with problems of both an ontological, epistemological and historical nature. That is, they address the questions: 'What are laws of nature?', 'What distinguishes laws of science?' and 'When did the laws of science emerge?'

The first part of this volume presents various philosophical views on the question. The second part looks at laws from the point of view of the scientist. And the third part is devoted to the development of the concept of laws of science from a historical perspective. (See also Eastwood [1967]; Ellis [1965]; Giere in this volume [Ch. II. pp. 120ff.: §2]; Home [1968]; Hund [1978]; Krajewski [1974]; Milton [1981]; Russell [1964]; Wilson [1969]; Zilsel [1942]).

Although discussions concerning laws of nature have never occupied centre-stage in the philosophy of science – in contrast to, for instance, the discovery of scientific revolutions or more recently the activities of experimental science – a considerable amount of literature has amassed over the years (see the Bibliography) and some of the central insights have been collected in this volume.

Law statements are of considerable importance in the sciences, since the discovery of regularities and their mathematical formulation present one of the pillars of scientific thinking. (In addition, science is based on theory and model construction as well as experimental and observational testing .) Their conceptual importance is reflected in the number of positions which have developed in response to the question, 'What is a law of nature?' Before a summary of these philosophical positions is given, some conceptual tools must be introduced without which this question cannot be properly discussed.

Firstly, then, various types of laws will be distinguished. The notions of laws and theories should be clearly kept apart. Although a separate chapter is devoted to this question (Scheibe [Ch. III. pp. 208-226.]), a preliminary distinction is to say that laws digitalise or conceptualise phenomena – they provide algorithmic compressibilty (Davies [Ch.III. pp. 248-267.] – , while theories conceptualise and explain laws – they provide coherence (Scheibe).

Secondly, various criteria are needed to distinguish genuine laws from mere accidental generalisations, since not every regularity is a law.

And finally the main features of the most important philosophical views on laws of nature will be sketched.

I. Classification of laws

Lit.: Bunge [1961]; Cartwright [1979]; [1983]; [1989]; Eddington [1958: 68-80]; Friedman [1992b]; Hull [1974: 70-100]; Nagel [1961: 73-75]; Sober [1984: §1.4], [1993: §1.5]; Stegmüller [1966: 650-1]).

1. *Types of Laws*

It is important to distinguish between *laws of science* and *laws of nature. Laws of nature* are those empirical regularities which govern the natural world around us, irrespective of whether or not intelligent beings possess knowledge of these regularities or whether or not an appropriate symbolic representation for at least some of these regularities has been developed. Thus, when Kepler formulated his three laws, he discovered regularities which had governed planetary motions long before his symbolic representations made them accessible to us.

Laws of science, by contrast, are those regularities of the natural world which are known to us and which have been cast in appropriate symbolic forms. Laws of science may be subject to various kinds of modifications. Refinements to their original formulations may be introduced: Maxwell's addition of displacement current to Ampère's law may serve as an example. Constants may be added to an equation: for instance the addition of the gravitational constant, G, to Newton's inverse-square law of gravitation. Or again limiting conditions may have to be formulated to save the validity of laws: the need to limit Newton's laws to macro-objects, such as lead balls and planets, and to velocities below the critical value of 30km/s, i.e. the average velocity of the earth in its orbit around the sun. Often these limitations become only known because when it is realized that limited laws are derivable from more general laws. Because science often deals with approximations, which it expresses in models of real physical systems, the laws of science are typically expressed in terms of these models.

Some philosophers have held that a philosophical account of laws must be given of laws of nature only, not of laws of science. The philosopher's task, on this view, is to deliver an ontological account of laws of nature, while it is left to the scientist to fill in the precise details. Differences in ontological commitments, as we shall see below, lie at the root of recent debates about the nature of laws. For present purposes, it is more appropriate to refer to laws of science rather than just laws of nature, because the laws of science provide us with important clues as to what a law of nature is. The idea is, roughly speaking, that the symbolic formulations of laws of science reveal important features of the structure of physical systems, which are governed by laws of nature. This can be maintained because the models of science display an appropriate 'fit' to the systems they model.

Both Coulomb's law and Newton's gravitational law, for instance, state that the force between two stationary charges or two material bodies is an inverse-square relationship. The symbolic formulation – $1/r^2$ – then reveals structural information about the force between two such bodies. Of course, this symbolization is conventional. But whatever symbols are chosen, they must refer to what Western culture happens to describe by an $1/r^2$ – relationship, because this is the constraint which the robustness of the physical world imposes on us. Other symbols of science – $\int$, ∂, $\bullet$, h, $\times$ – equally convey structural information about physical systems. (See Haken, Ch. III)

This decision is of immediate utility for the next distinction, namely that between *quantitative* and *qualitative* laws. Quantitative laws are the standard type of statement encountered in scientific textbooks, expressed in mathematical notation; these will be differentiated further in a moment. Qualitative laws are non-mathematical statements, for instance that all metals are good conductors. But such law statements are also used in physics, where their formulations make no reference to numbers or special symbols, but nevertheless use technical terms and often have mathematical implications. Thus, Newton's Third Law of Motion states:

If one object exerts a force *F* on a second, then the second object exerts an equal but opposite force $\neg F$ on the first.

(*F* is a vector.) Because of the validity of this law it is unconvincing that film heroes are seen to beat up whole armies without suffering the slightest adverse effects. And that is why Superman could not physically stop a speeding truck at arm's length without yielding a millimetre himself. For the force Superman exerts on the truck is, by this law, exerted on his body too. If frictional forces are taken into account, Newton's Third law can be seen to have mathematical consequences. The equation states that the kinetic friction force, f_k, is equal to the coefficient of kinetic friction, μ_k, times the normal force N:

$$f_k = \mu_k N.$$

According to this equation, it would take Superman, standing straight in front of the truck with his arm stretched out, a large distance of several kilometres to bring it to a halt! (Kane/Sternheim [1984: 55-6]). Other examples of qualitative laws in physics abound: Kepler's First Law, Newton's First

Law, the Zeroth Law of Thermodynamics, the Kelvin-Planck statement of the Second Law of Thermodynamics, Kirchhoff's rules, Lenz's Law, Huygens' s Principle, Fermat's Principle and Hund's rules in quantum mechanics. They all are stated in technical, though not directly mathematical language, and all have mathematical implications as just illustrated with respect to Newton's Third law. Biology, too, displays a number of qualitative laws: Mendel's laws of segregation and independent assortment, Dollo's law about the irreversibility of evolution and Mayr's founder principle may serve as examples.

The above equation is an example of a *functional law* or a law of association, in which the values of several parameters covary with each other. *Causal laws*, by contrast, are deterministic, asymmetric succession laws, in which one parameter functions as the cause of another. This, at least, is one understanding of causal laws in terms of universal correlations. Probabilities can be employed to measure causes, because, quite generally, a cause can be expected to increase the frequency of its effect. Causal claims can be further subdivided into singular causal claims and generic causal claims, where the latter are to be understood as ascriptions of capacities. Causal statements may also be employed by the physicist, for instance when he states that an inhomogeneous magnet (a Stern-Gerlach apparatus) causes a beam of silver atoms to spit into 2 components. But invariably physics will strive to turn such causal statements into quantitative laws; for example that the atom beam is split into 2j+1 components (where j is the quantum number characterizing total angular momentum).

Even if cause is not defined in terms of necessary and sufficient conditions, and hence the correlation between cause and effect is not universal, causal relationships have their place in certain areas of science. For instance, Hull suggests [1974: 74] that

> the connection between a particular gene and a particular gross phenotypic trait is on a par with that between smoking and lung cancer. Seldom is a particular gene necessary for a particular phenotypic trait and never is it sufficient.

By far the most common law statements in science are *quantitative laws* and quantification is indeed one of the hallmarks of modern science. These can be divided into *empirical* or *phenomenological* and *fundamental* laws, although it is probably best to think of this distinction as a continuum rather than a sharp division of types. Empirical laws are statements based on observed regularities which can often be derived from more fundamental statements (Presley [1960: 208]; Davies [Ch. III. pp. 248-267.]; Haken [Ch. III. pp. 227-247.]). Thus, Snell's law:

$$n_1 \sin\theta_1 = n_2 \sin\theta_2$$

relates the angle of incidence, θ_1, of a light beam to the angle of refraction, θ_2, via the refractive indices of the media (for instance, n_1 for air, n_2 for water) in which the light beam travels. Willebrord Snell (1591-1626) discovered this relationship experimentally but it is possible to derive it theoretically from Fermat's Principle, which states that a light ray travels the path of least time from A to B. It should be observed that such derivations can be complex so that a statement which is successfully derived from another statement may later be found to be derivable itself from an even more general statement. Thus the Dulong-Petit law, an empirical statement based on observed regularities, according to which the molar heat capacities of solids reach a maximum value ($C_v = 3R \approx 25 J/m.K$) at high temperatures can be derived – on a simple model of solids, involving degrees of freedom – from the equipartition theorem. But the equipartition theorem fails to explain why heat capacities vary with temperature and why rotational energy can be neglected in the internal energy balance at ordinary temperatures. A satisfactory explanation of the behaviour of solids is achieved by the Einstein model – or even better the Debye model – of solids, which involve the introduction of discrete energy levels required by quantum theory. Both the Dulong-Petit law and the equipartition theorem follow from these models for the limiting case of high temperatures. The derivation from these models is preferable, because an observed regularity is related to the internal structure of solids. Such multiple derivations are an important feature of science, involving both theoretical insights, models and the laws they comprise.

Not only are empirical laws based on observed regularities and thus derivable from higher-order laws (under the assumption of the validity of more complicated models), they also apply in most cases only to specific physical systems. There are usually a number of initial or boundary conditions which must hold before the empirical law applies to a system. Thus, the Dulong Petit law breaks down at low temperatures, while Snell's law only applies to isotropic media. In a certain sense, then, empirical laws are not truly universal – they only apply to physical systems which satisfy a number of boundary conditions.

Before the nature of truly fundamental laws is discussed, it is convenient to introduce the terms *physical system* more systematically. A physical system is an interrelated network of forces, fields and regularities into which individual particles may enter. Rather than saying that the world consists of individuals and relations between them, we shall say that the world consists of physical, biological, chemical and other systems. A system consists of a (cluster of) individuals and their interrelations. Physical systems are manifold: the earth and the moon form a physical system and so does the sun and its planet. A moving car constitutes a physical system and so does an elevator or a ball rolling down an inclined slope for the sake of an experiment; the hydrogen atom forms a physical system and of course all the more complicated atoms. All physical systems are subject to laws: Laws, it may be said, describe the structure or part of the structure of physical systems. Thus, two colliding cars form a physical system which can be completely described by Newton's laws. Newton's laws determine the structure of the event. Invariably, physical and in particular biological systems are so complex that models must be introduced to capture basic structures, then the laws are expressed by reference to these idealized systems. (Cf. d' Espagnat [1985: 104]; Scheibe [Ch. III. pp. 208-26.]; Rosenberg [1985: 214-6])

Although the present discussion concentrates on physical systems, much of what will be said about laws in such systems may be applicable, after suitable modifications, to chemical, biological and other systems. In particular, a position needs to be adopted regarding the much-debated question of whether there are true laws in biology, rather than just accidental generalizations. The strategy adopted is to treat laws in biology as limiting cases of laws in physics. That is to say that while it is possible to formulate lawful regularities in biology, the problem of isolating boundary conditions for biological systems is much more severe for the biologists than for the physicist. In a sense, the biologist will encounter many more reminders of the interrelatedness of nature than the physicist. Many of our results about laws in physical systems should also apply to laws in biological systems, but most likely in conjunction with a much greater idealization of and abstraction from boundary conditions. Or in Gould's convenient catchphrase: 'Laws in the background, contingency in the details.' (Gould [1989: 290]) Thus, against the ideas advanced by J. Smart and others (Smart [1963: Ch. 3]; Beatty [1981]; Helmer/Rescher [1959]; Mayr [1982], [1987a,b]; Rensch [1960]; Scriven [1959]; Simpson [1963]), we assume the existence of biological laws. (See Ghiselin [1989]; Hull [1974]; Rosenberg [1985]; Ruse [1970], [1973]; Sober [1984] [1993]; Steen/Kamminga [1991]). This does not mean, however, that biology is reducible to physics; as Ayala argues in this volume [Ch. III. pp. 268-86.], biology is distinguished from physics by employing teleological patterns of explanation.

While empirical or phenomenological laws are not valid for all physical systems, due to changing boundary conditions, truly *fundamental* laws can be expected to be part of the structure of any physical system. Thus, Newton's inverse-square law of gravitation will give us the strength of interaction in many different physical systems, with a minimum of boundary conditions. It basically only requires that the systems involved have material substance. In Einstein's theory of gravitation, even this requirement is abandoned since photons can be deflected by gravitational fields. A truly fundamental law, by this suggestion, will therefore be present in (almost) any physical system. A truly empirical law, at the other end of the scale, will only be found in a few physical systems, because a number of special boundary conditions need to be satisfied, before the law applies. Many laws will govern over a domain lying between these two extremes. It follows from this characterization that the distinction between empirical and fundamental laws should be regarded as a continuum, rather than a strict separation.

Staying with quantitative laws, a further differentiation suggests itself: that between *functional laws, deterministic* and *indeterministic* laws, *structural* laws and laws relating *fundamental constants* (Kneale [1949: §16]; Bunge [1961]; Nagel [1961: 77-90]; Stegmüller [1966]. A prominent example of the latter type is the so-called Rydberg constant which is important in the calculation of spectral lines. The Rydberg constant combines some of the most fundamental physical constants (the electron charge, e, the electron mass, m_e, the Planck constant, h, and the velocity of light, c) into one equation and allows the calculation of the values of some of these fundamental constants.(Weinert [forthcoming])

Functional laws, as was observed before, are very common in the sciences, since they state a functional relationship between various parameters (measurable quantities), which either may relate the rate of change of a parameter, y, with respect to time, t – for instance, in Galileo's law for free-falling bodies: $y = v_0t + 1/2at^2$, the distance, y, which the object falls, is related to its original velocity, v_0, as well as the time of fall and its acceleration a – or may be of a purely numerical kind, functionally relating various parameters: for instance, Boyle's law for ideal gases, PV=nRT, states a functional relationship between the product of pressure, P, and volume, V, on the one hand, and the temperature, T, the universal gas constant, R, and the mole number n, on the other. To state another example, the potential energy of an object thrown up into the air is functionally dependent on its mass, the height it reaches, h, and the gravitational acceleration near the surface of the earth, g:

$$PE = mgh$$

Thus any changes in the parameters 'm', 'h' changes the value of PE.

Statistical laws, unlike deterministic laws, only express the probability of an event. Typical examples are nuclear decay laws, like the exponential decay formula:

$$N/N_o = e^{-\lambda t}$$

which states what percentage of elements, N, of an original number of elements, N_o, of radio-active material remains after a certain time t (where λ is a decay constant). Quantum mechanics essentially contains statistical laws, corresponding to a large number of measurements on microphysical systems. Such laws may state the probability that an electron is to be found at a particular location, mathematically expressed by the square of the absolute value of the wavefunction $|\Psi|^2$, or the probability that a certain transition from, say, a higher-level energy state to a lower-level energy state (or vice versa) is likely to occur in an atom.

The employment of statistical laws in biology is well reflected in the Hardy-Weinberg law, which, as expected, works on a number of simplifying assumptions (Ruse [1970: §3]; Sober [1984: 32-8], [1993: 71-2]): we need a large random-mating population, consider only two different types of genes (A_1 and A_2) for one locus, assume that no external forces are acting on the population and 'that in the population at some particular time the proportion of A_1 genes to the total number at the locus is p and the proportion of A_2 is q, so that the ratio of A_1 genes to A_2 is p:q' (Ruse [1970: 240]). Because only two different types of genes are under consideration,

$$p + q = 1.$$

Then, the Hardy-Weinberg law states that in all successive generations the frequencies of the three genotypes A_1, A_1A_2, A_2 will be p^2, 2pq, and q^2:

$$p^2 A_1A_1 : 2pqA_1A_2 : q^2A_2A_2.$$

This is a universal statement about populations with certain characteristics such as sexuality which is not restricted to any place or time. It does not make reference to any particular individuals or populations. It can be derived from Mendelian genetics (Ruse [1970: 241-2], [1973: 32-7]). It allows reliable predictions and can form part of explanatory theories. It has all the hallmarks of a law, but it is based on a number of simplifying assumptions which may make approximations to real biological systems difficult.

A type of law whose significance has not been sufficiently stressed so far are so-called *structure laws*, introduced by Einstein and Infeld [1971: 143, 236ff] to emphasize the contrast between classical physics and the new field physics. (Cf. Harré/Madden [1975: Ch. 9]) With this term, Einstein and Infeld refer in particular to Maxwell's equations and the new gravitational laws developed in General Relativity Theory. They present, according to these authors, 'a new type of law' because they describe the structure of fields. There are two essential differences between, say, Newton's laws and Maxwell's laws:

> Maxwell's equations describe the structure of the electromagnetic field. All space is the scene of these laws and not, as for mechanical laws, only points in which matter or charges are present.

But there is a second difference which a comparison between Newton's gravitational law and Maxwell's laws will highlight:

> With the help of Newton's laws we can deduce the motion of the earth from the force acting between the sun and the earth. The laws connect the motion of the earth with the action of the far-off sun. The earth and the sun, though so far apart, are both actors in the play of forces.
>
> In Maxwell's theory there are no material actors. The mathematical equations of this theory express the laws governing the electromagnetic field. They do not, as in Newton's laws, connect two widely separated events; they do not connect the happening *here* with the conditions *there*. The field here and now depends on the field in the *immediate neighbourhood* at a time *just past*. The equations allow us to predict what will happen a little further in space and a little later in time, if we know what happens here and now. They allow us to increase our knowledge of the field by small steps. We can deduce what happens here from that which happened far away by the summation of these very small steps. In

Newton's theory, on the contrary, only big steps connecting distant events are permissible. (Einstein/Infeld [1971: 146-7]; italics in original).

It is clear that Einstein/Infeld envisage a new ontology in which the field concept replaces the matter concept as the primary vehicle of investigation. The purpose of shifting the ontological commitment 'would be the explanation of all events in nature by structure laws' (Einstein/Infeld [1971: 243]). The authors have in mind the trend of the physical sciences to go towards unification, and hence to arrive at fundamental relationships of a great level of abstraction, from which all less fundamental relationships can be derived.

The concept of structure laws may be of value to the philosophical analysis of laws, although the philosophical interest lies in the nature of laws in general, not just in those laws at which a unified physics will eventually arrive (which, according to von Weizsäcker [1971] will have a transcendental character). For the purposes of a philosophical interpretation of laws, it may be said that all genuine laws express structural features of physical systems. This interpretation suggests itself from the decision to bestow epistemological priority on physical systems, rather than individuals. (See Haken, Ch. III) Physical systems possess both core aspects and peripheral aspects encapsulated in various types of laws. Fundamental laws express the deepest level of structure in physical systems, while empirical laws combine with boundary conditions of various kinds to express structural aspects of specific physical systems.

2. *Epistemological Features of Laws of Science*

One further aspect, which can be derived from considerations of laws of science, may be termed *interrelatedness*. It is indispensable in any discussion of laws (as is, in fact, testified by several contributions to this volume). Interrelatedness signifies relations of mutual dependence of parameters which form part of a physical system. Kant called it 'the Principle of Continuity', according to which

All substances, as far as they coexist, stand in thoroughgoing community, that is, in mutual interaction. (Kant [1787: A221]).

Leading 19th century scientists, like von Helmholtz, Faraday and Darwin, emphasized the interrelatedness of physical systems, as they discovered the interactions of various physical forces or the interdependence between organisms and their environment. (Weinert [1993]; Weinert [1994]). If physical systems are interrelated, so are the laws, since laws encapsulate structural features of physical systems. Inferential relations are of great important in scientific theories. For instance, we can make the transition from the acceleration of a particle to its momentum in a small number of steps:

$$F = ma = mdv/dt = dmv/dt = dp/dt.$$

Hence, we go from the statement that the force a particle exerts is the product of its mass and acceleration to the statement that its force is also related to the rate of change of its momentum with respect to time. By a similar kind of argument, the slowing down of the earth's spin angular momentum, w, which is due to frictional forces on the earth's surface, can be related to the growing distance, r, between the earth and the moon. It is known that the moon recedes at a rate of 3cm per year from the earth – a fact which is due to an instance of one of the most fundamental laws of nature, the conservation of energy.

As already mentioned, fundamental physical constants figure prominently in laws of science and play a role in the fundamental interrelatedness of physical systems. For instance, the experimental confirmation of the velocity of light, c, as a constant in vacuum has had important implications for fundamental assumptions of classical physics: as it turns out, the acceptance of c as a constant of nature is incompatible with the ether concept (Weinert [forthcoming]).

In sum, laws can be understood as expressing structural aspects of interrelated physical systems. Hence, laws should not be seen as isolated regularities. (Skyrms/Lambert [Ch. II. pp. 139-156.]; Scheibe [Ch. III. pp. 208-226.]) The greatest scientists of the 19th century – Darwin, Faraday and von Helmholtz, discovering the interrelatedness of forces like electricity and magnetism and the interdependence of species and the environment – taught this lesson: the universe is a vast interactive system with a multitude of subsystems, which science, by a complicated process of idealization, abstraction and factualization, identifies, describes and explains. Michael Scriven expresses these features of laws in a succinct formula:

typical physical laws express a relationship between quantities or a property of systems which is the *simplest useful approximation* to the true physical behaviour and which appears to be *theoretically tractable* (Scriven [1961: 100]; italics in original; cf. Bunge [1961: 280])

The approximate and idealized character of laws of science is yet another feature discussed in various contributions to this volume (Davies, Giere, Scheibe). It is important enough to merit an illustration. The so-called simple pendulum is an idealized physical system which consists of a point mass, 'm', attached to a massless rod of length 'l', pivoting about the attachment of the rod. The period of this system is,

$$T = 2\pi\sqrt{l}/g$$

(where g is the gravitational acceleration near the surface of the earth). But massless rods, pivoting about the centre of the point mass, do not occur in nature so that a factualization of the equation should take us closer to a real pendulum. Such a system, taking into account some of the features of real physical systems, is the physical pendulum, which may be an object of arbitrary shape and mass 'm', attached to a rod of length 'd', swinging *frictionlessly* about an axis A. Its period, T, is then,

$$T = 2\pi\sqrt{I}/mgd$$

(where I is the moment of inertia, which differs for differently shaped objects, and 'd' is the distance from A to the centre of gravity). (For more on idealizations and factualizations, see

Brzezinski [1990, 1992]; Cartwright [1989]; Krajewski [1977]; Laymon [1989]). One of the most important features of science is reflected in this process of factualization, namely the testability of scientific theories and laws. Vehicles of testability are observation and experimentation. An important part of a consideration of laws is therefore a discussion of the role of experiments in the evaluation of laws. (Franklin [Ch. III. pp. 191-207.])

Various types of laws have been introduced so far and a number of characteristic features of laws of science have been mentioned. An attempt has been made to characterize laws in terms of structural features of physical systems. As mentioned before, a number of philosophical positions concerning laws of nature or laws of science have developed. Before these positions are outlined, it is important to turn to criteria by which true laws of nature are to be distinguished from mere accidental regularities.

II. Classification of Criteria

Lit.: Armstrong [1979], [1983]; Bunge [1961]; Chisholm [1955]; Earman [1978]; Kneale [1949], [1950, [1961]; Lange [1992], [1993], Molnar [1969], van Fraassen [1989], von Wright [1984]; also Duhem [1960].

It will be useful to reserve the term *lawful statements* for expressions of the true laws of nature and *lawlike statements* for expressions of mere accidental regularities. These two types of regularity may be marked by a logical distinction: accidental regularities are expressed by finite conjunctions of individual conjuncts (p, q, r), while laws of nature are typically expressed as 'If/then'-statements: If certain physical conditions are satisfied, then certain other conditions follow. Alternatively, it may be said that lawful (qualitative or quantitative) statements are (mathematical) expressions of structural features of physical systems. This statement has the advantage of capturing the idea that the very form in which laws of science are cast (for instance as integrals or second-order differential equations) tells us something about the structure of the physical world.

Reichenbach's classic example may serve to illustrate this logical point: It is a lawful statement that all solid spheres of enriched *uranium* have a diameter of less than one mile; this is true for structural reasons, independent of individual solid spheres and fundamentally interrelated with the existence of atoms in this universe. Hence this statement expresses much more than what the sum of individual observations on appropriate spheres can indicate. On the other hand, it is only a lawlike statement, hence an accidental generality, that all solid spheres of *gold* have a diameter of less than one mile, because nothing in the structure of gold forbids the existence of gold lumps with larger diameters. The lawlike statement only sums up the individual results (conjuncts) of a large number of findings. There is nothing structural about this statement.

Philosophers have worked out a number of criteria which should enable us to distinguish between statements expressing true laws from mere accidental generalities: Unrealized physical possibilities; counterfactual situations; inductive support, explanatory power, systemicity and universality. While there may be disagreement about the applicability of some individual criterion, there is general agreement that some cluster of these criteria is adequate for the distinction.(It is worth noting that at least some of these criteria were already employed by the founding fathers of modern science, see Steinle [Ch. IV. pp. 316-366.]).

a) *Unrealized physical possibilities.* It is often claimed that true laws must not only apply to the existing physical world but must also cover physical situations which, although nonexistent, are permitted by the laws of nature. Thus, according to thermodynamics, there are energetically improbable states which are not forbidden by the laws of thermodynamics, although these laws explain their improbability. For instance, a cake taken out of the oven cools down but will never, by the laws of probability, spontaneously and without the input of additional energy, warm up again. There is no law to forbid such an occurrence. But there is a law to explain why this is extremely unlikely: the improbability of such events is a consequence of the second law of thermodynamics (increase of entropy). Equally, if it is true that 90% of species which existed in the past are now extinct, then from the viewpoint of the distant past many of the species living today were unrealized physical possibilities.

Finally, many technological inventions, although subject to the laws of nature, were mere physical possibilities in the past. For instance, although the laser was only developed in the 1950s, the physical principle on which it is built, called 'stimulated emission', was predicted in Einstein's 1916 quantum theory of radiation.

Mere accidental generalities, however, do not extend to physically non-existing situations. They are conjunctive statements about some existing state of affairs. Let it be an empirical regularity that all moas – an extinct New Zealand bird – died before reaching the age of fifty and let it be stipulated that nothing in the constitution of the moa precipitated its untimely death. Due to some unfavourable conditions, however, – perhaps the presence of a virus or some other inhospitable feature – the moa simply did not reach its natural age. The universal statement, 'All moas die before the age of fifty', cannot be accepted as a law of nature, since, on the present criterion, this statement only sums up occurrent facts but does not tell us anything about the life expectancy of the moa in more favourable environmental conditions. (Popper [1959: 427])

b) *Counterfactual situations.* Talking about mere physical possibilities leads to statements about what would be the case, if a certain situation obtained. Thus, only true laws support counterfactuals, while accidental regularities do not. Astronomers have been searching, unsuccessfully, for a tenth planet in the solar system. Maybe there is no tenth planet. But, *If* there were a tenth planet, *it would* obey Kepler's laws. However, no such statement can be derived from the accidental generalization about moas: 'If there were still some moas in New Zealand, they would also die before the age of fifty', is unwarranted.

Such counterfactual-like statements actually play a significant role in physical research. For instance, counterfactual arguments are employed by physicists working on fundamental physical constants and null experiments. Thus, the observation of the strict validity of the Coulomb force law, an inverse-square relationship about stationary charges ($F_C = Kq_1q_2/r^2$), has confirmed physicists in their assumption of a zero rest mass of the photon. *If* there were a deviation from the law, the photon *would* have a non-zero rest mass. Equally, the constancy of the velocity of light (in vacuum) has led physicists to the rejection of the ether and the classical transformation rule (for, *if* the velocity of light were not constant ,in vacuum, there *would* be a preferred coordinate system, in violation of the relativity principle). (Weinert [forthcoming]).

c) *Inductive Support.* Only those regularities which can be inductively confirmed by their instances are to be regarded as expressions of true laws. An accidental generalization is not inductively confirmed by a further instance of the same regularity. As an accidental regularity has the logical structure of a conjunction of singular propositions,

p & r & s

the addition of a further conjunct does not inductively confirm the statement, for it is not possible to infer from this conjunction that a yet unexamined case can be added to it. If it is a fact that all men who ever enter this room wear wristwatches, then this regularity receives no support from the fact that John, on entering the room, is found to be wearing a wristwatch. If, however, laws express structural relations, of which a $1/r^2$ relationship is an example, then new instances do lend inductive support to this structural relationship. The concentration on a structure which individuals may share, makes the reference to these individuals unnecessary. It may be sufficient to examine a small number of well-chosen individuals to identify the structure; new, hitherto unexamined, individuals which exhibit this structure will then confirm the general statement made about the structure.

d) *Explanatory power.* There is a common conception that laws must be explanatory: laws explain the phenomena. This conception was promoted by Hempel's famous DN model of scientific explanation. According to this model, a phenomenon is explained if it can be subsumed under a universal statement. Thus, from the law that the period of a (simple) pendulum is related to its length, we can deduce that a particular pendulum, whose length is known, must have a particular period, determined by its length. Hence, its period is explained by this subsumption under a general law.

One may, however, entertain some doubts as to the reasonableness of this conception: at least stated without further qualifications, counterexamples can easily be found. Newton's law of gravitation, for instance, only states a functional relationship between the gravitational force obtaining between two material bodies, but does not explain why there is this particular force between these bodies. The Schrödinger equation, suitably modified, describes the behaviour of a particle in an electro-magnetic field, but it does not explain why the particle behaves like this. Snell's law states a functional relationship between angles of incidence and angles of refraction but does not explain why the angles are related in this way.

The difficulty with the above view is that explanation is a pragmatic, extendible concept. The subsumption of some empirical phenomenon under some general law may provide a first layer of explanation, but the law itself may be subject to further explanations, and the phenomenon may only be an aspect of a more general phenomenon. Empirical laws as such do not explain the phenomena. Fundamental laws may not explain the phenomena either. Explanations of phenomena are often obtained from laws *in conjunction with* scientific theories. Take the Newtonian question of why planets stay in their orbits. Neither the First law, nor the Second law alone can explain this fact. It is the conjunction of the two laws and their incorporation into Newtonian mechanics which give a first satisfactory explanation of this phenomenon.

The view implied in these remarks is that laws should not be regarded in isolation. Because of the fundamental interrelatedness of the physical world, laws, too, are interrelated. Laws form an interrelated network in a dual sense: as the physical world is, in the words of von Helmholtz, a *Kosmos*, that is, an interconnected system of forces and fields, the laws of nature, too, are interconnective; and as science describes and explains certain parts of this *Kosmos* – physical, biological, chemical systems – and penetrates deeper and deeper into their core, it too must describe and explain the interrelatedness of the physical world by appeal to laws, models and theories. By this process, physicists hope to arrive at a final theory. (See Barrow [1991]; Weinberg [1993] as well Morrison [Ch. II. pp. 157-188.] and Galison [Ch. VI. pp. 369-408.] in this volume.) This aspect of interconnectedness (systemicity) is often, though not always, emphasized as a further criterion of lawfulness:

e) *Systemicity.* The fundamental interrelatedness of the physical world, "its whole structural interconnection" (Feynman [1965: 125]), means that science has to look at interacting physical systems, although it must be realized that science is quite unable to embrace the total complexity of the natural world. Consequently, science operates with models of various kinds – increasingly computer models play a role – which concentrate on the most important aspects of physical systems and abstract from known negligible forces. Models invariably encapsulate idealizations and abstractions. The statements science makes about these various systems will have inferential relations to other statements: the more inferential relations there are, the more systemicity the theory will possess. One of the most important inferential relations in science is derivability. Many laws become derivable from a few fundamental laws. The above criterion then states that a statement counts as a law of science if one of its characteristics is systemicity: it bears a relation to other statements in the theory. It is embedded in a network of statements which are interrelated to each other.

It may be doubted that statements like 'All ravens are black', 'All polar bears are white' can lay any claim to the title of lawful statement. Such a doubt may have two sources: on the one hand, it may be due to the fact that such statements are not (yet) systematically interrelated with other statements in biological theory. In particular, it is required that if these statements are to acquire law status, they must express structural features about ravens and polar bears. (The statement must say more than: 'It so happens that all the ravens so far examined turned out to be black.') It may be found, for instance, that the genetic make-up of creatures like ravens is such that a change in their colour would seriously modify the very nature of ravenhood. If this could be shown, the above statement would be systematically related to other statements about ravenhood in particular and the genetics of feathered creatures in general. Such statements could therefore count as laws. But they would no longer specifically refer to ravens. They would refer to genetic structures. If, however, the study of ravens revealed that their colour is purely accidental, their blackness may just be the result of boundary conditions and would therefore not be systematically related to other statements about ravenhood. As a consequence, the above statement would reflect a mere accidental regularity with the logical structure of a conjunction of singular statements.

On the other hand, the doubt may also be derived from the observation that no law should refer to species like ravens and polar bears. Philosophers of biology and biologists have repeatedly pointed out that species are (spatially scattered) individuals, and that the laws of biology cannot refer to them. Biological laws have to be generalizations about classes of individuals. Unless this is recognized, Smart's doubts about biological laws cannot be effectively dispelled. (Ghiselin [1989: 53-5]; Hull [1974]; Rosenberg [1985: §7.6])

Accidental generalities cover only existing states of affairs. Exceptions may exist or may at least be possible. There are black swans and there may be white ravens. Accidental generalities may therefore not be truly universal. True laws, by contrast, cover both realized and unrealized states of affairs. Whether there are exceptions to true laws will depend on the existence of specific boundary conditions. But this only means that some structure may not be activated. Truly fundamental laws, as has been argued above, satisfy a minimum of boundary conditions.

Phenomenological laws require a greater number of boundary conditions. How universal, then, are the laws of nature?

f) *Universality.* It has already been noted that the existence of some uniformity is not necessarily an indication of the existence of a true law of nature, for uniformities may be accidental, and no accidental generality should count as a law of nature. Universality is therefore not sufficient to make lawhood. It may be doubted that truly universal, yet completely accidental regularities are ever to be found. For given the logical structure of accidental generalizations, it is doubtful that universality is conserved across a vast array of boundary conditions. But energetically possible (though highly unlikely) physical situations, not excluded by the laws of thermodynamics, may well be candidates to the title of accidental universalities. A viable philosophical account of laws should therefore be able to distinguish true laws from mere accidental generalities.

Thus, there may be evidence of a uniformity, conventionally described as,

$$\forall x(Fx \supset Gx)$$

which, however, is no more than an accidental regularity. Being a *de facto* generalization, it tells us nothing about counterfactual situations: If it were just an accidental generalization that all photons have zero rest mass, we could not possibly argue that as a consequence, Maxwell's equations would need to be changed and Coulomb's force law could not be strictly valid.

As we shall see below, basically all the philosophical positions developed, have endeavoured to mark the distinction between lawlike and lawful statements. One prominent recent suggestion centres on the introduction of the concept of natural or nomological necessity, qualifying true laws of nature. (Leckey/Bigelow [Ch. II. pp. 92-119.]) According to a version of the Regularity theory, however, the distinction is to be achieved by appeal to the epistemological criterion of systemicity: only statements which are properly incorporated into a network of other propositions should count as lawful statements.

But even if we leave accidental universalities aside, we may envisage the possibility of spatio-temporally limited laws, that is, laws which are restricted to a certain spatio-temporal period of the universe. Such considerations seem legitimate if recent models of the evolution of the universe are taken into account. (Davies [Ch. III. pp. 248-267.]; also Belashov [1992], Whitehead [1925]).

It is thus possible to distinguish two senses of universality: '(i) universality in the sense of unrestricted range over all space-times and (ii) for a given space-time S, universality in the sense of unrestricted range over S'. But in any case, law statements should make no reference to specific space-time coordinates. (Earman [1978: 173, 177]; Barrow [1992: Ch. 3]; Bunge [1961: 266-8, 277]; Hull [1974: Ch. 3]; Weinberg [1993: 26-9])

Most laws of science are stipulated without explicit reference to boundary conditions so that it looks as if most of these statements are actually violated in nature and hence not universally valid. For instance, it is often said that Newton's First law is never satisfied, because there are no bodies in the universe which are uninterfered with. (See Chalmers [1993]). Laws of science, however, must make abstractions from interfering forces and include idealizations from real physical features. The universality condition requires that law statements abstract from particular boundary conditions to capture the structure of physical systems. It is immaterial to Galileo's fall law whether air resistance slows down the fall of the feather and make it appear as if the lead ball behaved differently from the feather. The fall of the feather, and of the lead ball, in concrete situations is a component of many forces, not a cancellation of any particular force. Talk of the universal validity of laws in concrete physical systems can be restored, if component forces, often due to the presence of particular boundary conditions, are considered. Or at least if it is remembered that a statement of boundary conditions is always an appendix to the statement of laws. (Davies [Ch. III. pp. 248-267.]; Feynman [1965: 87]; Haken [Ch. III. pp. 227-247.]; Kneale [1949: §17]; Rosenberg [1985]: §§7.6-7.9]). Universality under the inclusion of boundary conditions means that whenever specific boundary conditions are satisfied, at any point in space time, the law determines the behaviour of particular physical systems in identical, predicable ways. This is not true for accidental generalities. Even if in the lifetime of the universe no energetically unlikely state actually materializes, there always remains the minute probability that under *identical* boundary conditions such a state does occur.

This concludes the discussion of the criteria in terms of which laws of science and laws of nature are usually discussed. We can now turn to a characterization of the most important philosophical theories of laws of nature and laws of science.

III. Classification of Philosophical Views

The most important development in the last decade has probably been the emergence of the so-called Necessitarian View of laws, since it poses a considerable challenge to the Regularity View which originates from Hume's philosophy. Modern versions of these two opposing views are represented in this volume. [Ch. II pp. 67-119] However, alternatives to these two main strands have also been tabled: the Model-Theoretic Approach (Giere, [Ch. II. pp. 120-138]) and what may be termed the Structural Approach, understood here as a synthesis grown out of the Necessitarian and the Regularity Views. (Cartwright [1983], [1989]; Chalmers [1993]; Harré/Madden [1975]; Weinert [1993]). The Instrumentalist View, which will be discussed first, has lost much of the attractiveness it once held under the influence of the early Wittgenstein. But it is instructive to see why its sway waned.

1. *Instrumentalism*

Lit.: Pearson [1892]; Mach [1976]; Wittgenstein [1922]; Rescher [1970]; Giere [1988]; van Fraassen [1989].

Instrumentalist theories about the nature of laws are attempts to explain laws as forms of descriptions which the human mind imposes on the realm of phenomena. The aim of scientific research is to organize these phenomena into a coherent network, which offers the most economic presentation. By its very nature, Instrumentalism is concerned with the laws of science, rather than the laws of nature, because the origin of laws, as statements of regularity, is seen in the constructive capacities of the human mind or of human language. The instrumentalist need not deny that these constructed law statements refer to some physical reality. In fact, he typically holds that there is some physical element, be it sense perceptions or even physical regularities, on which order is to be imposed by the human mind. According to Rescher [1970: 117], it is not 'the regularity claimed by a law but the lawfulness it builds into this claim that is mind-dependent.' The upshot is that there would be regularities in nature without the presence of human observers but no lawfulness: lawfulness is the result of a constructive act or an act of imputation, performed by intelligent observers of the physical world. This so-called law-idealism is reminiscent of Kant's transcendental philosophy: laws are seen as constructs of the human mind, functioning as a mental shorthand for sense impressions (Pearson) or restrictions on expectations (Mach).

In their denial of what Peirce dubbed 'the reality of the thirdness' – the realist idea that uniformity corresponds to a reality (Peirce [1934: 64]) – such views carry the strong nominalist commitment of treating both universality and lawfulness as relative to the human mind or in more modern parlance, as relative to scientific theories.

One of the most influential instrumentalist theories of laws in the 20th century is to be found in Wittgenstein's *Tractatus*, which helped shape the views of Watson [1938], Ryle [1949], Toulmin [1953], Harré [1960] and Hanson [1969]. (See Musgrave's discussion [1979-80]). Wittgenstein calls it an illusion 'that the so-called laws of nature are the explanations of natural phenomena' [T. 6.371]. He compares scientific theories, like Newton's mechanics, with conceptual networks, which bring 'the description of the universe to a unified form' [T. 6.341]. He adds that there can be different networks to which different systems describing the world correspond. And the fact that the world can be described by Newtonian mechanics 'asserts nothing about the world' [T. 6.342].

Wittgenstein's assertions about laws carry the hallmark of Instrumentalism: First, there is the view that causality and conservation laws are not laws at all, but law forms, i.e. 'possible forms of the propositions of science' [T. 6.32-6.34]. Second, there is the contention that scientific theories are only possible ways of describing the world and hence the basic instrumentalist idea that scientific theories are more or less efficient ways of making sense of the phenomena, without, however, acquiring truth values. (In her contribution to this volume [Ch. II. pp. 157-188.], Margaret Morrison discusses symmetries as methodological constraints on laws, hence as forms in which law statements are to be formulated. And Peter Galison [Ch. III. pp. 369-408.] looks at the problem of the empirical confirmation of string theory as one of the candidates for a final theory of nature.)

Amongst his followers (Ryle, Toulmin), Wittgenstein's assertions condensed into the so-called Inference-licence view of laws: 1) laws of science are interpreted as our method of representing nature; 2) they permit us to infer particular statements from other particular statements; (for instance, from the position and orbital velocity of a planet today it can be inferred where it was 300 years ago); 3) as they function as rules of inference, in accordance with which we infer particular statements from other statements, laws cease to be empirical statements: as mere inference tickets they do not inform us about the structure of the natural world.

A practising scientist will have difficulties accepting this position. The step from one particular set of data to another can only legitimately be made if the correct inference rule is used; and even if several rules seem to suggest themselves, it is often possible to specify the inferential conclusion to such an extent that only one particular rule will give the right answer. True, if our demands are not too stringent, it may be perfectly possible for the Ptolemaic system to give predictions of the positions of planets identical to those based on a Copernican-inspired astronomy. But the use of precision arguments in science is such that it is not only able to predict which results are expected from experiments and observations, but also which results would have to be regarded as spurious. (For arguments along this line and some enlightening examples see Franklin's contribution [Ch. III. pp. 191-207.]). The upshot is that our inference-licence tickets need to be improved in accordance with the constraints the physical world imposes on scientific theorizing. The question cannot be avoided why certain candidates for inference rules forsake us while others seem to perform particularly well.

A well-known argument against the instrumentalist view of laws is that instruments – if this is how laws should be treated – cannot be tested and refuted by experiments – they can only be shown to have limited applicability (Popper [1963: 112-4]; Musgrave [1979-80: 95-104]). The argument can be taken a step further. If a law is shown to have limited validity – for instance Newton's laws of motion begin to deviate from the correct, relativistic results at velocities greater than 30km/s – the legitimate question arises why a particular law becomes limited in its applicability. The classical answer, due to the work of Niels Bohr and Werner Heisenberg, has been that a domain-restricted law must be shown to be a limiting case of a more general law. This is known as the Correspondence Principle, according to which the limited validity of certain laws must be shown and be *justified* by a theory with wider scope. The restricted laws thus become a limiting case of a more general law under well defined conditions.

While the Inference-licence view of laws seems to have few followers these days, quasi-instrumentalist views of laws have recently emerged especially in the work of Bas van Fraassen [1989]. In this view, laws are not statements about the structure of the physical world, but attributes of the models used to describe and explain the physical phenomena. As was noted in the discussion of the earlier proponents of such views, law statement are at a remove from the physical world because *lawful-*

ness is an attribute of the relationship in which human beings stand to the physical world.

A strictly realist approach to the understanding of science is to show how the structure of scientific theories approximates or reflects the structure of the physical world. Van Fraassen's alternative, anti-realist approach, sees the study of the structure of science

as a product of the intellect that strives to order and unify the deliverances of experience. Both the notion of law and that of sufficient reason served as 'transcendental clues'. Departing from structural features of theory, they delineate the structure of any possible world allowed by physical theory – that is, the structure of its *models*. Both also were honoured or distorted (..) by a reification which accepts them as clues to the structure of the *world being modelled.* The alternative is to reject that reification (van Fraassen [1989: 13]).

The key terms of this *model-theoretic approach* to laws are 'structure' and 'model'. The models assign a structure to the physical systems they model. And the laws become features of these models. Unlike van Fraassen, however, Ronald Giere combines the language of models with the language of realism and considers that the models 'capture' the structure (or parts thereof) of physical systems. He calls his position *Constructive Realism*. (Giere [1988a: Ch. 3, 4])

Laws, on this account, are statements that characterize the structure of a theoretical model. Whenever the elements of a theoretical model are identified with physical properties, laws then describe constraints on the state or dynamic development of a physical system. The Boyle-Charles gas law, for example, is to be understood as asserting that the pressure, volume and temperature of an ideal gas are constrained to satisfy the relationship $PV = nRT$. Note that this statement is not to be understood as a universal generalization concerning real gases, but as part of the characterization of what it is to be an ideal gas (Giere [1988b: 42].

While in earlier publications, Giere used the notion of similarity to characterize the relation between the model and the physical systems it models [1988: 92-94], in his contribution to this volume, he sees this relation in terms of the notion of fit. [Ch. II. pp. 120ff.: §§4,5]. The alternatives to instrumentalist or quasi-instrumentalist interpretations of the laws of science are squarely rooted in the realist camp. The *Regularity View* has a long history focussing on the idea that laws are empirical universal regularities. The *Necessitarian View* is of more recent origin, but it gained its momentum from a dissatisfaction with the Regularity View, and developed into a fairly Platonistic view of the world, according to which, in its basic form, laws are necessary relations between universals. Both views have undergone several modifications.

2. *The Regularity View*

Lit.: Ayer [1963]; Braithwaite [1927]; Feynman [1965]; Hausman [1967]; Hempel/Oppenheim [1960]; Hitchcock [1992]; Hume [1973]; Jobe [1967]; Lauter [1970]; Molnar 1969; Urbach [1988]; Reichenbach [1954]; Swartz [1985]; Swartz's contribution to this volume, [Ch. II. A.]; Suchting [1968].

Criticism of the Regularity view: Armstrong [1979], [1983], [1984], [1988]; Bigelow/Pargetter [1990]; Carroll [1987]; Dretske [1977], [1978]; Forrest [1985]; Harré/Madden [1975]; Kneale [1949], [1950], [1961]; Molnar [1969]; Mundy [1986]; Pargetter [1984]; Swoyer [1982]; Tooley [1977], [1987]; Vallentyne [1988]; Suchting [1974].

In its original form this view simply identified the laws of nature with a Hume-inspired constant conjunction view of regularities. Statements of laws of nature satisfy the following criteria:

They are expressed in universal or statistical propositions,
they are true,
they are contingent,
they are purely descriptive in their terms, or in other words, they only contain non-local empirical predicates, apart from logical connectives and quantifiers.
(Armstrong [1983: 12]; Berofsky [1968: 318]; Molnar [1969: 79]; Swartz [1985: 28], and Swartz in this volume, [Ch. II. pp. 67-91])

While it is fairly typical of Regularists to view *laws of nature* as universal regularities, such a characterization of the laws of science faces a number of problems: the distinction between law-like and lawful statements, the existence of vacuous laws and functional laws are the most pressing. Vacuous laws state seemingly uninstantiated relationships, as for instance in Newton's first law, since there are no bodies in the universe which are not subject to any forces. Functional laws permit of values which a real parameter could not attain, for instance very high temperatures, as in the Ideas Gas Law: PV = nRT.

Regularity theorists are hardly oblivious to such problems. They are aware of the need for some further distinctions and criteria to single out, amongst the statements of science, those which are to be elevated to the status of laws, thereby also hoping to solve the other problems.

Their attitude to these questions reflects one of the main differences between proponents of the Regularity and the Necessitarian approaches. For while empiricists of all persuasion seek to distinguish between these two kinds of generalities by appeal to *epistemological attitudes*, proponents of Necessitarianism resort to *ontological criteria.* The question 'What there is' or, in Quinean parlance, 'What our theories say there is' lies therefore at the bottom of the dispute between Necessitarians and Regularists. With respect to the epistemological attitude, which empiricists adopt to distinguish properly between matters of fact and matters of law, and other distinctions consequent upon this one, it is convenient to distinguish a *personalist approach* from an *objectivist approach.* The basic ontological attitude is however shared by all empiricists.

Empiricist writers on laws of nature take ontologically independent events as the basic ontological units: 'laws are parasitic on singular occurrent facts, either of an observable or unobservable kind' (Earman [1984: 195]). For 'the Regularist, singular statements are ontologically primary. Physical laws are logically derivative.' ([Swartz 1985: 81], and in this volume [Ch. II. pp. 67-91]). Given such an ontological commitment, the empiricist finds it difficult to see any evidence in the natural world to distinguish between merely accidental (lawlike) and true (lawful) regularities. Indeed, as Ayer states,

> The factual information which is expressed by a statement of the form 'for all x, if x has Φ then it has Ψ', is the same whichever way it is interpreted (Ayer [1963: 230]).

At this stage, then, the need for a distinction between generalizations of fact and generalizations of law is needed, which cannot appeal to the way the word is. And it is typical for empiricists to invoke epistemological criteria, either of a personalist or objectivist kind, to draw the line between these two generalizations.

2.1. *The personalist approach to regularities*

Lit.: Ayer [1963]; Braithwaite [1927]; Hume [1973]; Lange [1993]; Urbach [1988].

The distinction is to be achieved, on this approach, by combining a constant-conjunction view of regularities with a customary- belief- theory of necessity. Necessity is often perceived as an accompaniment of causation and laws of nature. But the existence of physical necessity is rigorously denied, since there does not seem to be any evidence of the workings of physical necessity in the universe. The contradictory of any law, it is pointed out, is at least conceivable so that any notion of necessity is shifted either to the realm of psychology (as in Hume's psychological explanation of belief in the necessity of causation) or to the logico-linguistic level (as in modern-day empiricism), so that the notion of physical necessity is deprived of all meaning.

Ignoring complications, the structure of the argument, following Ayer, involves the requisite epistemological attitude a person, A, should adopt towards a regularity, deemed to be a law of nature. It should be such that

> a person A was treating a statement of the form 'for all x, if Φx then Ψx' as expressing a law of nature, if and only if there was no property X which was such that the information that a value α of x had X as well as Φ would weaken his belief that a had Y. (Ayer [1963: 232])

Of course, contemporary versions of this approach may highlight different aspects; for instance, it may be held that

> a claim's lawlikeness is not an ordinary fact but a fact about the way we ought to reason: to believe that "All Fs are G" is a law-statement is to hold that one

ought to reason by giving "...is F" as one's justification for believing "...is G". (Lange: [1993: 13]

Or again laws may be identified with subjunctive conditionals, $Aj \rightarrow Bj$, expressed by the probability term, $p(Bj/Aj)=1$, which is to be interpreted as a degree of belief (Urbach [1988]). The price of shifting the distinction between generalizations of fact and generalizations of law to the level of epistemological attitudes of rational agents seems to be that lawfulness as a real feature of the natural world is sacrificed. It should be noted at this stage that the empiricist's rejection of a *real* distinction between law-like and lawful generalizations is very much directed against the Necessitarian's identification of lawfulness with physical necessity. The introduction of such modal notions as possible worlds, essences and universals do not, at least according to one defence of the Regularity view, solve the problem of vacuous and functional laws either (Mellor [1980]). The empiricist's option for the realm of the actual should also determine his solution of the problem of vacuous and functional laws.

The unalterable empiricist constraint is that such laws must perform in the service of the actual; they must arise in the attempt to account for actual occurrent facts and regularities. (Earman [1984: 210]).

While this is no more than a recommendation, the objectivist approach actually does propose an answer to these concerns, which Earman shares. For Ayer, however, the problem posed by these two kinds of laws, makes it necessary to bring in possibilities: true laws cover both actual and possible cases (while accidental generalities do not). Hence, laws of nature cannot be just constant conjunctions. The evidence, however, does not tell us whether we are dealing with constant conjunctions only or with true laws of nature. Hence, the evidence must be 'ordered' by human epistemological attitudes, as outlined above. (See Ayer [1963: 220-34]).

The alternative to personalist attitudes towards how to classify the evidence, still embedded in the empiricist tradition, is an objectivist approach to the problem of laws which appeals, not to degrees of belief, but to objective features of scientific systems to single out, amongst the observed regularities, those which we should consider as true laws of nature. The approach is, of course, via the (known) laws of science. But empiricism (to be understood here in a very broad sense) has no difficulty in adding the realist assumption that these laws of science are reliable indicators of at least the regularities which the evidence reveals. The aim is to achieve a distinction without reference to modal notions. From the empiricist's point of view terms like 'nomonological necessity' or 'possible world' are fraught with difficulties: the empiricist denies that there are irreducible modal facts and finds the notion of physical necessity epistemically inaccessible.

As a transition to the objectivist approach to laws, in terms of systemicity, it is useful to take a look at an analysis of nomic necessity in terms of statistical invariance.

2.2. *The Resiliency Solution*

Lit.: Skyrms [1980], [1984]; Skyrms/Lambert in this volume [Ch. II. pp. 139-156.].

The key to an understanding of nomic necessity, according to these authors, lies in the notion of probabilistic invariance or *resiliency*. It is hoped that this notion will make reliance on the metaphysics of possible worlds unnecessary. But the epistemological attitude, so typical of the regularity approach, is unmistakably present:

> Certain patterns of phenomena recur throughout the world. It is economical for a thinking animal to form habits of thought connecting various features of a pattern. The habits corresponding to invariant patterns are refined into laws of nature, the residue of the phenomena being regarded as de facto conditions. (...) The distinction between belief in a statement *qua* law and belief in a statement *qua* matter of fact is, then, to be found in the domain of invariance of that belief. Confirmation *qua* law should carry not only the requirement that the evidence warrant high probability, but also that it warrant *invariantly* high probability. (Skyrms [1980: xi])

Skyrms agrees with Mill's understanding of necessity as *unconditionalness* such that 'high resiliency can be thought of as a statistical notion of *necessity*' [1980: 12]. The disputed notion of physical necessity, taken by some Necessitarians as a primitive, unanalyzable concept, is resolved here into more tractable features:

> Necessity = Unconditionality = Invariance under Conditionalization = Statistical Independence.

A universal law becomes a limiting case of a statistical law: everything in its scope has a propensity of one not to be a counterexample. The universal proposition, p, 'All ravens are black', is interpreted as 'Every raven has a propensity of one to be black.' In the language of resiliency, this means that the resiliency of Pr(p) is 1. Thus we expect, as noted earlier, that a law, unlike an accidental generalization, can be inductively confirmed, which in terms of resiliency means:

> 'All ravens are black' to be well confirmed, 'If it's a raven, it's black' should be resilient over 'It's a raven.' (Skyrms [1980: 35-6; 59-7]).

It was observed above that one may have legitimate reservations as to the law status of this type of statement. The Resiliency approach, it seems, must accept empirical generalizations as lawful statements, even if reasonable doubts exist. The Resiliency account also makes room for the aspect of systemicity, for which the term *network of laws* is used: 'Laws do not stand in isolation. Rather, they operate as part of a network, or theory, containing other laws.' (Skyrms [1980: 66]; Skyrms/Lambert [Ch. II. pp. 139-156]). The resiliency condition is then transposed from the instantial language used to evaluate individual laws to networks:

> For a network of laws, 'All F_1s are G_1s,'...'All F_ns are G_ns,' take its confirmation as the resiliency of the conjunction of its instances: $R[(F_1a...G_1a)]$ & $(F_2a...G_2a)$ & ...$(F_na...G_na)]$. (Skyrms [1980: 67]).

It can be gathered from this brief discussion of the Resiliency approach that the Regularist's insistence on epistemological attitudes to single out the laws readily leads to a discussion of important features of laws of science. Necessitarians, however, grounding the distinction between lawlike and lawful generalities in the ontology of the world, have so far had little to say about the epistemological features of laws. (For a step in this direction, see Leckey/Bigelow in this volume, [Ch. II. pp. 92-119.] and Mundy [1987]). This insistence on the features of laws of science is further emphasized by the objectivist approach.

2.3. *The objectivist approach to regularities*

Lit.: Berofsky [1968]; Braithwaite [1953]; Earman [1984]; Hesse [1980]; Hempel [1966]; Hochberg [1981]; Job [1967]; Lauter [1970]; Lewis [1973], [1983]; Mellor [1980]; Mill [1906]; Nagel [1961]; Niiniluoto [1978]; Papineau [1986]; Ramsey [1931]; Reichenbach [1954]; Smart [1985].

Harking back to von Helmholtz's notion of a *Kosmos,* according to which the natural world has to be seen as a network of interlocking physical systems, which scientific theories must reflect, the empiricist may try to develop his account of laws of nature on the most economical metaphysical picture he can muster. Thus, the natural world displays regularities which are sorted into accidental and non-accidental regularities by an appropriate epistemic attitude. Referring to the feature of systemicity, he can gain two advantages over the subjectivist approach to regularities by treating 'regularities not one at a time, but rather as candidates to enter into integrated systems.' (Lewis [1983: 367]) Regularities are then seen as interdependent, with inferential links to other regularities, *and* symbolically represented in conceptual systems which may be called scientific theories. A further requirement is, of course, that such conceptual systems must be empirically confirmed. Accidental regularities are those which fail to entertain systematic connections with other known regularities. The appeal to systemicity also serves to account for the fact that only lawful regularities support counterfactuals, for whether a given law statement supports a counterfactual may depend on our knowledge on how the regularities expressed in the statement is related to other known regularities in a conceptual system. Thus our indecision with respect to the status of a statement reporting the blackness of ravens may be due to our ignorance as to how the colour of feathered creatures is related to their genetic make-up. By contrast, it is clear that the diameter of a gold lump (say of less than 1m) is purely accidental because nothing in the atomic structure of gold forbids the existence of gold-lump of a greater diameter. Vacuous and functional laws may also cease to pose problems, as indicated above, because they are now part of a system which is supported by empirical evidence. Such laws are consequences of other statements, as part of conceptual networks. Thus they enjoy theoretical support. (Ayer [1963: 224-5]; Braithwaite [1953: Ch. 9]; Broad [1935]; Earman [1984: 210]; Hempel [1966: Ch. 4]; Nagel [1961: 60-1]).

It is worth mentioning that the notion of conceptual network, employed by this approach, permits of different characterizations. In a well-known passage, D. Lewis reformulates Ramsey's theory of lawhood (1928) by reference to the term 'deductive system':

> A contingent generalization is a *law of nature* if and only if it appears as a theorem (or axiom) in each of the true deductive systems that achieves a best combination of simplicity and strength. (Lewis [1973: 73]; cf. Earman [1984: 196ff]; Armstrong [1983: §5]and Bigelow/Pargetter [1990: §5.3] for criticism).

It is not required that we have knowledge of such a deductive system; it is sufficient that 'deductively closed, axiomatizable sets of true sentences' exist (Lewis [1973: 73]). By contrast, Braithwaite only requires that an established hypothesis, *h*, is to be regarded as a law of nature when

> the hypothesis either occurs in an established scientific deductive system as a higher-level hypothesis containing theoretical concepts or that it occurs in an established scientific deductive system as a deduction from higher-level hypotheses which are supported by empirical evidence which is not direct evidence for *h* itself. (Braithwaite [1953: 301-2])

Apart from objections which have been raised with respect to the notion of a deductive system and the condition of derivability – it seems possible to derive statement from conceptual systems which we would not regard as candidates for lawhood (Suchting [1974]) – the main concern of Necessitarians is that such an approach seems to fail to distinguish between the (known and fallible) laws of science and the true laws of nature. Lawhood, it appears, has nothing to do, or at least should have nothing to do, with membership of a statement in a deductive system (Armstrong [1979: II]; Papineau [1986: §3]; Vallentyne [1988]; Bigelow/Pargetter [1990: §5.3]; Forrest [1985]).

At this stage in the argument we are faced with a difficult choice, namely either to accept the empiricist's world of regularities, sorted into lawlike and lawful components by an adequate epistemological attitude or to embrace the Necessitarian's ontologically richer world, inhabited by universals and necessary relations. As John Earman so aptly observes:

> What the universalists seek to achieve through ontological ascent, the empiricist achieves by ascent of explanatory level. (Earman [1984: 215])

The question is whether the empiricist's ontology of singular occurrent facts cannot be replaced by an appeal to structures in order to free his approach from its present theory-dependence of lawhood. This consideration will lead us to the Structural approach. However, the Regularist's immediate opponent is Necessitarianism.

3. *The Necessitarian View*

As with the Regularity approach, the Necessitarian view has undergone several modifications and refinements, focussing, in particular, on the notion of physical necessity which plays a prominent role in this approach. Few Necessitarians hold that laws involve logically necessary relations (for a discussion of this position see Nagel [1960: 52-6] and Armstrong [1983: Ch. 11]); the predominant view is that laws involve physical necessities. While logical necessity can be characterized as true in all possible worlds, physical necessities are only true in some possible worlds which stand in a suitably qualified relation with the actual world. Laws, then, involve a much stronger relation than the mere conjunction of events: not all universal generalities qualify as laws of nature. The ontology, required by Necessitarians, also exceeds that needed by the Regularists, since the selection of certain generalities as laws of nature does not proceed via epistemological mechanisms but is held to be grounded in features of the physical world itself. Apart from the occurrent facts, the Necessitarian admits both universals and relations between them into his ontology and distinguishes the true laws of nature from accidental generalities by appeal to these features. So while the Necessitarian's approach is ontological, his motivation is epistemological: the existence of universals and their relations is stipulated in order to distinguish properly between matters of law and matters of fact. The introduction of a Platonist ontology is justified by an Inference to the Best Explanation of the fundamental distinction between lawlike and lawful regularities. While these features are common to all Necessitarian approaches, three developments of Necessitarianism should be distinguished.

3.1. The Origins of Necessitarianism

Lit.: Armstrong [1979], [1983], [1984], [1988]; Carroll [1987]; Dretske [1977], [1978]; Forrest [1985]; Kneale [1949], [1950], [1961]; Molnar [1969]; Mundy [1986]; Pargetter [1984]; Swoyer [1982]; Tooley [1977], [1987]; Vallentyne [1988]).

For criticism of Necessitarianism, Hesse [1980]; Hochberg [1981]; Mellor [1980]; Niiniluoto [1978]; Papineau [1986]; Swartz [1985] and his contribution this volume [Ch. II. pp. 67-91.]; van Fraassen [1981], [1987], [1989]; Weinert [1993].

The approach, which originated with Kneale, Armstrong, Dretske and Tooley, was motivated by what these authors saw as the failure of the Regularity view to deal adequately with the distinction between accidental and true regularities and attendant problems, such as counterfactuals, unrealized physical possibilities, uninstantiated and functional laws, the problem of induction and probabilistic laws.

It was therefore suggested (Armstrong [1983: 77]) that the sentence:

It is a law that Fs are Gs

should not be analysed as,

All Fs are Gs

but as,

It is physically necessary that Fs are Gs,

where physical or nomic necessity is a relation with a force intermediate between logical necessity and contingency. On this scheme, a law is not a universal truth about the class of material objects in the natural world; rather, it expresses a relationship between universal qualities (F-ness and G-ness) or quantities (Dretske [1977: 253-4]). This move reveals the ontological ascent (Earman) to second-order quantification over properties and marks the difference between a Regularity and a Necessitarian standpoint. (Cf. Leckey/Bigelow [Ch. II.B.]) Individual objects fall under or instantiate these universal qualities or quantities. A clear distinction must be drawn between particulars and universals: universals can be instantiated by particulars, but particulars can only instantiate, never be instantiated by universals. The most pressing concerns for Necessitarianism are a specification of the relation of necessitation between universals, an explication of the nature of universals and a demonstration of their relevance for scientific theorizing.

While some defenders of this approach admit uninstantiated universals into their ontology (Tooley [1977]; Bigelow/Pargetter [1990]), Armstrong has defended an a posteriori attitude towards universals, to be understood 'either (as) properties or relations of some or holding between some real particulars' [1983: 93-6]. That is, universals are just 'the repeatable features of the spatio-temporal world' and it is the task of 'total science', not of philosophy, to determine which universals there are ([1983: 82-83; Ellis, [forthcoming]).

The admittance of universals into their ontology is one respect which distinguishes Necessitarians from Regularists; the acceptance of relations between universals is another. The appropriate relation between universal properties is one of necessitation, symbolized as

N(F,G),

which means that 'x's being F necessitates x's being G'. This relation provides the means of distinguishing *ontologically, not epistemologically,* between lawful regularities and merely lawlike generalities, for while a nomic relation entails a corresponding cosmic uniformity, the converse relation does not always hold, since not every cosmic uniformity is also a law of nature. Symbolically (Armstrong [1983: 85-6]),

$$N(F,G) \rightarrow \forall x(Fx \supset Gx)$$

but we cannot always infer N(F,G) from $\forall x(Fx \supset Gx)$.

While laws of nature, on this view, are understood as principles of necessitation, whereby it is impossible for an F-thing not to be a G-thing (Kneale [1949: §17]), the status of this principle or relation, on a cosmic scale, is one of contingency, at least according to most Necessitarians. That is, *if* there is a world, in which a certain law of nature holds, *then* F-ness necessitates G-ness, but there may well be worlds in which this law does not hold at all. Only in those worlds which contain the universal F is it a law that all Fs are Gs, but this may not be true in all possible worlds. (See Leckey/Bigelow, [Ch. II. pp. 92-119.])

Regularists have an obvious complaint: that there is no epistemological access to physical necessities. Hence, even if there are physical necessities, they cannot be known. Thus, their stipulation does no real work. David Lewis, in a well-known passage, complains:

> Whatever N may be, I cannot see how it could be absolutely impossible to have *N(F,G)* and *Fa* without *Ga*. (..) The mystery is somewhat hidden by Armstrong's terminology. He uses 'necessitates' as a name for the lawmaking universal *N;* and who would be surprised to hear that if *F* 'necessitates' *G* and *a* has *F*, then *a* must have *G?* But I say that N deserves the name of 'necessitation' only if, somehow, it really can enter into the requisite necessary connections. It can't enter into them just by bearing a name, any more than one can have mighty biceps just by being called 'Armstrong'. (Lewis [1983: 366])

To this charge the Necessitarian will often reply that the nomic relation, N, is a theoretical postulation on a par with other theoretical postulations in science. But this reply does not really alleviate the concern, shared by later Necessitarians, that the original approach treats N, the relation of nomic necessity, as a *primitive* which allows of no further explanation. 'At the end of all explanation, this factor of necessitation remains unexplained.' (Armstrong [1983: 92]; Dretske [1977: 264]; Kneale [1949], [1961]; Tooley [1977]). A curious disanalogy, however, creeps into the picture: theoretical postulations in science, which carry the explanatory weight, assigned to the nomic relation, N, by the Necessitarians, would hardly be treated as primitives. The history of science seems to teach us that theories with such theoretical entities are subject to a number of stringent tests. It is possible, however, that Necessitarians at this stage thought of the relation of necessitation as a metaphysical belief, on a par with such basic beliefs as determinism.

Nevertheless, Necessitarians themselves have felt dissatisfied with this state of things (see Carroll [1987]; Forrest [1985]; Molnar [1969]; Mundy [1986]; Swoyer [1982]; Vallentyne [1988]) and have made two notable attempts to clarify the notion of nomic necessity.

3.2. A realist possible worlds account of laws of nature

Lit.: Bigelow [1990]; Bigelow/Pargetter [1990]; Pargetter [1984].

As in the original approach, this explanation of laws of nature endorses a strong realism about physical universals (properties, relations, patterns and structures). In *Science and Necessity*, Bigelow and Pargetter conceive Platonism as Mathematical Realism, with a three-tiered structure: 1) Individuals; 2) determinate relationships among individuals; 3) relations of proportion among these determinate relationships ([1990: §2.5; see also Leckey/Bigelow, [Ch. II. pp. 92-119]). Hence, this position embraces a full-fledged modal realism, in which (physical) universals have as much reality as particulars.

Within this realist framework, laws of nature, which are a subclass of natural necessities, are explicated by reference to a special relation – nomic accessibility – which holds between possible worlds. This relation in turn is explained by appeal to the essential natures of worlds. The theory requires of laws both that they describe regularities but also, importantly, that they ascribe natural necessity to these regularities. Laws, it claims, are natural necessities and natural necessities are true in all appropriately accessible worlds. Unlike accidental generalisations, laws must be regularities in all accessible worlds, barring all exceptions. It is not possible to go from a world in which certain laws, or at least regularities hold, to a world in which some of these regularities do not hold. But accidental generalizations in a world w_0 may be false in some worlds which are nevertheless accessible from w_0. So something is a law of nature if it is true in all possible worlds which resemble each other in their essential properties, including higher-order universals, i.e. which are accessible from each other.

> A generalisation is a law in world w_0 if and only if the generalisation is true in worlds w_0 and in all possible worlds accessible from world w_0. (Pargetter [1984: 337]; Bigelow [1990: 216])
>
> A statement is logically necessary in any given world if it is true in all worlds. It is physically or nomically necessary, or a law of nature, in a given world if it is true in all worlds which are accessible from that world. (Bigelow/Pargetter [1990: 245])

A law can be regarded as a statement of the form:

$$\text{Necessarily, } \forall x(Fx \supset Gx).$$

Natural necessities are features of a world which make this world the way it is – they make up the *nature* or *essence* of such a world. This world, for instance, could not be the way it is without possessing its essential features. 'Something is a "natural" necessity in a given world, when it is true in all possible worlds with the same essence, or "nature"' (Bigelow [1990: 214]). Worlds which possess the same essence are accessible from each other. The accessibility relation, in this approach, is supervenient on the intrinsic natures of the worlds involved;

depending on how 'similar' the intrinsic natures are between worlds, the accessibility relation comes in degrees. If only minor changes are required to transform one world into another, these worlds are easily accessible from each other. On the other hand, two worlds could not differ in their accessibility relations, unless there were a difference in the universals which are instantiated in these two worlds.

The degree of accessibility between two worlds should be determined by the intrinsic natures of those worlds. Each world arises from a recombination of individuals in the actual world with the various quantitative properties and relation in the actual world, such as mass, charge, relative velocity, and force. More specifically, each world is a structural universal, standing in a host of internal relations to its own constituent universals and to other possible worlds. The degree of accessibility between worlds will be a function of the proportions holding between the different quantities assigned to the same individuals in these different worlds. (…) Thus, accessibility will be a quantity, a determinable, and each determinate degree of accessibility will be a structural universal standing in a host of internal relations to the worlds it relates and to various other universals. (Bigelow/Pargetter [1990: 263; cf. pp. 122, 239]; Pargetter [1984]).

What has been said so far is strictly speaking true only of fundamental laws. Derived laws are not true in all accessible worlds, for they depend on boundary conditions which may be specific to one particular world. Still, it is possible to explain the lawful status of derived laws along the lines of fundamental laws, 'since any proposition which is a derived law in world w_0 will be entailed by the fundamental laws of w_0 together with the particular facts $q_1, \ldots q_m$.' (Bigelow/Pargetter [1990: 257]; Pargetter [1984: 345-6]).

Of course, this programme involves a heavy ontological commitment and has, as yet, paid scarce attention to the epistemological features of the laws of science. And even among Necessitarians there are those who remain sceptical about the stipulation of possible worlds. They have sought an explanation of laws of nature in terms of real existents in the actual world, but within the framework of nomic necessity.

3.3. A natural kinds account of laws of nature

Lit.: Bigelow/Ellis/Lierse [1992]; Ellis [forthcoming], Leckey/Bigelow [this volume, p. 114f]

The domain on which the analysis is to be carried out is that of real existent things. The notion of nomic necessity is not to be taken as a primitive but as supervenient on the essential nature of things. In particular, laws of nature, which are a subclass of nomic necessities, are grounded in the essential nature of things.

Such an approach offers an advantage over the original idea that laws of nature are contingent relations of necessitation between universals. This relation of necessitation is no longer an unanalysable primitive but a consequence of the essential nature of things. If, for an individual, or rather for a type of individual, to be, say, A, it must possess a certain nature, then this nature is essential to its identity, and everything which possesses this nature must be A. For instance, the atomic number of gold is 79: nothing could be gold unless it possessed this atomic number. It is also part of the nature of a metal that it conducts electricity. Hence, possessing a certain nature necessitates some of the physical behaviour of things, other things being equal. Ultimately, the laws of nature can be derived from the essential constituents of the things and structures in the physical world.

In particular, the world consists of natural kinds – things like copper, fields and tigers. And it is natural kinds, with their essential properties which should be used to explain the laws of nature. Laws of nature, it is claimed, 'derive from the attribution of essential properties to things.' (Bigelow/Ellis/Lierse [1992: 373]) Hence the nomic necessity, which essential properties of natural kinds possess, is bestowed on the laws of nature.

Laws of nature, we argue, are truths whose necessity is grounded in the essential properties of this world and the things in it. Hence, it is not the relation between universals that constitutes the necessity of laws, but rather, their necessity results from the essential natures of the properties on which the nomological relation supervenes. (...)

How is it possible for the laws of nature to arise from essential properties? This can be explained by direct appeal to the nature of a property itself. We argue that included among the essential properties *of a property* is the *propensity* or disposition of whatever possesses it to display a particular kind of behaviour in a specific kind of context. What science observes and codifies are the manifestations of these dispositions. (...)

This means that, given the essential natures of the kinds of things said to be nomically connected, the supervenient relation (law) must *necessarily* follow. Hence, there is no possible world where things of these kind exist, but the supervening (nomic) relation fails to hold. (Bigelow/Ellis/Lierse [1992: 378]).

As it is altogether contingent which natural kinds exist in a world, the laws of nature, too, are contingent on which kinds of things exist. The necessity possessed by the laws is conditional on the natural kinds from which they derive.

The concept of conditional physical necessity which this account implies – necessity in terms of entailments – is not identical with a sense of relative necessity which the Regularity theorist may want to embrace:

The proposition *p* is physically necessary if and only if *p* is implied (entailed) by the set of physical laws. (Swartz [1985: 46]; Lewis [1979]; [1986]).

Thus, the Regularist explains natural necessity by reference to laws. If occurrent facts fall under a set of laws, then what happens to them is entailed by these laws. Hence, the Regularist is willing to accept that this stone, subject to a set of laws, must break this window, given appropriate boundary conditions. What is physically necessary is only the entailment of singular facts acting under the constraint of laws.

Necessitarian accounts explain the laws by reference to natural necessities. The natural laws are entailments of essential properties of worlds or natural kinds or of natural states. What is related by the natural necessities are second-order things, like properties, not the particular to the general. (Cf. Bigelow/Pargetter [1990: § 5.2]; Bigelow/Ellis/Lierse [1992: 379]; Ellis [forthcoming]; Leckley/Bigelow [Ch. II. pp. 92-119])

Ellis's approach marks a significant departure from the traditional empiricist ontology in that it makes laws dependent, not on occurrent facts, but on natural kinds, including fields and structures. One problem with making the existence of laws supervenient on the existence of essential properties (of existing natural kinds) is that many laws (of science) deal with accidental properties. For instance, the distance an object will travel under a certain acceleration is not part of its essential properties. In addition, the concept of natural kind is subject to different interpretations and science also deals with things we would not normally accept as natural kinds.

The Ellis approach brings us closer to what seems to be a neglected explanation of laws of nature, one which will be dubbed 'Structural Approach' and is understood here as a middle course between the Regularity View and Necessitarianism.

4. *The Structural Approach*

Lit.: Bhaskar [1978]; Cartwright [1983], [1989], [1992]; Chalmers [1993]; Harré/Madden [1975]; Laymon [1989]; Nehrlich/Suchting [1967]; Popper [1959: Appendix X]; Popper [1967]; Suchting [1969]; Weinert [1993].

As has been emphasized throughout the preceding discussion, ontological attitudes are inseparable from considerations of laws of nature. Often, deficiencies in one account of laws will lead to an enrichment of the ontology to overcome the defect. Thus in a revealing passage, John Earman admits that the empiricist ontology of singular occurrent facts may have to undergo a radical modification if the quantum picture of reality is correct, i.e.,

> if the familiar classical ontology has to give way to an ontology of shimmering potentialities which from time to time actualize themselves in patterns of occurrent events which can only be described statistically. (Earman [1984: 202]).

However, short of evoking the problem of the reduction of the wave packet, it must be recognized that the thrust of the Necessitarian objection against the Regularity view is that the accepted ontology of empiricism is not sufficiently rich to account for the existence of laws of nature in the real world. This concern is shared by the Structural approach. But there is a striking difference between the two: while the former interprets laws of nature in a strong sense of natural necessity, either as a relation between universals, or as natural entailments by appeal to possible worlds, essential properties or natural states, the latter operates at best with a weak sense of natural necessity as an entailment of the observable by underlying structures.

Thus, the first move of the Structural approach is to enrich the ontology by what may generically be termed *structures* (although in the literature cited above other terms, like 'generative mechanisms', 'powerful' or 'active particulars', 'capacities' and 'tendencies' are employed). A second important feature of this approach is the recognition of the interrelatedness or connectivity of things in the natural world, which an appropriate scientific theory must capture. Structures, not events, are seen as ontologically primary; structures generate the regular patterns and phenomena which are observable in the world.

One consequence of the stipulation of structures is an agreement between the Structural and the Necessitarian viewpoints that law statements do not refer to existing classes of objects. Law statements are not be understood as descriptive of classes of particular objects, although they are used, of course, to describe the behaviour of such objects under specific boundary conditions. But the important point is that law statements say *more* than that for a well-specified set of objects [A], if the set A has members {a,b,c}, then each member of A obeys the structural constraints laid down in the law statement.

Rather, the laws of science can be characterized as expressing (aspects of) the structure of interrelated physical systems, i.e. networks of interrelated regularities, involving forces, fields and particles. Thus, Bernard d'Espagnat observes that

> a general agreement seems nowadays to exist among physicists that the aim of their scientific investigations is to discover structural relationships between individual "happenings". (d'Espagnat [1971: 372]; cf. Dürr [1990: 57]; Roman [1969: 363])

Fundamental laws will express the structure of many physical systems, quite independent of specific boundary conditions, while phenomenological laws will be specific to certain physical systems which are distinguished from others by a number of boundary conditions. The distinction between lawlike and lawful generalities will be made by reference to the physical structures and their interrelatedness.

The purely accidental generality, for instance, that all moas died at the age of fifty, is unrelated to any of the lawful regularities which may characterize the behaviour of living systems. Equally, the blackness of ravens, as we have seen, may be totally unrelated to the genetic structure of ravens. These accidental regularities would not be admissible as laws of nature. The accidental feature of these examples is well expressed in the logical character of accidental generalities as a conjunction of events: (p & q & r). Lawful relationships, it was argued, cannot be thus expressed, because they convey structural information. In symbolic language, this structural information is expressed in mathematical relationships, as for instance, in Gauss's law:

$$\nabla \bullet B = 0$$

i.e., the divergence of the magnetic field is always zero. This is a mathematically precise statement of the well-known fact that magnetic monopoles do not exist. According to the Structural approach, the Humean characterization of laws as conjunction of events is too weak, because events or facts do not possess ontological primacy.

Given the proposal that laws are to be seen as structural features of physical systems, the introduction of unactualized physical possibilities poses no problems, nor does the fact that laws are commonly seen as supporting counterfactuals. For a physical system may have a structure the effects of which are never manifested; still, we would be entitled to say of such a system that 'if it were actualized it would manifest certain types of behaviour.' Equally, a system with a certain structure may see its boundary conditions manipulated in such a way that it no longer manifests a certain type of behaviour observed before. For example, even if atoms were never again exposed to homogeneous magnetic fields, so that the (normal) Zeeman effect would never be observed again, the Schrödinger equation, suitably modified, would still describe the motion of such particles in homogenous magnetic fields.

One consequence of these considerations is that the distinction between fundamental and phenomenological laws becomes a continuum, and so does the distinction between accidental and true regularities. For in spite of what was claimed earlier about the accidental universality of energetically improbable states of affairs, these states are a consequence of the second law of thermodynamics and therefore do not stand in isolation. Their improbability has a reason. Should it not be said that they participate in the lawfulness of the second law? In view of the interrelatedness of the physical world this uncertainty with respect to some propositions is not a disturbing consequence of the Structural approach.

One further point is worth pointing out. We have spoken of laws of nature and laws of science without invoking classes of objects. Laws have merely been characterized in terms of structural features of physical systems. But material objects move within physical systems and are subject to their laws. We can redescribe what has been said about laws of nature from the perspective of individual objects. From the perspective of an object within the framework of a physical system (being accelerated from rest to a velocity v, crossing an electric field, making a transition from a p-state to an s-state in an atom), a law of nature can be seen as imposing structural constraints on the physical behaviour it is permitted to display. Depending on whether these structural constraints are expressed in fundamental or phenomenological laws, the object may not fall under the law. Just as the mathematician sees equations as constraints on the possibilities of solutions, the Structural Approach sees laws of nature as constraints on the physical options for the behaviour of a physical entity.

From the point of view of individual particles, laws of nature, then, are principles of impossibility (Popper [1980: 430]) or physical constraints. But from the point of view of the existence of physical systems, laws of nature are highly contingent, for, in Popper's words, 'there may be *structurally different*

ent worlds - worlds with different natural laws.' (Popper [1980: 480])

This discussion was intended as a general introduction to the problems raised by laws of nature and laws of science. It is hoped that it will make it easier for the reader to assess the arguments in the articles which follow. It is also hoped that those interested in the philosophy of science will become more aware that, quite apart from theory construction and experimental strategies, the symbolic formulation of laws is one of the central tasks of science.

(The author would like to thank Edwin Mares for critical comments on an earlier draft of this Introduction.)

Bibliography

Although this Bibliography is not comprehensive, it should give any serious student of Laws of Nature a useful starting-point. Thank you to Paul Davies, Edwin Mares and Brian Skyrms for suggesting some items for inclusion.

Achinstein, P. [1971]: *Law and Explanation.* London: Oxford UP.

Amsterdamski, St. [1975]: *Between Science and Philosophy.*Boston Studies in the Philosophy of Science XXXV. Dordrecht: Reidel.

Armstrong, D. [1979]: 'Laws of Nature'. *Proceedings of the Russellian Society* (Sydney), 4: 46-61.

Armstrong, D. [1983]: *What is a Law of Nature.* Cambridge:Cambridge UP

Armstrong, D. [1984]: 'Reply to Earman'. In: R.J. Bogdan(Ed.), *D.M. Armstrong.* Dordrecht: Reidel, pp. 262-69.

Armstrong, D. [1988]: 'Reply to van Fraassen.' *Australasian Journal of Philosophy, 66/2*, pp. 224-9.

Ayer, A.J. [1963]: 'What is a Law of Nature?' In: *The Concept of a Person.*London: Macmillan, pp. 209-34. Reprinted from 'What is a Law of Nature?' *Revue International de Philosophie, 10/2* (1956), pp. 144-65.

Baird, D. [1983]: 'Conceptions of Scientific Law'. In: Rescher, N. (Ed.) [1983],pp. 33-41.

Balashov, Y.V. [1992]: 'On the Evolution of Natural Laws.' *Brit.J.Phil.Sc., 43/3*, pp. 343-71.

Barrow, J. [1988]: *The World Within the World.* Oxford: Clarendon Press.

Barrow, J. [1991]: *Theories of Everything.* Oxford: Oxford UP.

Barrow, J. [1992]: *Pi in the Sky.* Oxford: Oxford UP.

Beatty, J. [1981]: 'What's wrong with the received view of evolutionary theory?' In: P. Asquith/R. Giere(Eds.): *PSA 1980*, Vol. 2, Philosophy of Science Association, pp. 397-426.
Berofsky, B. [1966]: 'Causality and General Laws'. *Journal of Philosophy, 63*,pp. 148-57.
Berofsky, B. [1968]: 'The Regularity Theory'. *Noûs, 2*, pp. 315-40.
Bhaskar, R. [1978]: *A Realist Theory of Science.* Sussex: Harvester.
Bigelow, J. [1990]: 'The World Essence'. *Dialogue, XXLX*,pp. 205-17.
Bigelow, J./R. Pargetter [1990]: *Science and Necessity.* Cambridge:Cambridge UP.
Bigelow, J./B/ Ellis/C. Lierse [1992]: 'The World as One of a Kind: Natural Necessity and Laws of Nature.' *Brit.J.Phil.Sc., 43/3*, pp. 371-89.
Boutroux, E. [1894/1991]: *De la Contingence des Lois de la Nature.* Paris:Presses Universitaires de France. [Engl. Translation: *The Contingency of the Laws of Nature.* Chicago: Open Court 1976].
Boyd, R. [1972]: 'Determinism, Laws and Predictability in Principle.'*Philosophy of Science, 39*, pp. 431-50.
Braithwaite, R.B. [1927]: 'The Idea of a Necessary Connexion I.' *Mind, 36*,pp. 467-77.
Braithwaite, R.B. [1928]: 'The Idea of a Necessary Connexion II.' *Mind, 37*, pp. 62-72.
Braithwaite, R.B. [1935]: 'Experience and the Laws of Nature.' *Actes du Congrès International de Philosophie Scientifique*, Sorbone, Paris 1935. *Actualités Scientifiques et Industrielles* 388-95, pp. V. 57-64.
Braithwaite, R.B. [1953]: *Scientific Explanation.* Cambridge: Cambridge UP.
Brzezinski, J. et al. [1990]: *Idealization I: General Problems.* Amsterdam/Atlanta: Rodopi.
Brzezinski, J. et al. [1990]: *Idealization II: Forms and Applications.* Amsterdam/Atlanta: Rodopi.
Brzezinski, J. et al. [1992]: *Idealization III: Approximation and Truth.* Amsterdam/Atlanta: Rodopi.
Brittan, G.G. Jr. [1978]: *Kant's Theory of Science.* Princeton (N.J.): Princeton UP.
Broad, C.D. [1935]: 'Mechanical and Teleological Causation.' *Proceedings of the Aristotelian Society,* Supplementary Volumes XIV.
Bronowski, J. [1978]: *The Origins of Knowledge and Imagination.* Yale UP.
Brown, J.R. [1991]: *The Laboratory of the Mind.* Thought Experiments in the Natural Science. London:Routledge.
Buchdahl, G. [1951]: 'Science and Logic: Some Thoughts on Newton's Second Law of Motion.' *Brit.J.Phil.Sc., 2*, pp. 217-35.
Buchdahl, G. [1967]: 'Semantic Sources of the Concept of Law.' In:Cohen/Wartofsky [1967], pp. 272-92.
Bunge, M. [1961]: 'Kinds and Criteria of Scientific Laws.' *Philosophy of Science, 28*, pp. 260-81.

Campbell, N. [1953]: *What is Science?*. New York: Dover.
Carnap, R. [1966]: *Philosophical Foundations of Physics*. New York: Basic Books. Reprinted as *An Introduction to the Philosophy of Science*. New York: Basic Books 1974. [Dt. *Einführung in die Philosophie der Naturwissenschaft*. München: Nymphenburger Verlagshandung 1969.
Carroll, J. [1987]: 'Ontology and the Laws of Nature.' *Australasian Journal of Philosophy, 65*, pp. 261-76.
Carroll, J. [1994]: *Laws of Nature*. Cambridge: Cambridge University Press.
Cartwright, N. [1979]: 'Causal Laws and Effective Strategies.' *Noûs, 13*, pp. 419-37.
Cartwright, N. [1983]: *How the Laws of Physics Lie*. Oxford: Clarendon Press.
Cartwright, N. [1989]: *Nature's Capacities and their Measurement*. Oxford: Clarendon Press.
Cartwright, N. [1992]: 'Aristotelian Natures and the Modern Experimental Method.' In: J. Earman (Ed.): *Inference, Explanation and other Frustrations*. Essays in the Philosophy of Science. Berkeley: University of California Press, pp. 44-71.
Chalmers, A. [1993]: 'So the Laws of Physics Needn't Lie.' *Australasian Journal of Philosophy 71*, pp.196-205.
Chisholm, R. [1955]: 'Law statements and counterfactual inference.' *Analysis 15*, pp. 92-105.
Cohen, L.J. [1980]: 'The Problem of Natural Laws.' In: D.Mellor (Ed.): *Prospects for Pragmatism*. Cambridge:Cambridge UP, pp. 211-28.
Cohen, R.S./M.W. Wartofsky (Eds.) [1967]: *Boston Studies in the Philosophy of Science*. Vol. III. (1964-66). In Memory of N.R.Hanson. Dordrecht: Reidel.
Cohen, R.S./M.W. Wartofsky (Eds.) [1969]: *Boston Studies in the Philosophy of Science*. Vol. V. (1966-88). Dordrecht: Reidel.
Cohen, R.S./M.W. Wartofsky (Eds.) [1974]: *Methodological and Historical Essays in the Natural and Social Sciences*. Boston Studies in the Philosophy of Science. Vol. XIV.. Dordrecht: Reidel.
Colodny, R. (Ed.) [1965]: *Beyond the Edge of Certainty*. Englewood Cliffs(N.J.): Prentice-Hall.
Colodny, R. (Ed.) [1977]: *Logic, Laws and Life*. Pittsburgh: University of Pittsburgh Press.
Comte, A. [1830-42]: *Cours de Philosophie Positive*. Paris.
Couturat, L. [1969]: *La Logique de Leibniz*. Hildesheim: Olms.
Danto, A./S. Morgenbesser (Eds.) [1960]: *Philosophy of Science*. New York: Meridian Books.
Davies, P. [1988]: 'Law and Order in the Universe.' *New Scientist*,(October 15), pp. 58-60.
Davies, P. [1989]: 'What are the Laws of Nature?' In: J. Brockman (Ed.): *The Reality Club No. 2*. Lynx Communications.
Davies, P. [1990]: 'Chaos frees the Universe'. *New Scientist* (6 October), pp. 36-9.

Davies, P. [1992]: *The Mind of God.* New York/London: Simon & Schuster.
d'Espagnat, B. [1971]: *Conceptual Foundations of Quantum Mechanics.* Menlo Park (California): W.A. Benjamin,Inc.
d'Espagnat, B. [1985]: *Une Incertaine Réalité.* Paris: Gauthier-Villars.
Deutsch, D. [1988]: 'On Wheeler's notion of "Law Without Law" in physics'. In: W./A. van der Merwe/W.A. Miller (Eds.): *Between Quantum and Cosmos.* Princeton UP, pp. 583-92.
Dretske, F.I. [1977]: 'Laws of Nature'. *Philosophy of Science, 44*, pp. 248-68.
Dretske, F.I. [1978]: 'Reply to Niiniluoto'. *Philosophy of Science, 45*, pp. 440-444.
Dürr, P. [1990]: *Das Netz des Physikers.* München: Deutscher Taschenbuch Verlag.
Duhem, P. [1954]: *The Aim and Structure of Physical Theory.* Transl. by Ph.Wiener. Princeton: Princeton UP. [Dt. *Ziel und Struktur der physikalischen Theorien.* Übers. F. Adler. Hamburg: F. Meiner 1978].
Duhem, P. [1960]: 'Physical Law'. In: Danto/Morgenbesser [1960], pp. 182-97.(Pt, II/Ch. 9 of Duhem [1954]).
Earman, J. [1984]: 'Laws of Nature. The Empiricist Challenge.' In: R.J. Bogdan (Ed.): *D.H. Armstrong.* Dordrecht: Reidel, pp. 191-224.
Earman, J. [1978]: 'The Universality of Laws.' *Philosophy of Science, 45*, pp.173-81.
Eastwood, B.S. [1967]: 'Grosseteste's "Quantitative" Law of Refraction: A Chapter in Non-Experimental Science. *Journal of the History of Ideas, 28*, pp. 403-14.
Eddington, A. [1958]: *The Nature of the Physical World.* Ann Arbor: University of Michigan Press.
Eigen, M./R. Winkler [1975]: *Das Spiel: Naturgesetze steuern den Zufall.* München: Piper. [Engl.: *Laws of the Game.* Transl. by Robert and Rita Kimber.Penguin 1983.]
Einstein, A. [1949]: 'Autobiographical Notes.' In: P.A. Schilpp (Ed.), *Albert Einstein. Philosopher-Scientist.* Vol. I. La Salle (Ill.): Open Court 1949, pp. 2-94.
Einstein, A. [1953]: 'The Laws of Science and the Laws of Ethics.' In: H. Feigl/ M. Brodbeck (Eds.): *Readings in the Philosophy of Science.* New York: Appleton-Century-Crofts, 1953, pp. 779-80.
Einstein, A./L. Infeld [1938/1971]: *The Evolution of Physics.* Cambridge: Cambridge UP.
Ellis, B. [1965]: 'The Origin and Nature of Newton's Laws of Motion.' In: R. Colodny [1965], pp. 29-68.
Ellis, B. [forthcoming]: 'Natural Kinds and Natural Kind Reasoning.'
Exner, F. [1919]: 'Über Naturgesetze' *Vorlesungen über die physikalischen Grundlagen der Physik.* Vienna: Deutike.
Feigl, H./G. Maxwell (Eds.): *Current Issues in the Philosophy of Science.* New York: Holt, Rinehart & Winston.

Fetzer, J.H. [1983]: 'Transcendent Laws and Empirical Procedures.' In: N. Rescher [1983], pp. 25-32.
Fetzer, J.H. [1981]: *Scientific Knowledge.* Dordrecht: Reidel.
Feynman, R. [1965/1992]: *The Character of Physical Law.* Penguin Books.
Fisk, M. [1970]: 'Are there necessary connections in nature?' *Philosophy of Science, 37*, pp. 385-404.
Fodor, J.: [1991]: 'You can fool some of the people all of the time, everything else being equal.' *Mind, 100*,pp. 19-34.
Forge, J. [1986]: 'David Armstrong on Functional Laws.' *Philosophy of Science, 53*, pp. 584-7.
Forrest, P. [1985]: 'What Reasons do we have for Believing there are Laws of Nature?' *Philosophical Inquiry, 7*, pp. 1-12.
Frege, G. [1891/1967]: 'Über das Trägheitsgesetz.' In: *Kleine Schriften,* hrsg. v. I. Angelelli. Hildesheim 1967. [Engl.: 'On the Law of Inertia.' Transl. and Notes by H. Jackson and E. Levy. In: R.S. Cohen/M.W.Wartofsky [1974], pp. 257-76.]
Friedman, M. [1992a]: *Kant and the Exact Sciences.* Harvard UP.
Friedman, M. [1992b]: 'Causal Laws and the Foundations of Natural Science.' In: P. Guyter (Ed.): *The Cambridge Companion to Kant.* Cambridge: Cambridge UP [1992], pp. 161-99.
Gavin, A.H. [1958]: 'General Statements as Rules of Inference.' In: H.Feigl/ G. Maxwell/M. Scriven (Eds.): *Concepts, Theories, and the Mind-Body-Problem.* Minnesota Studies in the Philosophy of Science, II. Minneapolis: University of Minesota Press, 1958.
Ghiselin, M.T. [1989]: 'Individuality, History and Laws of Nature.' In: M. Ruse (Ed.): *What the Philosophy of Biology Is.* Dordrecht: Kluwer, pp. 53-66.
Giere, R. [1979]: 'Propensity and Necessity.' *Synthese, 40*, pp. 439-51.
Giere, R. [1988a]: *Explaining Science.* Chicago/London: University of Chicago Press.
Giere, R. [1988b]: 'Laws, Theories and Generalizations.' In: Grünbaum/Salmon [1988], pp. 37-46.
Goodman, N. [1955]: *Facts, Fiction and Forecast.* New York: Bobbs-Merrill.
Gould, St.J. [1989]: *Wonderful Life.* London: Hutchinson.
Grünbaum, A. [1961]: 'Law and Convention in Physical Theory.' In: Feigl/ Maxwell [1961], pp. 140-55.
Grünbaum, A./W.C. Salmon (Eds.) [1988]: *The Limitations of Deductivism.* Berkeley: University of California Press.
Haken, H./A. Wunderlin [1990]: 'Le Chaos Déterministe.' *La Recherche, 225*, (October), pp. 1248-55.
Hall, M.B. [1970]: *Nature and Nature's Laws.* Documents of the Scientific Revolution. New York: Walker.
Hanson, N. [1958/1965]: *Patterns of Discovery.* Cambridge:Cambridge UP.

Hanson, N. [1965]: 'Newton's First Law: A Philosopher's Door into Natural Philosophy.' In: R. Colodny (Ed.):*Beyond the Edge of Certainty.* Englewood Cliffs(N.J.): Prentice-Hall, pp. 6-28.
Hanson, N. [1969]: *Perception and Discovery: An Introduction to Scientific Inquiry.* San Francisco: Freeman, Cooper & Co.
Harré, R. [1960]: *Introduction to the Logic of the Sciences.* London: Macmillan.
Harré, R./E.M. Madden [1975]: *Causal Powers.* A Theory of Natural Necessity. Totowa (N.J.): Rowman and Littlefield.
Hausman, A. [1967]: 'Hume's Theory of Relations.' *Noûs, 1/3*, pp. 255-83.
Heisenberg, W. [1970]: *Natural Law and the Structure of Matter.* London: Rebel Press.
Helmer, O./N. Rescher [1959]: 'On the epistemology of the inexact sciences.' *Management Science, 6*, pp. 25-52.
Hempel, C. [1965]: *Aspects of Scientific Explanation.* New York: Free Press.
Hempel, C. [1966]: *Philosophy of Natural Science.* Englewood Cliffs (N.J.): Prentice Hall. [Dt. *Philosophie der Naturwissenschaften.* Übers. W. Lenzen. München: Deutscher Taschenbuchverlag. 1974]
Hempel, C./P. Oppenheim [1960]: 'Problems of the Concept of General Law.' In: Danto/Morgenbesser [1960], pp. 198-204.
Hempel, C. [1988]: 'Provisos: A Problem Concerning the Inferential Function of Scientific Theories.' In: Grünbaum/Salmon [1988], pp. 19-36
Hesse, M. [1980]: 'A Revised Regularity View of Scientific Laws.' In: D. Mellor [1980], pp. 87-103.
Hetherington, S. [1983]: 'Tooley's Theory of Laws of Nature.' *Canadian Journal of Philosophy, 13*, pp. 101-6.
Hitchcock, Ch. [1992]: 'Urbach on Laws of Nature.' *Analysis, 52/2*, pp. 61-4.
Hochberg, H. [1981]: 'Natural Necessity and Laws of Nature.' *Philosophy of Science, 48*, pp. 386-99.
Holton, G. [²1973]: *Introduction to Concepts and Theories in Physical Science.*(Revised and with new material by St. G. Brush). Addison Wesley.
Holton, G./D.H. Roller [1958]: *Foundations of Modern Physical Science.* Addison Wesley.
Home, R. [1968]: 'The Third Law in Newton's Mechanics.' *The British Journal for the History of Science, 4*, pp. 39-51.
Hull, D.L. [1974]: *Philosophy of Biological Science.* Englewood Cliffs (N.J.) Prentice-Hall.
Hume, D. [1739/1973]: *A Treatise of Human Nature.* (Edited and with an Analytical Index by L.A. Selby-Bigge.). Oxford: Clarendon Press.
Hund, F. [²1978]: *Geschichte der physikalischen Begriffe.* 2 Vols.. Mannheim/Wien/Zürich: Bibliographisches Institut.
Hunt, G.M.K. [1992]: 'Laws: projectibility and uniformity.' *International Studies in the Philosophy of Science, 6.*
Jobe, E.K. [1967]: 'Some Recent Work on the Problem of Law.' *Philosophy of Science, 34*, pp. 363-81.

Kane, J.W./M.M. Sternheim [[2]1984]: *Physics.* New York: John Wiley & Sons.
Kant, I, [1787/1971]: *Kritik der reinen Vernunft.* Hamburg: Felix Meiner [Engl. *Critique of Pure Reason.* Transl. by Norman Kemp Smith. London: Macmillan 1976].
Kitcher, Ph./W.C. Salmon (Eds.) [1989]: *Scientific Explanation.* Minnesota Studies in the Philosophy of Science. Vol. 13. Minneapolis: University of Minnesota Press.
Kneale, W. [1949]: *Probability and Induction.* Oxford: Clarendon Press.
Kneale, W. [1950]: 'Natural Laws and Contrary-to-Fact Conditionals.' *Analysis, 10*, pp. 121-5. Reprinted in M. MacDonald (Ed.) [1954]: *Philosophy and Analysis.* Oxford: Blackwell.
Kneale, W. [1961]: 'Universality and Necessity.' *Brit.J.Ph.Sc., 12*, pp. 89-102.
Körner, St. [1963]: 'On Laws of Nature.' *Mind, 62*, pp. 218-29.
Krajewski, W. [1968]: 'Das Naturgesetz als notwendiger Zusammenhang.' In: Kröber [1968].
Krajewski, W. [1969]: *Concepts of Laws of Science at the End of the XIXth Century.* Wroclaw (In Polish).
Krajewki, W. [1974]: 'The Idea of Statistical Law in 19th Century Science.' In: R.S. Cohen/M.W. Wartofsky [1974], pp. 397-405.
Krajewski, W. [1977]: *Correspondence Principle and Growth of Knowledge.* Dordrecht: Reidel.
Kröber, G. (Ed.) [1968]: *Der Gesetzesbegriff in der Philosophie und den Einzelwissenschaften.* Berlin.
Lambert, K./G. Brittan [1970]: *Introduction to the Philosophy of Science.* Englewood Cliffs (N.J.): Prentice Hall. [3]1987 Atascadero (Ca.): Ridgeview.
Lange, M. [1992]: 'Armstrong and Dretske on the Explanatory Power of Regularities.' *Analysis, 52/3*, pp. 154-9.
Lange, M. [1993]: 'Lawlikeness.' *Noûs, 27/1*, pp. 1-21.
Lauter, H.A. [1970]: 'An Examination of Reichenbach on Laws.' *Philosophy of Science, 37*, pp. 131-45.
Laymon, R. [1989]: 'Cartwright and the Lying Laws of Physics.' *The Journal of Philosophy, 86/7*, pp. 353-72.
Lewis, D. [1973]: *Counterfactuals.* Oxford: Basil Blackwell.
Lewis, D. [1979]: 'Counterfactual Dependence and Time's Arrow.' *Noûs, 13*, pp. 455-76. Reprinted in *Philosophical Papers II.* Oxford: Blackwell 1986, pp. 32-52.
Lewis, D. [1983]: 'New Work for the Theory of Universals.' *Australasian Journal of Philosophy, 61/4*, pp. 343-77.
Lewis, D. [1986]: *On the Plurality of Worlds.* Oxford: Basil Blackwell.
Mach, E. [1908]: 'Sinn und Wert der Naturgesetze.' In: *Erkenntnis und Irrtum.* Leipzig: Barth, pp. 449-63. Abridged Transl.: 'The Significance and Purpose of Natural Laws.' In: A. Danto/M. Morgenbesser [1960], pp. 266-73.
Mackie, J.L. [1962]: 'Counterfactuals and Causal Laws.' In: R.S. Butler (Ed.): *Analytic Philosophy.* New York: Barnes & Noble (1962), pp. 66-80.

Mackie, J.L. [1973]: *Truth, Probability and Paradox.* Oxford: Clarendon Press.
Maxwell, N. [1968]: 'Can there be Necessary Connections between Events?' *Brit. J. Phil. Sc., 19*, pp. 1-25.
Mayr, E. [1982]: *The Growth of Biological Thought.* Cambridge (Mass.): Harvard UP.
Mayr, E. [1987a]: 'The ontological status of species: scientific progress and philosophical terminology.' *Biology and Philosophy, 2*, pp. 145-66.
Mayr, E. [1987b]: 'Answers to these Comments.' *Biology and Philosophy, 2*,pp. 212-20.
Mellor, D.H. [1971]: *The Matter of Chance.* Cambridge: Cambridge UP.
Mellor, D.H. [1980]: 'Necessities and Universals in Natural Laws.' In: D. Mellor [1980], pp. 105-26.
Mellor, D.H. (Ed.) [1980]: *Science, Belief and Behaviour.* Essays in Honour of R.B. Braithwaite. Cambridge: Cambridge UP.
Menger, C. [1893/1933]: *The Collected Works of Carl Menger.* Vol. II: *Untersuchungen über die Methode der Socialwissenschaften, und der Politischen Ökonomie insbesondere.* Leipzig: Duncker & Humboldt (London School of Economics Reprints. No. 18)
Mill, J.S. [1868]: *A System of Logic.* London: Longmans, Green, Reader & Dyer.
Milton, J.R. [1981]: 'The Origin and Development of the Concepts of the Laws of Nature.' *Arch. Europ. Sociol., XXII*, pp. 173-95.
Milne, E.A. [1937]: 'On the Origin of Laws of Nature.' *Nature, 139* (June), pp. 997-99.
Molnar, G. [1969]: 'Kneale's Argument Revisited.' *Philosophical Review, 78*, pp. 79-86.
Mundy, B. [1987]: 'The Metaphysics of Quantity.' *Philosophical Studies, 51*, pp. 29-54.
Musgrave, A. [1979-80]: 'Wittgensteinian Instrumentalism.' *Theoria, 45/6*, pp. 65-105.
Musgrave, A. [forthcoming]: 'Realism and Idealisation.' In: J. Mizieck (Ed.): *Philosophy of Science.* The Polish Conferences. (Boston Studies in the Philosophy of Science). Dordrecht: Kluwer.
Nagel, E. [1961]: *The Structure of Science.* New York: Harcourt, Brace & World.
Nehrlich, G.C./W.A. Suchting [1967]: 'Popper on Law and Natural Necessity. *Brit. J. Phil. Sc., 17*, pp. 233-5.
Nernst, W. [1893]: *Theoretische Chemie.* Stuttgart: Enke.[Engl.: *Theoretical Chemistry.* London 1911].
Nernst, E. [1922]: 'Zum Gültigkeitsbereich der Naturgesetze.' *Die Naturwissenschaften, 10*, pp. 489-95.
Niiniluoto, I. [1978]: 'Dretske on Laws of Nature.' *Philosophy of Science, 45*, 431-39.
Nozick, R. [1981]: *Philosophical Explanations.* Oxford: Clarendon.
Nugayev, R. [1992]: 'The fundamental laws of physics can tell the truth.' *International Studies in the Philosophy of Science, 6.*

Öhlschläger, G. [1974]: 'Einige Unterschiede zwischen Naturgesetzen und sozialen Regeln.' In: H.J. Heringer (Ed.): *Der Regelbegriff in der praktischen Semantik.* Frankfurt a./M.: Suhrkamp, pp. 88-110.

Pap, A. [1962]: *An Introduction to the Philosophy of Science.* Glencoe (Ill.): The Free Press of Glencoe.

Papineau, D. [1986]: 'Laws and Accidents.' In: G. Macdonald/C. Wright(Eds.): *Fact, Science and Morality.* Essays on A.J. Ayer's *Language, Truth and Logic.* Oxford: Basil Blackwell, pp. 189-217.

Pargetter, R. [1984]: 'Laws and Modal Realism.' *Philosophical Studies, 46,* pp. 335-47.

Passmore, J. [1946]: 'Prediction and Scientific Law.' *Australasian Journal of Philosophy, 24,* pp. 1-33.

Pearson, K. [1892]: *The Grammar of Science.* London: Walter Scott.

Peierls, R.E. [1955]: *The Laws of Nature.* London: George Allen & Unwin.

Peirce, Ch.S. [1934]: 'The Reality of Thirdness.' In: *Collected Papers of Charles Sanders Peirce,* ed. by Ch. Hartshorne/P. Weiss. Vol. V: *Pragmatism and Pragmaticism.* Cambridge: Harvard UP, pp. 64-76.

Peirce, Ch.S. [1958]: 'The Laws of Nature and Hume's Argument Against Miracles.' *Values in a Universe of Chance,* ed. by P.P. Wiener Stanford: Stanford UP.

Penrose, R. [1989]: *The Emperor's New Mind.* Oxford: Oxford UP.

Pietarinen, J. [1972]: *Lawlikeness, Analogy and Inductive Logic.* Amsterdam: North Holland.

Pietroski, P./G. Rey [1995]: 'when Other Things Aren't Equal: Saving *Ceteris Paribus* Laws from Vacuity.' *Brit. J. Phil. Sc.* , 46, pp. 81-110.

Planck, M. [1933]: *Where is Science Going?* London: George Allen & Unwin.

Planck, M. [1949/1968]: *Scientific Autobiography, and other papers.* New York: Greenwood Press.

Planck, M. [1965]: *Vorträge und Erinnerungen.* Darmstadt: Wissenschaftliche Buchgesellschaft.

Poincaré, H. [1952]: *Science and Hypothesis.* New York: Dover [*La Science et L'Hypothèse.* Paris 1902.]

Poincaré, H. [1963]: 'The Evolution of Laws.' In: *Mathematics and Science: Last Essays (Dernière Pensée,* 1913), transl. by J.W. Bolduc. New York: Dover, pp. 1-14.

Popper, K. [1949]: 'A note on natural laws and so-called "contrary-to-fact conditionals".' *Mind, 58,* pp. 62-6.

Popper, K. [1957]: *The Poverty of Historicism.* London: Routledge & Kegan Paul.

Popper, K. [1959]: 'Universals, Dispositions and Natural or Physical Necessity.' In: *The Logic of Discovery.* London: Hutchinson, pp. 420-41. [Dt.: 'Universalien, Dispositionen und Naturnotwendigkeit.' In: *Logik der Forschung.* Tübingen: J.C.B. Mohr (Paul Siebeck) ^{4}1971, pp. 375-96.]

Popper, K. [1967]: 'A Revised Definition of Natural Necessity.' *Brit. J. Phil.Sc., 18*, pp. 316-21.

Presley, C.F. [1954]: 'Laws and Theories in the Physical Sciences.' *Australasian Journal of Philosophy, XXXII*, pp. 79-103. Reprinted in Danto/Morgenbesser [1960], pp. 205-25.

Prigogine, I./I. Stengers [1979]: *La Nouvelle Alliance.* Métamorphose de la science. Paris: Édtions Gallimard. [Engl. *Order Out of Chaos.* Foreword by Alvin Toffler. London: Heinemann 1984]

Putnam, H. [1970]: 'On Properties.' In: N. Rescher et al. (Ed.) [1970]: *Essays in Honor of Carl G. Hempel.* Dordrecht: Reidel. Reprinted in H. Putnam: *Mathematics, Matter and Method.* Philosophical Papers Vol. 1. Cambridge: Cambridge UP 1975, pp. 305-22.

Ramsey, F.P. [1931]: *Foundations of Mathematics and other Logical Essays.* (Ed. by R.B. Braithwaite) London: Routledge & Kegan Paul.

Reichenbach, H. [1947]: *Elements of Symbolic Logic.* New York:Macmillan.

Reichenbach, H. [1954]: *Nomological Statements and Admissible Operations.* Amsterdam: North-Holland Publishing Company. Reissued as *Laws, Modalities and Counterfactuals.* Berkeley: University of California Press 1976.

Rensch, B. [1960]: 'The Laws of Evolution.' In: *Evolution after Darwin.* Vol. I., Sol Tax. Chicago: Chicago UP, pp. 95-116

Rescher, N. [1970]: *Scientific Explanation.* New York: Free Press.

Rescher, N. [1973]: *Conceptual Idealism.* Oxford: Basil Blackwell.

Rescher, N. (Ed.) [1983]: *The Limits of Lawfulness.* Studies in the Scope and Nature of Scientific Knowledge. University of America Press.

Rosenberg, A. [1985]: *The Structure of Biological Science.* Cambridge: Cambridge UP.

Roman, P. [1969]: 'Symmetry in Physics.' In: Cohen/Wartofsky [1969], pp. 363-9.

Rothman, M.A. [1963]: *The Laws of Physics.* New York: Basic Books.

Rothman, M. A. [1972]: *Discovering the Natural Laws.* The Experimental Basis of Physics. New York: Doubleday & Co.

Rudolph, E. [1991]: 'Naturgesetze und gegenständliche Realität.' *Philosophia Naturalis, 28/1.*

Ruse, M. [1970]: 'Are there Laws in Biology?' *Australasian Journal of Philosophy, 48*, pp. 234-46.

Ruse, M. [1973]: *The Philosophy of Biology.* London: Hutchinson University Library.

Ruse, M. [1977]: 'Is biology different from physics?' In: R. Colodny [1977], pp. 89-127.

Russell, J. [1964]: 'Kepler's Laws of Planetary Motion, 1609-66.' *British Journal for the History of Science, 1*, pp. 1-24.

Ryle, G. [1949]: *The Concept of Mind.* London: Hutchinson.

Salmon, W.C. [1976]: 'Laws, Modalities and Counterfactuals.' *Synthese, 35*, pp. 191-229. Reprinted in Salmon [1979], pp. 655-96.
Salmon, W.C. (Ed.)[1979]: *Hans Reichenbach.* Logical Empiricist. Dordrecht: Reidel.
Salmon, W.C. [1984]: *Scientific Explanation and the Causal Structure of the World.* Princeton: Princeton UP.
Scheibe, E. [1988]: 'The Physicist's Conception of Progress.' *Studies in the History and Philosophy of Science, 19/2*, pp. 141-59.
Schiffer, S. [1991]: 'Ceteris Paribus Laws.' *Mind, 100*, pp. 1-17.
Schlick, M. [1931]: 'Die Kausalität in der gegenwärtigen Physik.' *Die Naturwissenschaften, 19.*[Engl. Transl.: 'Causality in Contemporary Physics.' *Brit. J. Phil. Sc., 12*, 1962, pp. 177-93.]
Schlick, M. [1953]: 'Are Laws Conventions?' In: H. Feigl/M. Brodbeck (Eds.): *Readings in the Philosophy of Science.* New York: Appleton-Century-Crofts. Inc., 1953, pp. 181-8.
Schrödinger, E. [1967]: *What Is Life?* (1944) Cambridge: Cambridge UP.
Schrödinger, E. [1987]: 'Was ist ein Naturgesetz?' (1922) In: Schrödinger, E.: *Was ist ein Naturgesetz*, München: Odenbourg, pp. 9-17.
Scriven, M. [1959]: 'Explanation and Prediction and Evolutionary Theory.' *Science, 130*, pp. 477-82.
Scriven, M. [1961]: 'The Key Property of Physical Law – Inaccuracy.' In: Feigl/ Maxwell [1961], pp. 91-104.
Scriven, M. [1962]: 'Explanations, Predictions and Laws.' In: H. Feigl/ G. Maxwell (Eds.): *Scientific Explanation, Space and Time.* Minnesota Studies in the Philosophy of Science, Vol. 3. Minneapolis, 1962, pp. 170-230.
Sellars, W. [1948]: 'Concepts as Involving Laws and Inconceivable without them.' *Philosophy of Science, 15*, pp. 287-315.
Sellars, W. [1958]: 'Counterfactuals, Dispositions, and the Causal Modalities.' In: H. Feigl/M. Scriven/G. Maxwell (Eds.): Minnesota Studies in the Philosophy of Science. University of Minnesota Press, 1958, pp. 225-308.
Simpson,G.G. [1963]: 'Historical Science.' In: C.C. Albritton Jr. (Ed.):*The Fabric of Geology.* Stanford (Cal.): Freeman, Cooper & Co., pp. 24-48.
Skyrms, B. [1966]: 'Nomological Necessity and the Paradoxes of Confirmation.' *Philosophy of Science, 33*, pp. 230-49.
Skyrms, B. [1975]: 'Physical Law and the Nature of Physical Reduction.' In: G. Maxwell/Anderson (Eds.): Minnesota Studies in the Philosophy of Science, Vol. 7. Minneapolis: University of Minnesota Press, pp. 496-529.
Skyrms, B. [1977]: 'Resiliency, Propensity and Causal Necessity.' *Journal of Philosophy, 74*, pp. 704-13
Skyrms, B. [1980]: *Causal Necessity.* New Haven/London: Yale UP.
Skyrms, B. [1984]: *Pragmatics and Empiricism.* New Haven/London: Yale UP.
Smart, J.J.C. [1963]: *Philosophy and Scientific Realism.* London: Routledge & Kegan Paul.

Smart, J.J.C. [1968]: *Between Science and Philosophy.* New York: Random House.
Smart, J.J.C. [1977]: 'Cosmic Coincidence.' In: *Proceedings of the Russellian Society,* (Sydney), 2.
Smart, J.J.C. [1985]: 'Laws of Nature and Cosmic Coincidences.' *Philosophical Quarterly, 35,* pp. 272-80.
Sober, E. [1984]: *The Nature of Selection.* Cambridge (Mass.)/London (England): MIT Press.
Sober, E. [1993]: *Philosophy of Biology.* Boulder/San Francisco: Westview Press.
Steen van der, W.J./H. Kamminga [1991]: 'Laws and Natural History in Biology.' *Brit. J. Phil. Sc., 42,* pp. 445-67.
Stegmüller, W. [1958/9]: 'Conditio Irrealis, Dispositionen, Naturgesetze und Induktion.' *Kant-Studien, 50,* pp. 363-90.
Stegmüller, W. [1966]: 'Der Begriff des Naturgesetzes.' *Studium Generale, 19,* pp. 649-57.
Strawson, P. [1952]: *Introduction to Logical Theory.* London: Methuen & Co.
Suchting, W.A. [1968]: 'Functional Laws and the Regularity Theory.' *Analysis, 29,* pp. 50-1.
Suchting, W.A. [1969]: 'Popper's Revised Definition of Natural Necessity.' *Brit. J. Phil. Sc., 20,* pp. 349-52.
Suchting, W.A. [1974]: 'Regularity and Laws.' In: R.S. Cohen/M. W.Wartofsky [1974], pp. 73-90.
Suppe, F. [1989]: *The Semantic Conception of Theories and Scientific Realism.* Chicago: University of Illinois Press.
Swartz, N. [1985]: *The Concept of Physical Law.* Cambridge: Cambridge UP.
Swoyer, C. [1982]: 'The Nature of Natural Laws.' *Australasian Journal of Philosophy, 60,* pp. 202-23.
Swoyer, C. [1987]: 'The Metaphysics of Measurement.' In: John Forge (Ed.): *Measurement, Realism and Objectivity.* Dordrecht: Reidel, pp. 235-90.
Temple, D. [1983]: 'Can Science Know What's Necessary?' In: N. Rescher [1983], pp. 43-50.
Thöle, B. [1990]: *Kant und das Problem der Gesetzmäßigkeit der Natur.* Berlin: Walter de Gruyter.
Tooley, M. [1977]: 'The Nature of Laws.' *Canadian Journal of Philosophy, VII/4,* pp. 667-98.
Tooley, M. [1987]: *Causation.* A Realist Approach. Oxford: Oxford UP.
Tondl, L. [1973]: *Scientific Procedures.* Boston Studies in the Philosophy of Science, X. Dordrecht: Reidel.
Toulmin, St. [1953]: *The Philosophy of Science.* An Introduction. London: Hutchinson. [Dt. *Einführung in die Philosophie der Wissenschaften.* Göttingen: Vandenhoeck & Ruprecht. O.J.]
Urbach, P. [1988]: 'What is a Law of Nature? A Humean Answer.' *Brit. J. Phil. Sc., 39,* pp. 193-210.
Urbach, P. [1992]: 'Reply to Hitchcock.' *Analysis, 52/2,* pp. 65-8.

Vallentyne, P. [1988]: 'Explicating Lawhood.' *Philosophy of Science, 55*, pp. 598-613.
van Inwagen, P. [1979]: 'Laws and Counterfactuals.' *Noûs, 13/4*, pp. 439-53.
van Fraassen, B.C. [1977]: 'The Only Necessity Is Verbal Necessity.' *Journal of Philosophy, 74*, pp. 71-85.
van Fraassen, B.C. [1981]: 'Essences and Laws'. In: R. Healey (Ed.): *Reduction, Time and Reality.* Cambridge: Cambridge UP, pp. 189-200.
van Fraassen, B.C. [1987]: 'Armstrong on Laws and Probabilities.' *Australasian Journal of Philosophy, 65*, pp. 243-60.
van Fraassen, B.C. [1989]: *Laws and Symmetry.* Oxford: Clarendon Press.
von Weizsäcker, C.F. [1971]: *Die Einheit der Natur.* München: Hanser.[Engl. *The Unity of Nature.* New York: Farrar, Strauss, Giroux, 1980].
von Wright, G.H. [1984]: 'Laws of Nature.' In: *Truth, Knowledge and Modality.* Philosophical Papers. Volume III. Oxford: Basil Blackwell, pp. 134-49.
Walters, R.S. [1968]: 'Laws of Science and Lawlike Statements.' In: P.Edwards (Ed.): *Encyclopedia of Philosophy.* Volume 4. New York: Macmillan.
Wartofsky, M. [1968]: *Conceptual Foundations of Scientific Thought.* New York: Macmillan.
Watson, W.H. [1938]: *On Understanding Physics.* London
Weinberg, St. [1993]: *Dreams of a Final Theory.* New York: Pantheon Books.
Weinert, F. [1993]: 'Laws of Nature. A Structural Approach.' *Philosophia Naturalis, 30/2*, pp. 147-71.
Weinert, F. [1994]: 'The Correspondence Principle and the Closure of Theories: Two Incompatible Aspects of Heisenberg's Philosophy of Science.' *Erkenntnis* 40, pp. 303-23.
Weinert, F. [forthcoming]: 'Constraints on Constructibility: A Role for Fundamental Constants and Null Experiments.'
Wheeler, J.A. [1983]: 'On Recognizing "Law without Law"'. *Am. J. Phys., 51/5* (May), pp. 398-404.
Whitehead, N. [1925]: *Science and the Modern World.* London: Macmillan.
Wigner, E.P. [1964]: 'Laws of Nature and Invariance Principles.' *Science, 145*, pp. 995-8.
Wilson, C. [1969]: 'From Kepler's Laws, So-Called, to Universal Gravitation: Empirical Facts. *Archive for History of Exact Science, 6*, pp. 89-120.
Wittgenstein, L. [1922]: *Tractatus.* Oxford: Basil Blackwell.
Yolton, J.W.[1958]: 'Locke on the Law of Nature.' *Philosophical Review, 67*, pp. 477-98.
Zilsel, E. [1942]: 'The Genesis of the Concept of Scientific Law.' *Philosophical Review, 51*, pp. 245-67. [Dt.: 'Die Entstehung des Begriffs des physikalischen Gesetzes.' In: E. Zilsel: *Die Sozialen Ursprünge der neuzeitlichen Wissenschaft.* Frankfurt: Suhrkamp 1976, pp. 66-97]

II. Philosophical Views

Norman Swartz

A Neo-Humean Perspective: Laws as Regularities

Introduction[1]

I was seven or eight years old. In Hebrew school we had just learned the Aleph-Bet and were, haltingly, beginning to sound out words. As we spoke the ancient text, our teacher translated: "... And God said: 'Let there be light.' And there was light. ..."[2] Here was magic; here was the supernatural; here was the creation of the universe. I resonated to the story. I was filled with wonder, far more than had ever been elicited by any fairy tale my parents had read to me. I pictured in my imagination the majesty of it: God speaks and Nature obeys.

But I did not *believe* the story that was unfolding.[3] I was certainly enraptured by it; but I was not seduced by it. I reveled

1 It has been ten years since I wrote *The Concept of Physical Law*. In the meantime I have been working on other research. Friedel Weinert's kind invitation to me to contribute this paper affords me a welcome opportunity to produce what is really a Postscript to that book. The principal arguments for the Regularity Theory, developed more leisurely and in greater detail, are to be found therein. Here I will try to present the theory briefly, using the occasion to air some new arguments and (without mentioning their names explicitly) reply to some of my critics.

2 Genesis 1:3, *The Holy Scriptures According to the Masoretic Text*, 1955. (Philadelphia: The Jewish Publication Society of America)

3 Our Hebrew school teachers – throughout a formal religious training that lasted another eight years or so – never once asked me or my classmates whether we believed what we read in the Bible. The goal was always *understanding* the text, not believing it (which, of course, is not to say that we were expected to *dis*believe it); in short, it was never an issue whether or not we believed what we were reading. (It comes as something of a shock for some Christians to learn that Judaism is a religion whose criteria for membership and standing do not *mandate* theological beliefs.)

in the imagery while harboring a conviction, one that would strengthen in the ensuing years, that it was false.

The fable of Genesis is one of the most enduring of Western Civilization. It is so much a part of the fabric of our collective thinking that in its modern-day guises it is all but invisible. But parts of the biblical Creation-myth do persist. The fable survives, even in the thinking of many persons who believe that they have divested themselves of any shred of a theistic or supernatural cosmology, persons who are certain that they have adopted a thoroughgoing materialistic naturalism.

Sometimes ancient mythology is not vanquished: it merely accommodates itself to the prevailing temper of the times and 'goes underground' to form the unarticulated presuppositions of one's metaphysics.[4]

What exactly is the 'physics' (so to speak) of the ancient biblical Myth? It is that God's words alone – just His voice, His thoughts, His willing that something occur – are sufficient to bring things (e.g. light, the sun and moon, the creatures of the waters and of the dry land) into existence. And these words – these thoughts, these proclamations – determine the eternal destiny of (some at least of) His creation: "And God set them [the sun and moon] in the firmament of the heaven to give light upon the earth, and to rule over day and over the night, ...".[5] It is, in short, the theory of the sovereign power of The Word. God's Words do not describe Nature; God is no Chronicler. God's Words move and shake Nature; God is Creator, God is Prime Mover, and God is Legislator.[6]

The myth has been secularized for consumption in the twentieth century. Few scientists ever invoke God in their explana-

4 See Swartz 1993b.

5 Genesis 1:17-18

6 It was, we do well to recall, fewer than 250 years ago, well within the Modern Period, that Montesquieu in his influential *The Spirit of the Laws* (1748) devoted several paragraphs of Chapter 1, Book 1, to explaining that the Laws of Nature (what we would call physical laws) are the handiwork of God.

tions of events in the world.[7] But if God Himself has dropped out of the picture, His Words remain and retain their primeval role. The principal difference is that now those Words are called "physical laws". Just as the background radio hiss in the universe is the remnant of an ancient Big Bang, the theory that physical laws *govern* the world is the remnant of the ancient supernatural theory that God spake and Nature obeyed.

Here is a typical presentation of the secularized version. This is the introductory paragraph in the section "Looking for Laws: The Scientific Approach to Behavior" in the textbook *Psychology: Themes and Variations*, by Wayne Weiten:

> Whether the object of study is gravitational forces or people's behavior under stress, *the scientific approach assumes that events are governed by some lawful order*. As scientists, psychologists assume that behavior is governed by discernible laws or principles, just as the movement of the earth around the sun is governed by the laws of gravity. The behavior of living creatures may not seem as lawful and predictable as the "behavior" of the planets. However, the scientific enterprise is based on the belief that there *are* consistencies or laws that can be uncovered. Fortunately, the plausibility of applying this fundamental assumption to psychology has been supported by the discovery of a great many such consistencies in behavior, some of which provide the subject matter for this text. (Weiten 1992, p. 34)

And the mathematician/physicist P.C.W. Davies writes:

> ... subatomic physics is not complete anarchy. Many conceivable transmutations and reactions simply do not occur at all. We do not see protons changing into positrons, or electrons into neutrinos. ... Why not? What rules are there that bring at least partial discipline and order to the tangle?
>
> Classical physics has rules governing the interactions of matter [laws of conservation of energy and momentum]. ... When relativity is taken into account, we must include mass with energy because of the equivalence $E = mc^2$. Less familiar is the conservation of angular momentum or spin.
>
> ... There is also a universal *asymmetrical* law which regulates the *organization* of the activity [the second law of thermodynamics]. (Davies 1980, pp. 155-157. Parenthetical glosses summarize omitted portions.)

The subject matter changes from psychology to physics. But the underlying metaphysical view (we see in these two exam-

7 By "world" I mean the entire universe, not just the planet Earth.

ples) remains constant. The vocabulary describing the relationship between laws on the one hand and events (/states, /properties, etc.) of the world on the other is the same. Both Weiten and Davies write of laws 'governing' the world; and Davies, note, invokes near-synonyms when he goes on to write of 'rules that bring discipline and order [to Nature]' and of '[laws] regulating the activity [of Nature]'. There is nothing unusual or unfamiliar in these passages. Indeed they were purposely selected just because they are so very ordinary: they express the prevailing view. Indeed it would be a trivial matter to multiply indefinitely examples from other authors in every other scientific field: economics, pharmacology, geography, epidemiology, archaeology, neurophysiology, etc. (And I have recounted in Swartz 1985, p. 120, similar views expressed, not by scientists [as here], but by philosophers as well.)

This standard, or received, view – that physical laws (or Laws of Nature) govern or regulate Nature – is so commonplace, so entrenched, so widely promoted, that it is taken to be truistic. But for all that, this iconic view is challengeable. Indeed – I want to argue – it is misleading, radically unempirical, and overdue for replacement. However, before we turn to that matter, we must pause to make an important distinction.

Physical Laws and Scientific Laws

Physical laws are the 'real' laws of Nature. These laws are true independent of human beings coming to learn their truth. The law of the constancy of the speed of light, for example, has nothing to do with human beings having learned it to be true. It is true – and would have been true – even if human beings had never discovered its truth, indeed even if human beings (or any other sentient creatures for that matter) had never existed in this universe.

Physical laws have at least five properties. They are: (1) true (for all time and all place[8]); (2) universal or statistical generalizations; (3) purely descriptive (i.e. free of any terms naming specific items in the universe); (4) conditional; and (5) contingent (i.e. not necessary [logical] truths). Being believed or being known is *not* a defining property of physical laws. What properties beyond these minimal five properties are necessary for a proposition's being a physical law is the area of contention between the Regularity Theory and its rivals. According to the Regularity Theory these five properties are individually necessary and jointly sufficient for a proposition's being a physical law. Rival theories – which I here wish to reprove – argue that additional properties are necessary for a proposition's being a physical law, that is, these competing theories argue that the list, just given, of five properties is not sufficient.

Scientific laws, in contrast, are human creations. These are statements adopted in our collective effort to explain, predict and control the world. Some of these laws are more-or-less 'read off' nature directly. (Many statistical generalizations are of this sort. Such generalizations may be exceedingly useful tools for prediction. For example, managers of public transportation systems are skillful, using such empirically ascertained generalizations, at hiring the 'right' number of overtime drivers to handle increased demands on New Year's Eve. No sociologist could derive such statistical laws from any extant theory, but the laws are available, standing isolated from any overarching theory, as useful tools. Similarly engineers will determine, through extensive testing, the mean time between failures [MTBF] for various devices – copying machines, fluorescent bulbs, silicon memory chips, etc. – such results being unobtainable from basic theory.)

8 This latter feature – viz. being true for all time and place – is the central theme of James Trefil's *Reading the Mind of God: In Search of the Principle of Universality*, 1989.

Other laws, we know, are guessed at (hypothesized) and then tested. While still others 'fall out' of higher-order theories.[9]

Scientific laws – with few exceptions – are understood (1) to be not literally true, i.e. to be false. Nearly all scientific laws – as pointed out by Scriven and, far more extensively, by Cartwright – are false: not just possibly false, but actually false, and moreover *known* to be false.[10] (Cartwright writes colorfully that the 'laws of physics *lie*'. Literary license understood, we know that she means that the [scientific] laws of physics are *false*.) In addition, scientific laws are understood (2) to be approximations to the truth, 'idealized' reconstructions, or instrumental tools; and (3) to be held only tentatively, always subject to the possibility, and in many instances the actuality, of refutation, abandonment, and replacement.

Scientific laws – except for a tiny handful at most – are not physical laws, i.e. they are not the laws of Nature itself, but of human beings trying to understand and subdue Nature. For some proposition (statement) to be regarded as a *scientific law*, it must satisfy a certain subset of a very complex set of properties. Trying to determine what these properties are – e.g. conferring predictive abilities, providing explanatory premises, being interconnected in a net of theoretical liaisons, being tractable, being testable (even if only very indirectly), etc. – has occupied a very great deal of time among philosophers of science. In pursuing such a philosophical activity one is probably well advised to keep Wittgenstein's model of 'family resemblances' at the fore. There almost certainly is no single set of properties shared by all

9 I use the imprecise term "fall out" because I am loath to describe the process as that of their being 'deduced'. The deriving of Kepler's Laws from Newton's, or specific heats [of particular elements] from statistical thermodynamics, when actually examined in the writings of physicists, is nothing like the mathematical proofs or derivations in logic textbooks – these 'derivations' by physicists are far more natural-language prose than they are symbol manipulation in accord with deductively valid inference rules.

10 Although Scriven adopts the term "physical laws", it is clear from his discussion that he is referring to what I am here calling "scientific laws".

scientific laws. What it is to be a scientific law is to have a certain subset of a varied mix of 'law-conferring' properties *and* to have won the approbation "scientific law" from a number of prominent practicing scientists. In short, it is exceedingly difficult to say precisely what a scientific law *is*, and this for the reason that there is no single defining set of necessary and sufficient conditions for a statement's being a scientific law.

To explicate the concept of *scientific law* calls upon a great deal of *empirical* research, specifically concerning how scientists actually use the concept of law and what criteria they use in designating something a scientific law. This philosophical task is as daunting as any. While it has a significant, perhaps preponderant, empirical dimension, it also has a normative one: How – given the empirical data about how scientists actually use the concept of scientific law – *ought* we to reconstruct (/modify /conceive of) this concept so that it captures actual practice while at the same time can inform that practice and allow us to 'make sense of' (best understand) the scientific enterprise?

In contrast, the question as to what *physical laws* are, or better, how they are to be conceived, has little, if anything, to do with actual scientific practice. What are to be the defining characteristics of physical laws is ultimately to be decided by our metaphysics, not by our epistemology, and not by an empirical survey of the writings of scientists. (To be sure, metaphysics and epistemology are not wholly distinct disciplines, but in this instance I am talking about relative balance, and on whole, the question of how we are to conceive of physical laws is far more a matter of metaphysics than it is of epistemology or the philosophy of science, especially when the latter is conceived to be a branch of epistemology.)

The terminology we are forced to use is unfortunate. All of the expressions, "physical law", "law of Nature", and "Natural Law", connote – in the first instance at least – the context of physics and chemistry. It would be useful to have a more neutral, more inclusive, terminology, a name for any and all (real) laws whatsoever: those pertaining to the subject matter of physics as well as the subject matter of economics, sociology, geography, sociobiology, anthropology, genetics, etc. Lacking such a term, we will have to make do with "physical law". (The term "Natural Law" carries still other baggage, even more problematic than does "physical law". In some theories in ethics and in theology, the expression "natural law" has been used to denote the warrant in nature or in reason for ethical and moral precepts.[11] Thus I eschew the use of "natural law".) In short, I use the term "physical law" broadly: to encompass the laws (i.e. the 'real' laws, not the scientific laws) of all phenomena whatsoever, including, for example, economic laws, sociological laws, etc. Even though it is not usual to regard (the real) laws of economics, or (the real) laws of sociology, as "physical laws", they will here – for lack of any better term – be so regarded.

Throughout this century, much philosophical writing about physical laws (the laws of Nature) has been infected with a serious confusion, or more accurately, a conflation. Many writers fail to take any account whatever of the difference between *physical laws* on the one hand and *scientific laws* on the other. (You can detect this conflation in the paragraph by Weiten,

---

11 See, for example, Wollheim 1967.

The concept of *natural law* features prominently and frequently in Pope John Paul II's encyclical *Veritatis Splendor*, 1993, where he writes (quoting Aquinas): "the natural law 'is nothing other than the light of understanding infused in us by God, whereby we understand what must be done and what must be avoided. God gave this light and this law to man at creation'" (section 40).

In the 1993 federal election in Canada, the Natural Law Party fielded a candidate in every riding (constituency). Their 44-page advertising brochure includes these explanations: "The most fundamental level of Natural Law is the Unified Field of Natural Law, the Constitution of the Universe. Both modern science and ancient Vedic Science locate the source of Nature's perfect order in a single self-interacting Unified Field of pure intelligence. This field sequentially creates, from within itself, all the diverse Laws of Nature governing life at every level of the manifest universe" and "The technology to enliven the Unified Field of Natural Law is the group practice of Maharishi's Transcendental Meditation and TM-Sidhi program, including Yogic Flying" (p. 5).

quoted a moment ago.) These writers often use the terms "laws of nature" and "scientific laws" interchangeably. John Hospers, for example, in *An Introduction to Philosophical Analysis* (first edition, 1953) – the text through which doubtless some of the readers of this present volume will have been introduced to analytic philosophy – tacitly treated the two concepts as equivalent: "Scientific laws are *generalizations*; that is to say, from observing particular examples of nature's uniformities we *generalize* and assert that these uniformities hold in all cases. ... All laws of nature, then, in what they assert, go beyond the evidence that is available for them at any given time" (p. 168). The confusion remains, and is indeed exacerbated, in the latest (3rd) edition (1988). Writing of laws of nature, Hospers says "An important feature of laws is one that may come as a surprise: no single observation is sufficient to undermine a law" (p. 160). Such a claim may be apt for *scientific laws*; it is utterly misconceived in the context of *physical laws*, where the issue of acceptance or rejection cannot intelligibly arise.

In this chapter, I treat the two concepts – physical law and scientific law – as distinct. As I prepare this chapter, having before me only a list of chapter titles and the authors invited to contribute to this volume, it would appear that the majority of the papers in this anthology are given over to examining *scientific* laws. This chapter may help to even the balance slightly. This chapter will dwell principally on *physical* laws.

Metaphysics in Science

Many scientists naively believe that they practice a craft free of metaphysics. I have had frequent discussions with scientists who have told me that science – particularly as practiced by them personally – is free of metaphysics that infected science before the modern period, that they themselves carefully and deliberately eschew *all* metaphysics in their own science.

Such opinions are self-delusional. No science is, or for that matter could be, free of a very substantial component of metaphysical presupposition. (For an extended defense of this latter claim, see Swartz 1991, esp. Chapters 2-8.) The question is not whether our science has metaphysical components, but rather whether we learn to detect those aspects and whether we learn, or want, to examine them critically.[12]

What I have been calling the 'standard' or 'received' view of physical laws – viz. that physical laws 'govern' or 'regulate' Nature itself – is so familiar, so widely promoted, that it has taken on the appearance of being an (indubitable) 'fact'. We have been told that this was one of the great discoveries of the sixteenth and seventeenth centuries, that this is one of – and even perhaps the most important of – the cornerstones upon which the edifice of modern science rises (again see the quotation from Weiten above). We have been told that the belief in this claim has proven so successful, and so often confirmed, that it is – now – quite beyond dispute. In short, the standard view is presented as if it were itself a (virtually incontestable) datum of science.

That manner of regarding the epistemic status of the standard view of physical laws is a mistake. The standard view of physical laws is not a scientific datum, but a *metaphysical theory*, one which needs to be seen as such and to be appraised as such.

No one can reasonably (rationally) claim that there is not considerable order in Nature. There is. Such order occurs in the subatomic through to the galactic; in nonliving through to living matter; in protein fragments through to viruses; and in

12 Here are just a very few examples (taken from a virtually inexhaustible list): (1) that there is (/is not) a real external world; (2) that there is (/is not) such a thing as 'the' causal relation; (3) that truth is (/is not) manifest; (4) that there are (/are not) purposes in Nature; (5) that material objects are (/are not) 'amalgams' of instances of properties *in* an underlying substance; (6) that matter 'preserves' (/does not preserve) itself in existence; (7) that all aspects of the world are (/are not) quantifiable [measurable]; (8) that relations are (/are not) 'real'; etc. (On the last-mentioned topic, viz. whether relations are 'real', I have recently corresponded at length with an American sociologist who has conveyed to me an ongoing dispute he has had with professional colleagues on the issue of the reality of 'social' relationships.)

single-cells through to multi-celled organisms; in small groups, through to complex societies and to entire civilizations; etc. Most, if not all, of Nature 'falls under' physical laws. We know this. It would be irrational to deny this. The (imperfect) laws of science daily add further data pointing to an (underlying) orderliness in Nature itself, that is, pointing to the existence of 'real' (i.e. what I have called 'physical') laws. In challenging the standard view, I do not want to challenge the thesis of the orderliness of Nature or of the existence of 'real' laws. On the contrary, as should be obvious, I want to insist on the existence of such 'real' laws.

What I am intent, however, to challenge are certain metaphysical claims made about the nature of these underlying (or 'real' or 'physical') laws. In particular, I want to challenge the theory (i) which portrays physical laws as *governing* the world and (ii) which attributes to physical laws a special property of 'lawfulness', usually called by philosophers "nomicity" and taken to be a kind of 'physical necessity'.

Truth Reported *versus* Truth Imposed

It is only in the Twentieth Century, more exactly since the 1930s, that philosophy can finally claim to have the makings of a viable theory of truth. (In Tarski's theory, the vehicles of truth and falsity were sentences. Here I will assume that the vehicles of truth and falsity are *propositions*[13] and will adapt his theory accordingly.) On Tarski's semantic theory of truth (modified as just explained), the proposition expressed by the sentence "Snow is white" is true if and only if snow is white. (The theory says nothing about our being able to find out whether a given proposition is true, but that is not the intent of his theory. Tarski's theory is meant to explain, or at least remove some of the historical seeming mystery surrounding, the nature of truth itself, i.e. it sets out to explicate the semantic concept of truth.

13 See Bradley and Swartz 1979, pp. 65-86.

The theory does not even address, let alone answer, how to discern actual truths from actual falsities. Indeed one of the great merits of his theory is that it thoroughly dissociates questions of semantics from questions of epistemology.)

Propositions do not 'become' true. A person may become the treasurer of a company; her becoming so is an event which is datable. The truth of a proposition is not a datable event or occurrence; propositions bear their truth omnitemporally. "John F. Kennedy is (/was /will be) assassinated on November 22, 1963" is true, has always been true, and will always be true. It did not 'become' true on November 22, 1963. The proposition's having been true for all time prior to November 22, 1963 was not the *cause* of Kennedy's death. (So-called 'logical determinism' is a crass confusion.) The cause of Kennedy's death was an *occurrence* (event) in Dallas, Texas, on the fateful day; it was not a proposition or its truth.

More generally, and perhaps somewhat colloquially but at the same time more perspicuously, the Tarskian theory of truth has it that propositions (originally "sentences") 'take their truth from the way the world is'. The semantic 'truth-making' relation – if I may be permitted to put it that way – proceeds *from* the world *to* propositions (or sentences).[14] If the world is a certain way, then any proposition which *says* (reports) that the world is that way is true; and any proposition which *says* that the world is otherwise is false. The way the world *is*, *was* and *will be*, accounts for certain propositions being true (and of course for others being false). (The technical details are, to be sure, formidable; but enough has been said for current purposes.)

In a bizarre quirk of history, the received account of physical laws, deriving from a medieval view that physical laws were dictates of God – the account which has it that the world is governed by physical laws – turns the Tarskian account of truth precisely 'on its head'.

14 Tarski, himself, put it this way: "The truth of a sentence consists in its agreement with (or correspondence to) reality" (Tarski 1949, p. 54).

On a Tarskian theory of truth, the proposition expressed by the sentence "The charge on each electron is -1.6×10^{-19} coulombs" is true if and only if the charge on each electron is -1.6×10^{-19} coulombs.[15] However, according to the 'standard' account of the nature of physical laws, electrons bear the charge they do because they are governed by the 'law' "The charge on each electron is -1.6×10^{-19} coulombs". The Moon and the Earth dance (to first approximation) in elliptical orbits around a common focus *because* they are governed by the Einsteinian laws of Space-time (General Relativity). Cats learning to escape from puzzle boxes do so in accord with (what Thorndike called) 'the law of effect': "If a response in the presence of a stimulus leads to satisfying effects, the association between the stimulus and the response is strengthened".[16] And, acting in accord with the laws of sociobiology, a person is more likely to come to the aid of a grandchild than of a cousin. In other words, on the standard account, there are certain, special, propositions, viz. physical laws, which do not take their truth from the way the world is, but instead 'dictate' to Nature, forcing it, governing it, regulating it, to 'behave' in certain ways and not in others.

There are, then, in the 'standard' account *two* radically different kinds of propositions: ordinary, garden variety ones (such as expressed by "Snow is white" and "Kennedy was assassinated on November 22, 1963") which 'take their truth' from the way the world is; and 'lawful' propositions (e.g. the second law of thermodynamics) which 'reverse the truth-making semantic relationship', which do not take their truth from the way the world is, but rather 'govern', 'regulate', 'discipline', 'rule', and 'impose order on' Nature itself. These latter propositions are said to be *nomologically* (or *nomically* [occasionally *physically*, *onticly*, or *contingently*]) *necessary*. Philosophers who plump for the existence of nomological necessity are called "Necessitarians".

15 More exactly: $-1.6021892 \times 10^{19}$ coulombs, with an uncertainty of 2.9 parts per million. See Taylor 1983.

16 Weiten 1992, p. 201.

It is a mark of just how peculiar the Necessitarian account of physical laws is that the theory of truth it presupposes has been so little examined in the philosophical literature that it does not even bear an accepted name. In the *Concept of Physical Law*, I tried to fill this void by dubbing that theory of truth 'the Autonomy Theory', the idea being that nomological propositions bear their truth 'autonomously': their truth does not arise out of the way the world is but rather occurs 'primitively' or 'independently' (or more aptly, *sui generis*).[17]

In greater detail: on the standard, Necessitarian, account, physical laws bear their truth, not indicatively, but subjunctively (or counterfactually). Physical laws are not only true, they would remain true (and their contraries would remain false) even if the world were different, in particular, even if those laws had no actual instances. Nomological laws do not bear their truth on the basis of correctly reporting (perhaps future) states of the world. For example, nomological laws of economics, of psychology, of biology, of sociology, etc., are regarded as being true from time immemorial *even if life had never appeared in the universe*. On the Necessitarian view, the way the universe

17 Necessitarianism comes in a variety of 'flavors'. In recent years, some philosophers have argued that physical necessity is not imposed on nature by laws but instead arises (or is grounded) directly in nature itself. For example, Armstrong – along with some others – argues that nomological necessity is a relation between universals. However, he admits "... the following complaint [against any such theory] may be made. At the end of all of our explanations, this factor of necessitation remains unexplained. This reproach is just, I think, but the inexplicability of necessitation just has to be accepted. Necessitation, the way that one Form (universal) brings another along with it as Plato puts it in the *Phaedo* (104d-105), is a primitive, or near primitive, which we are forced to postulate. ... We must admit it in the spirit of natural piety, to adopt Samuel Alexander's phrase" (Armstrong 1983, p. 92). Some of my objections below are not appropriate for this particular version of necessitarianism; however, most of them – e.g. concerning the lack of an empirical test; the dispensability of the notion; its mischievousness; etc. – remain intact.

unfolds depends on two 'factors': on the 'initial state' of the universe and on the nomological laws of the universe.[18]

This latter view of physical laws – that they bear their truth subjunctively and autonomously and that they function like inviolable prescriptions – is the single most important difference between the Standard (i.e. Necessitarian) Account of physical laws and the Regularity Theory. On the Regularity Theory, physical laws are descriptions, nothing more. They do not govern Nature; they do not bear their truth subjunctively or autonomously. On the Regularity Theory, there is but a single theory of truth.

The Regularity Theory, or, Being More Humean than Hume

Until fairly recently, the received reconstruction of Hume's theory of physical laws had it that Hume advanced a regularity theory, that for Hume, physical laws were nothing more or less than 'constant conjunctions'. Hume was alleged to have argued that physical laws are simply (a proper subset of) universally true generalizations and, more especially, were not 'necessary' – neither logically necessary nor physically (i.e. nomologically) necessary.

Thus, for example, in the first edition (1952) of his widely-read and respected *A History of Western Philosophy*, we find W.T. Jones offering this reconstruction of Hume's theory:

> Like identity [through time], necessary connection is something in us, not something in the object; like identity, it is grounded in a custom or habit of the imagination rather than in a rationale in the universe. (Jones 1952, p. 780)

Or again, a few years later, we find Ernest Nagel writing in *The Structure of Science*:

> ... Hume proposed an analysis of causal statements in terms of constant conjunctions and *de facto* uniformities. Ignoring important details in Hume's ac-

---

18 For an even more extravagant view of nomicity, see Rescher 1984. For a critique, see Swartz 1993a.

count of the spatiotemporal relations of events which are said to be causally connected, the substance of the Humean position is briefly as follows. The objective content of the statement that a given event *c* is the cause of another event *e*, is simply that *c* is an instance of a property *C*, *e* is an instance of a property *E* (these properties may be quite complex), and any *C* is as a matter of fact also *E*. On this analysis, the "necessity" allegedly characterizing the relation of *c* to *e* does not reside in the objective relations of the events themselves. The necessity has its locus elsewhere – according to Hume, in certain habits of expectation that have been developed as a consequence of the uniform but *de facto* conjunctions of *C* and *E*. (Nagel 1961, pp. 55-56)

It is not clear that either Jones or Nagel subscribed to the reconstructions they offered of Hume's theory of physical laws, indeed they probably did not. But the point, for our purposes, is that they believed that they had accurately reported Hume's view.

More recent scholarship refutes this earlier 'standard' reconstruction of Hume.[19] For an even more extravagant view of nomicity, see Rescher 1984. For a critique, see Swartz 1993a. What Hume denied was that there was any *empirical* evidence for necessity. His so-called skepticism concerned our finding out that physical laws are nomological. But his skepticism about our ability to find evidence did not carry over to a skepticism about the existence of such necessity. He had a belief in such necessity; his problem was to justify that belief rationally and he found it difficult to do so.

The regularity and the necessitarian theories of physical laws seesaw. For much of this century, the regularity theory – the supposedly Humean theory – was, I think, predominant in the thinking and writing of many, perhaps most, philosophers. But in recent years a number of developments have reversed that balance. Certainly the concept of *modality* is far more familiar and attractive than it had been in the first half of the century – possible-world semantics and related techniques have made various concepts of necessity respectable. Philosophers have turned to a number of new, related, problems, e.g. the investi-

19 See, e.g., Wright 1983, esp. pp. 143-144; Beauchamp and Rosenberg 1981; and Strawson 1989.

gation of the truth-conditions for counterfactuals. And modern scholarship has, as I have said, substantially revised the received view of the historical Hume. The latter reversal is sometimes summed-up in the aphorism "Hume was no Humean" (i.e. the actual, historical, Hume did not advance the theories traditionally attributed to him).

I am heir to some of Hume's skepticism.[20] But where his own worry about physical necessity was only epistemological and was not mirrored in counterpart metaphysical beliefs, my own is far more corrosive and pervasive. For my own skepticism about physical necessity permeates both my epistemology and my metaphysics. While I agree with Hume that there is no empirical *evidence* for such a thing, I do not share his belief that metaphysics, nonetheless, requires the positing of physical necessity. It is possible to advance a cogent metaphysics free of the concept of physical necessity.

The Nomicity of Scientific Laws and the Nomicity of Physical Laws

There are two, wholly different, accounts of nomicity.

As I have said, philosophers of science have expended much effort in trying to uncover (better 'reconstruct') by what criteria scientists distinguish between 'mere accidentally true generalizations' and those which are designated as scientific laws. Many philosophers have come to regard the crucial difference under the label of "nomicity", i.e. on this account of nomicity, physical necessity is a property accruing to *scientific laws* out of the *praxis* of actual science. (An analogy may be helpful: the distinction is not terribly unlike the difference between a musical performance's being judged prize-winning while another

20 But only to some. For example, I do not share Hume's skepticism, or even strong worries, about the existence of the external world or of the endurance of unperceived objects.

is deemed merely competent.) The nomicity of *scientific laws* is strictly a human artifact.

This first kind of nomicity is entirely innocuous, one to which I have no objection whatsoever. (It is, of course, a gross confusion – to which many have succumbed – to ascribe or transfer this sort of nomicity to *physical* laws.)

Of far more, indeed central, concern to this chapter is the 'deeper' necessity conceived to attach – quite independent of human knowledge or scientific practice – to the real, i.e. *physical*, laws of nature. Is there, or could there be, any empirical evidence for such a thing?

Hume believed that there could not be, and I think he was right. However, a number of philosophers, particularly in recent years, have tried to argue that there can be such evidence. Elsewhere I have reviewed and tried to rebut (in Swartz 1985 and Swartz 1988) some of these attempts. On this occasion, I will have to content myself with a single later example. Martin Gerwin has written (in what I would describe to be 'a Lockean fashion'):

> It is my contention that, contrary to what Hume has claimed, there is a very familiar type of experience which perfectly well could, and most probably does, give rise to the idea of necessary connection between phenomena. It is the experience of trying to do something and failing, so that one concludes, "I can't do it". Such an experience is quite fundamental to our awareness of ourselves as agents, and in the most intimate way. This, it seems to me, *is* what Hume was searching for, and he *did* overlook it. (Gerwin 1987, p. 7)

Gerwin and I have corresponded about this paragraph. Here is an edited version of my reply:

> I doubt that 'trying and failing' will do the trick. After all, many times we try to do something, fail at it, and yet – for whatever reasons – do *not* conclude "I can't do it." (It happened to me recently. I was trying to install a new pump on a dishwasher and was having no success, indeed I was failing miserably. I stopped for a while, tried again, and succeeded.) I am absolutely convinced that there is *no* phenomenological, introspective, felt (call it what you will) difference whatsoever between failing to do something which is possible (e.g. installing a dishwasher pump) and failing to do something which necessitarians call 'nomically impossible', e.g. flapping my arms and flying.

> What do I feel when I find that I repeatedly fail to do something? Disappointment, remorse, anger, sadness, annoyance, irritability, fury, etc. Do I experience (physical or nomological) impossibility? Not that I can tell. I would not know how to recognize it if I did. I can experience that I have not done what I wanted; that I have tried especially hard; etc. But I do not see that I have experienced that I *cannot* do it. I may *say*, "I can't do it." But I have not experienced anything *more* than failure. (personal correspondence, March 1, 1989)

In short, I have never found any good argument to the effect that there is *empirical evidence* for the existence of nomicity (physical necessity).

The Need for a 'Gestalt Switch'

The standard, necessitarian, account of physical laws is so much a part of the contemporary metaphysics in which modern science is pursued that it is difficult to discern its existence and perhaps even more difficult to imagine abandoning that view.

To do so requires the philosophical equivalent of a 'gestalt switch': believing that the world unfolds, certain patterns emerge, certain 'orderliness' prevails, but that none of this is 'governed'; it simply occurs.

To abandon necessitarianism means to elevate – and to live with – contingency: the world does not *have to be* the way it is; it just *is*. The charge on the electron does not *have to be* -1.6×10^{-19} coulombs; it just *is*. Light does not *have to* have a constant, finite, velocity; it just *does*. To invoke nomological necessities to 'account' for such constancies (order, etc.) is to engage in explanatory hand-waving. Is it really any more informative to be told that light has a constant velocity because there is a law of nature to that effect than to be told that opium is sleep-inducing because it has a 'dormative power'? The *form* of an explanation has been given, but the content is chimerical.

There is orderliness is Nature. That's the way Nature *is*. There are no secret, sublime, mystical laws forcing Nature to be that way. Or at least, there is no good rational reason to believe that there are such queer entities. Physical laws are *descriptions*, they neither are, nor function like, *prescriptions*.

Advantages of the Regularity Theory

There are three advantages the Regularity Theory has over the Necessitarian Theory.

First of all, the Regularity Theory falls within the Empiricist worldview which has managed to discard a significant amount of historical baggage now regarded as explanatorily superfluous. No scientists in the late twentieth century invoke such metaphysical concepts as substance, vital forces, or natural places. These metaphysical arcana have fallen out of use because we have come to reject the kind of explanations offered which utilized them. Such concepts are not like the empirical concepts of phlogiston, caloric, and electric fluids, for example. These latter concepts came to be rejected because the theories in which they occurred were superseded by better empirical theories. But no empirical theory, no scientific theory, competed with the theory of substance, for example. The theory of substance fell into disuse because philosophers and scientists came to believe that its putative explanatory power was illusory: there was no empirical test possible for the existence of substance, and positing its existence merely deferred, without genuinely solving, the very problems it was invoked to explain.[21] Similarly, there is no empirical test for nomicity, in spite of claims to the contrary by some Necessitarians. Nomological necessity is purely a metaphysical posit, an anachronism that has outlived its usefulness. It is not that we must regard the theory as false, only as superfluous. Speaking of God, Laplace is alleged to have said to Napoleon "I have no need of such an hypothesis." Speaking now of nomicity, I would echo Laplace's words.

Second, the Regularity theory provides a metaphysical underpinning to a philosophy of science which legitimizes the so-

21 See Swartz 1991, chapters 10-12.

cial sciences as readily as it does the so-called more basic sciences of physics and chemistry. One infrequently articulated, but clearly central, cluster of presuppositions of the Necessitarian theory is that there are only a finite number of bona fide physical laws, that these laws are in fact relatively few in number, that they concern the subject matter of physics and chemistry, and that all 'other' laws are supposed to be consequences (or implications) of these basic laws. The trouble with this latter view is that it promotes the idea that the social sciences are, in some sense, 'second-class' sciences: that the scientific laws of the latter are less 'authentic' than the scientific laws of physics and chemistry which are 'closer to' the real laws of Nature. No such attitude is warranted in the Regularity Theory. For the Regularity Theory rejects all of the presuppositions just mentioned. On the Regularity Theory, there is no warranted claim that the number of physical laws is small (or even finite for that matter) and there is no claim – for there are no good grounds upon which to make such a claim – that the laws of physics are more basic than those of, let's say, economics (see again note 9). Furthermore, the Regularity Theory is more congenial to the existence of statistical laws. Accommodating the existence of statistical laws was an historical wrench for the Necessitarian theory and remains an awkwardness in a theory which sees laws as governing nature and as bearing their truth autonomously: "How could statistical truths be *laws*? What sort of underlying 'mechanism' could one imagine that would account for the permanency and constancy of certain statistical truths (not only those of quantum mechanics, but of economics, psychology, sociology, etc.)? How, more precisely, could a 'necessity' which 'drives' Nature operate stochastically? Is there 'partial' nomological necessity?" There are no counterpart worries in the Regularity Theory. Inasmuch as physical laws, in the latter theory, 'take their truth from the way the world is', i.e. are descriptions, there can be no problem about statistical laws. If there are certain stochastic regularities in Nature, then there are true (probabilistic) descriptions of those regularities. The metaphysical conundrum concerning 'partial necessities' does not arise.

Third, adopting the Regularity Theory solves one of the perennial problems, not just of academic philosophy, but of the wider Western worldview. For generations, thinkers have struggled with 'the problem of determinism and free will'. "How could human beings possibly have, and exercise, free will in a world governed (in every detail) by physical laws?" The number of attempted solutions fills volumes, but few of these solutions manage to wring much conviction from their readers. The very persistence, the obduracy, of the problem suggests that there is something seriously wrong with its *presuppositions*. (It is not unlike the puzzle "What is the last digit in the decimal expansion of π?" None of the answers "0", "1", ... "9" is acceptable: instead one must reject one of the presuppositions of the question itself, viz. one must argue that there is no last digit in the decimal expansion of π.) The solution to the perdurable problem of determinism and free will lies ready at hand if one adopts the Regularity Theory. For one is then in a position to reject the principal presupposition of the puzzle, namely the belief that the world is 'governed' by physical law. Give up that belief – more exactly, continue to believe that there are physical laws, but abandon the Necessitarian account of physical laws which makes them nomologically necessary, regard physical laws simply as true descriptions of what occurs in the world – and the problem of free will and determinism is solved so completely that it cannot even coherently be stated so as to appear to be a problem. (The argument in its full comprises chapters 10-11 in Swartz 1985.)

Technical Appendix – The 'Paradox' of Regularity

This anthology is expressly designed for an interdisciplinary audience. I have, therefore, tried above to keep technical details to a minimum. But there is one technical aspect of the Regularity Theory which has attracted attention and deserves discussion here.

Consider any omnitemporally false existential statement. George Molnar's example (1963) was "There is a river of Coca Cola" [in symbols "$(\exists x)(Rx \ \& \ Cx)$"]. Each such statement is the logical contradictory of (and hence inconsistent with) a true negative universal statement [e.g., in the case at issue, "$(x)(Rx \supset \sim Cx)$"]. On the Regularity Theory, the latter (true [negative] universal) statement turns out to be a physical law. Add now the usual definition of "physical impossibility", viz. that anything logically inconsistent with a physical law is physically impossible, and one can generate an unanticipated 'paradox': Every omnitemporally false existential statement (e.g. that there is a river of Coca Cola) is not just false *but is physically impossible.*

Many writers have taken the conclusion of the argument in the preceding paragraph to be the reductio ad absurdum of the Regularity Theory. The Regularity Theory cannot possibility be a correct account of physical laws – it is alleged by these critics – inasmuch as it has egregiously counterintuitive consequences: it would make every omnitemporally false existential statement a physical impossibility.

To this I would reply that the critics' objection is guilty of an unwitting equivocation. What it does is to carry over the baggage associated with the concept of "physical impossibility" in the Necessitarian account and to use it (illicitly) to tar the Regularity Theory.[22] On the Necessitarian account – i.e. on the received theory – for something to be 'physically impossible' means (roughly) that it is 'disallowed' or 'forbidden' by the nomological laws of nature. On the Necessitarian account, being 'physically impossible' is to have encountered an insurmountable metaphysical obstacle.

22 The mistake is not unlike that many persons make when they invoke 'intuitions', arising out of acquaintance with finite mathematics, to reject as paradoxical the theses of transfinite mathematics. If one is to consider a new theory, one must adopt (even if only tentatively) all its unique contextual definitions, and not selectively import or retain key concepts from earlier or competing theories.

On the Regularity Theory, an existential proposition's "being physically impossible" connotes something far more benign: simply that proposition's being omnitemporally *false*. If (ex hypothesi) there never is (past, present, or future) a river of Coca Cola, then it is a universal truth (physical law) that no rivers are (constituted of) Coca Cola. Accordingly, in the sense in which "physical impossibility" is understood *in the Regularity Theory*, it is physically impossible for there to be a river of Coca Cola. Physical impossibility in this latter theory is simply omnitemporal falsity: it is *not* the confronting of an insurmountable metaphysical obstacle. Rather than being a fatal flaw in the theory, the physical impossibility of all omnitemporally false existential propositions rightly ought to be seen to be nothing more than an innocuous logical triviality.

References

Armstrong, David [1983]: *What is a Law of Nature?* (Cambridge: Cambridge University Press).

Beauchamp, Tom L., and Alexander Rosenberg [1981]: *Hume and the Problem of Causation.* (New York: Oxford University Press).

Bradley, Raymond, and Norman Swartz [1979]: *Possible Worlds: An Introduction to Logic and Its Philosophy.* (Indianapolis: Hackett).

Cartwright, Nancy [1983]: *How the Laws of Physics Lie.* (Oxford: Oxford University Press).

Davies, Paul.C.W. [1980]: *The Forces of Nature.* (Cambridge: Cambridge University Press).

Gerwin, Martin [1987]: "Causality and Agency: A Refutation of Hume", in *Dialogue* (Canada), *XXVI*, pp. 3-17.

Hospers, John [1953]: *An Introduction to Philosophical Analysis.* (New York: Prentice-Hall).

Hospers, John [1988]: *An Introduction to Philosophical Analysis*, 3rd ed. (Englewood Cliffs, NJ: Prentice-Hall).

Jones, W.T. [1952]: *A History of Western Philosophy.* (New York: Harcourt, Brace and Company).

Molnar, George [1963]: "Kneale's Argument Revisited", in *Philosophical Review, 78/1*, pp. 79-89.

Montesquieu, Baron de [1988]: *The Spirit of the Laws* [1748, 1st ed.; 1757, last edition, posth.], translated and edited by Anne M. Cohler, Basia Carolyn Miller, and Harold Samuel Stone. (Cambridge: Cambridge University Press).

Nagel, Ernest [1961]: *The Structure of Science: Problems in the Logic of Scientific Explanation.* (New York: Harcourt, Brace & World).

Rescher, Nicholas [1984]:*The Riddle of Existence: An Essay in Idealistic Metaphysics.* (Lanham, MD: University Press of America).

Scriven, Michael [1961]: "The Key Property of Physical Laws – Inaccuracy", in *Current Issues in the Philosophy of Science – Proceedings of Section L of the American Association for the Advancement of Science*, ed. by Herbert Feigl and Grover Maxwell, pp. 91-104. (New York: Holt, Rinehart and Winston).

Strawson, Galen [1989]: *The Secret Connexion: Causation, Realism, and David Hume.* (Oxford: Oxford University Press).

Swartz, Norman [1985]: *The Concept of Physical Law.* (New York: Cambridge University Press).

Swartz, Norman [1988]: "Reply to Ruse", in *Dialogue* (Canada), *XXVII*, pp. 529-532.

Swartz, Norman [1991]: *Beyond Experience: Metaphysical Theories and Philosophical Constraints.* (Toronto: University of Toronto Press).

Swartz, Norman [1993a]: "Can Existence and Nomicity Devolve from Axiological Principles?", in *Electronic Journal of Analytic Philosophy, 1/1* (Aug. 1993). (Available by anonymous FTP from phil.indiana.edu.)

Swartz, Norman [1993b]: "The Obdurate Persistence of Rationalism", in *Vicinae Deviae: Essays in Honour of Raymond Earl Jennings*, ed. Martin Hahn, pp. 125-141. (Burnaby, B.C.: Simon Fraser University).

Tarski, Alfred [1949]: "The Semantic Conception of Truth", in *Readings in Philosophical Analysis*, ed. Herbert Feigl and Wilfred Sellars, pp. 52-84. (New York: Appleton-Century-Crofts, Inc.). Reprinted from "Symposium on Meaning and Truth", in *Philosophy and Phenomenological Research*, vol. IV (1944).

Taylor, Barry N. [1983]: "Atomic Constants", in *McGraw-Hill Encyclopedia of Physics*, ed. Sybil P. Parker, pp. 47-50. (New York: McGraw-Hill Book Company).

Trefil, James [1989]: *Reading the Mind of God: In Search of the Principle of Universality.* (New York: Charles Scribner's Sons).

Weiten, Wayne [1992]: *Psychology: Themes and Variations*, 2nd ed. (Pacific Grove, Calif. Brooks/Cole Publishing Company).

Wollheim, Richard [1967]: "Natural Law" in *The Encyclopedia of Philosophy*, ed. Paul Edwards, vol. 5, pp. 450-454. (New York: Macmillan Publishing Co. and The Free Press).

Wright, John P. [1983]: *The Skeptical Realism of David Hume.* (Manchester: Manchester University Press).

Martin Leckey and John Bigelow

The Necessitarian Perspective: Laws as Natural Entailments

1. Introduction

"It's a very interesting thing in physics," said Mr. X, "that the laws tell us about permissible universes, whereas we only have one universe to describe." – Richard Feynman (Gleick 1992, p.126).

We maintain that there is something called *natural necessity* that is involved in the laws of nature -laws are concerned with what must happen, and what could not possibly happen, rather than merely what does and does not happen. Some recent believers in natural necessity, such as Dretske [1977], Tooley [1977, 1987] and Armstrong [1978, 1983], have argued that this natural necessity arises from certain relations among the properties of things in our world – they argue that there are relations of "necessitation" holding among universals. We endorse this theory that laws involve relations among properties of things. However, we think that more needs to be said about what underlies this relation of necessitation. In this paper we assume that there exist certain complex properties, which we call states of systems, or just states, and we postulate that there are some states that are *natural* for our world and other states that are *non-natural* for our world. Given that there does exist a real distinction between natural and non-natural states for our world, there will be a definable kind of natural necessity – there will be a significant sense in which laws really do tell us what must happen and what could not possibly happen.

The account of laws adopted here is presented in a framework that is adaptable to a range of distinct versions of a necessitarian theory of laws, by fine-tuning certain elements in the framework in various different ways. Among the distinct theories which can be derived within the broad framework, there will be articulations of several existing necessitarian theories, such as that of Harré and Madden, or that of Dretske, Tooley and Armstrong, or that of Ellis and Lierse, and so on. We will advocate a fine-tuning of our own. Both the broad framework under which all these theories fall, and the particular theory we present within this framework, are original to this paper, we believe.

2. Climbing the Hierarchy of Laws

We claim that laws of nature exist and that human beings have discovered many laws of nature. We suggest that William Whewell's methodology, involving the *consilience of inductions,* furnishes one way in which claims of the discovery of laws of nature can begin to be justified. Before developing the metaphysics of laws in detail we will introduce some examples of laws by means of a Whewellian methodology. Whewell's methodology is less well known than the more restrictive empiricism of his contemporary, J.S. Mill, but we have been impressed with work by Malcolm Forster [1988] which shows Whewell to have been more in touch with the practice of real scientists. (Leckey [1991] takes issue with some aspects of Forster's interpretation of Whewell, but nevertheless shows that the Whewell/Forster epistemology is well suited even to quantum mechanics.)

When we begin a scientific investigation we may find a certain relation among observed facts, but we may be unsure whether this relation amounts to a law of nature. Whewell calls this relation between experimentally determined facts a *colligation of facts.* In this paper we will consider mainly laws that are expressed in mathematical terms, although we trust that our metaphysical account is applicable to all laws. In mathematical sciences, Whewell points out that a colligation of facts can often be divided into three steps, which he calls the determination of the independent variable, the formula, and the coefficients (Butts [1968, pp.210-211]).

As an example, consider a body dropped from rest near the surface of the Earth. We wish to find a law that describes the way its velocity changes as it falls towards the ground. The velocity is labelled the *dependent variable.* The first step is to choose another quantity that varies with the velocity, labelled the *independent variable.* In this case we will take time as the independent variable. The next step is to choose a general *formula* linking the two variables. In this case the formula is as follows:

$$v = at,$$

where v is the velocity at time t of an object in free fall near the surface of the Earth, given that it began from rest. The formula is generally characterised by one or more coefficients – here the coefficient is labelled a. This coefficient can be determined by experiment, which is the final step in forming a colligation of facts. Historically, it is a myth that this law was first discovered by performing such experiments and tabulating results, without any preconceptions about what kind of law was likely to pop out of the data. Nevertheless, once discovered, the laws do stand in need of justification by appeal to supporting evidence, and careful experimental work is done both to justify the law and to determine the precise value of the coefficient, which in this case is called the *acceleration due to gravity*.

The equation $v = at$ describes a relation between velocity and time that will hold to a reasonable degree of approximation in circumstances where the effect of air resistance is small. In all actual cases there will be some contribution from air resistance; but the velocity can be split into two components – the component it has due to gravity and the component lost due to air resistance. The relation $v = at$ will then apply with high precision to the velocity component due to gravity.

The above relation between time of fall and the velocity due to gravity is asserted to hold under particular circumstances: namely, for medium-sized bodies falling from rest near the surface of the Earth. Other conditions of an experiment can be varied a great deal, while keeping these circumstances fixed. For example the mass, size and composition of the object can be varied, the place on Earth where the experiment is done can be varied and the experimenters themselves can be varied. Thus if many experiments on free falling bodies were carried out in many locations on the surface of the Earth, varying as many of these other factors as possible, then approximately the same value of the acceleration, *a*, would be determined in each case. This value is called g, the acceleration due to gravity. (If we eliminate air resistance and other interfering factors sufficiently, the exact value measured would vary between the equator and the poles, due to the fact that the Earth is not spherical, its diameter being greater at the equator, and due to the centrifugal effect of the Earth's rotation; but we are discussing here a law that involves a low degree of precision.) Discovering that approximately the same value obtains for a coefficient like this under different conditions is an simple instance of what Whewell calls a *consilience of inductions*.

After discovering that this consilience of inductions holds for a very wide range of conditions, provided the circumstance of being a medium sized body falling from rest near the surface of the Earth is kept fixed, we become confident that this relation between velocity and time would hold under a much wider range of cases than those we have observed so far. Under these circumstances we become confident in saying, "if any body were to be in free fall anywhere near the Earth's surface then this relation would hold." The prediction that this relation would hold is inductively extended to other cases, whether these cases ever actually occur or not – the prediction is equally extended to both factual and counterfactual cases. We do not merely predict what will happen but also what would happen under the given circumstances. Popper notwithstanding, it is reasonable to draw such inferences from observed actual cases both to unobserved actual and to merely possible cases. Thus is reasonable to infer that there is a sense in the which a contrary state of affairs to that predicted *could not possibly* occur under those circumstances. And it is then that we infer that the predicted state of affairs obtaining under those circumstances is not an accident but a law of nature.

Now we need to clarify what we mean when we say that a contrary state of affairs "could not" obtain under those circumstances. Although we become very confident that in some sense the state of affairs could not obtain, it is of course not *logically* impossible for such a state of affairs to obtain. It seems in some sense a necessity that the state of affairs will not obtain, but it is not a logical necessity. The reason it is natural to call this necessity a *natural* necessity, is that it arises out of something we might call the "nature" of our world, or at least of some significant part or aspect of the world. When a state is one that "could not" obtain under certain circumstances, then for that state of affairs to obtain under those circumstances would be contrary to the nature of our world – such a state of affairs would be *non-natural* under those circumstances. In other words the combined state of being in those circumstances and in a contrary state of affairs is a *non-natural state*, for our world. Then on condition that things only take on natural states, states that are natural for our world, it is logically necessary that they take on the given state of affairs in those circumstances. The key to our theory will lie in drawing a distinction between natural and non-natural states. Natural necessity can then be defined as logical necessity conditional on systems taking on natural states only. We will draw out this idea later in the paper. First, however, we will look a little longer at the kinds of evidence that support laws, and the logical forms that laws take.

Consider now a broader range of circumstances in which we examine the velocity of a falling body. Imagine that free fall experiments were carried out at different heights above the Earth's surface. In this case an experimenter would find that the acceleration due to gravity would no longer be a constant; instead the magnitude of the acceleration would fall off as the inverse square of the distance from the centre of the Earth:

$$a = k/r^2,$$

where k is constant. Here we have formed a new colligation of facts. In this case the dependent variable is a, the acceleration due to gravity, which was a constant coefficient in the earlier colligation of facts. The independent variable r is the distance from the centre of the Earth, and k is a new coefficient.

Now suppose the same experiment were carried out on the moon and on other planets, then we would find that k is not constant but varies according to the mass M of the heavenly body:

$$k = GM,$$

where G is the universal gravitational constant. G is a constant under all circumstances, as far as we know. Having reached a universal constant, we have reached the end of this particular hierarchy of laws. Putting the last two equations together we see that

$$a = GM/r^2.$$

This is a law for the acceleration due to gravity for bodies attracted to a body of mass M. It is an approximate Newtonian law; but properly construed it should still be regarded as an actual law of nature, provided it is understood as asserting a relationship which holds only within some margin of approximation.

Thus, epistemologically, we can move up a hierarchy of colligations of facts, uncovering laws that hold in a wider and wider set of circumstances. The coefficients of the laws will hold constant over wider and wider ranges of situations, until we reach a law that holds, at least approximately, over a very wide range of circumstances. The laws towards the top of the hierarchy are the more fundamental laws, and those towards the bottom are the more phenomenological laws. The phenomenological laws can be thought of as derived laws, laws that follow from the fundamental laws by deriving the logical consequences which would follow from them within a restricted range of circumstances. Some philosophers argue that phenomenological laws, such as Galileo's law $v = at$, are not laws at all, because they lack sufficient generality, being restricted in this case to free fall near to a particular body, namely the Earth. However we do regard this as a genuine law, even though it is valid over a relatively limited range of circumstances. As we shall explain, we take the defining feature of laws to be the fact that a contrary state of affairs to that specified by the law under the given circumstances would be a non-natural state of affairs, and that condition is met in this case.

Note that even the most "fundamental" of the laws considered here is subject to limitations over its range of application. For one thing, it is expressed in terms of the distance to the centre of the planet rather than the distance from the centre of mass of the pair of objects taken together. Thus, for the law to be accurate, there will be limitations on the mass of the object concerned, because if the object is very massive then the centre of mass for the planet and the object taken together will not be approximately at the centre of the planet. Another limitation on the range of circumstances over which the law holds is that the distance r has to be greater than the radius of the planet. There are other reasons why this law does not hold precisely. The Earth is not a perfect sphere nor is it of perfectly spherically symmetric density, and the same would be true of any actual body in the universe. There will also be small deviations from the inverse square law due to general relativistic effects.

It is possible that all laws hold only approximately and only over a fairly extensive range of circumstances but not absolutely universally. We do not assume that a law will always enter into a hierarchy of laws of the type discussed, or that all the separate hierarchies will be connected in one huge hierarchy. There exist many separate hierarchies of laws in science, and the explicit reduction of one set of laws to another may be the exception rather than the rule. Thus we are going to make no assumptions about reductionism or about the ultimate unification of laws – there may be many separate sets of laws. We will give an account of laws that is neutral on these matters, and applies to both fundamental and phenomenological laws.

Some people think that Einstein's revolution in physics revealed the fact that Newton's supposed laws of nature were not laws at all; this stern verdict, however, flows from an unrealistic expectation of what laws can be expected to deliver for us. We prefer to affirm approximate laws, such as Newton's, as genuine laws of our world, as long as it is made clear the degree of approximation that is involved and the limitations on the range of circumstances under which this approximation holds. We now know of some more accurate laws than Newton's, but most likely they also hold only to a certain degree of precision under certain circumstances. In a sense we do not know these new laws as completely as we know Newton's, because we are as yet unclear about their limitations. If laws were as precise and unified and universal as tradition assumed, our theory of laws would have no trouble encompassing that possibility; but our theory is tailored to allow the alternative possibility that there is only a loose confederation of approximate laws.

Consider the relation between variables v and t

$$v = gt.$$

In philosophical language variables like this are known as *determinables*, and they take on an infinite number of *determinate* values. The relation between determinables is equivalent to an infinite conjunction of relations between *determinates*. Each relation between determinates is a relation of proportion – each velocity and time stand in a proportion equal to g, which is a constant quantity. In fact determinate velocities and times do not stand *directly* in any relation of proportion at all. Velocities only stand in direct relations of proportion to other velocities, and similarly times only stand in relations of proportion to other times. In general for all determinables, some determinate is chosen as the 'unit' and the other determinates can be identified by the proportion in which they stand to the unit. It is these relations of proportion, of the velocities and times to their respective units, that in turn stand in the relation of proportion given by g. Thus this indirect relation of proportion between velocities and times can only be expressed once units of measurement have been specified. In SI units g has a value of about 9.8 $m\ s^{-2}$.

These relations of proportion between quantities are referred to by means of real numbers. On the philosophy of mathematics we support, real numbers are in fact *identical with* relations of proportion such as these – on other philosophies of mathematics the relation between numbers and these proportions will be more indirect. See Bigelow [1988] and Bigelow and Pargetter [1990] for more on the theory that numbers simply are proportions.

Thus, the upshot of this Whewellian climb up a hierarchy of laws is that laws emerge as involving relations among properties. In the case used for illustration, the law concerns a relation of proportion holding among properties. What we arrive at, by consilience of inductions, is a confidence that within real states of systems there are components which stand in proportions which lie within some interval of approximation around some precise value. The holding of these relations has a special status, which we call natural necessity, and which we will elucidate by reference to a distinction between natural and non-natural states for our world.

3. Formal Expression of Laws

When discussing laws of nature in thoroughly general terms, philosophers need a brief and schematic representation that can stand proxy for any particular law of nature, shedding all the idiosyncrasies that distinguish each law from the others. The standard scheme that philosophers have adopted for this purpose is that of a universal generalisation, framed within the predicate calculus:

$$\forall x(Fx \supset Gx),$$

"For any thing, x, if it is an F then it is a G", or "All F's are G's". For our purposes, however, a different schema will be needed.

One problem with the "all F's are G's" schema is that it quantifies over individuals, whereas the laws we have been dealing with express relations between properties, like the properties of acceleration and distance from the centre of the Earth. Therefore we suggest that a law should involve quantification over properties rather than individuals, or at least be expressed as a relation among properties which in turn entails quantification over properties.

The theory that laws fit the standard schemas that quantify over individuals, rather than properties, is the core of what we call a regularity theory. An objection which can be raised against any regularity theory takes off from the plausible premise that there could be laws that are never instantiated. Arguably, such a law could not be captured by quantification over individuals. Consider for instance a very early stage of the universe, just after the alleged Big Bang. On our theory there may be many laws of nature that hold of the universe at that time, even if there are as yet very few regularities which manifest those laws. Indeed, imagine a universe just like ours, except that it has a God that decides to cut the universe short a few seconds after it begins. We maintain that if its God had allowed it to evolve according to its own nature it would have evolved in a similar way to ours. But most of the entities that make up our world never actually exist in that world. In our world, the laws are displayed in regularities which are absent in that world. Yet both worlds share the very same laws. Thus the laws cannot be identified with the regularities which they explain.

Furthermore, there are some theories that attempt to explain the origin of the universe in terms of the laws of physics. (See Halliwell [1991], Davies [1992, Chapters 2 & 3], and Davies's contribution to this volume.) In this case, these laws are taken to explain the origin of matter, so the existence of these laws cannot depend on the existence of matter. While these explanations of the origin of the universe are very speculative and may not be correct, we would not want to adopt a philosophical account of laws that rules these theories out, and the regularity account does do this.

On our theory therefore, laws may hold at an early stage of the universe whether or not the future is destined to reveal complex regularities which will manifest these laws. If suitable regularities do emerge, they do so because of the laws that held in the world already. The laws are not laws at the early stage of the universe because of the regularities that the future holds in store. The laws are not constituted by the future regularities. Rather, we maintain, the laws are relations among properties, and it is the obtaining of these relations among properties which will bring about the emergence of various regularities whenever the circumstances shape up in the right way for the laws to manifest themselves among the contingently existing individuals in this world. As a consequence, we construe a law as a relation between properties, and we urge that a law should be expressed by quantification over properties rather than over individuals.

Consider the equation

$$a = k/r^2,$$

which expresses a relation between the determinables a and r that holds to a certain degree of approximation in circumstances C. By quantifying over determinate values of these determinables a and r, this law can be re-expressed in terms of determinate quantities, and at the same time indicating that the relation holds only in circumstances C, as follows:

$$\forall_C a' \forall_C r' \; a' = k/r'^2.$$

The expression $\forall_C r'$ is read "for all r' in circumstances C". We have written dashes on the quantities a' and r', to indicate that these are determinate values rather than the determinables which feature in the original equation. The key thing to note about this equation relating determinates is that it quantifies over properties, not over individuals. It says that any property of a certain sort (acceleration in circumstances of a given kind) and any other property of another sort (distance from a certain sort of body) will stand in a certain relation.

More can be said to express the relation between the circumstances *C* and the relation of proportion between acceleration and distance. To do this we return to the law expressed in terms of determinables. Since we wish to express the law in terms of properties, we will refer to the property of "being in circumstances *C*". The key idea is that there may be some sort of connection, an "entailment", between the property of "being in circumstances *C*", and the property of "being such that the specified relation between acceleration and distance holds". Strict logical entailment will not, in the end, be quite what we want; but for the moment, just consider the theory that would emerge if we were to take there to be a simple logical entailment between these properties. The theory we would arrive at could then be formulated by saying that the above law of nature is a relation of entailment that holds just in case:

$$\hat{x}(Cx) \text{ entails } \hat{x}(a_x = k/r_x^2).$$

The symbol $\hat{x}$ ("*x* hat") can be read as "is a system (call it *x*) such that". The expression *Cx* means that *x* is in the circumstances *C*. The inset proposition is thus to be read: "the property 'is an *x* such that *x* is in circumstances *C*' entails the property 'is an *x* such that the acceleration of *x* stands in proportion *k* to the inverse square of its distance from the centre of a body of mass *M*'". There is a logical entailment between properties just in case it is logically impossible for a thing to have one of these properties without also possessing the other.

Consider now the "consequent" of this law:

$$\hat{x}(a_x = k/r_x^2).$$

We have written the subscript *x* on the determinables *a* and *r* to remind us that the acceleration and distance relate to the same object which we have labelled *x*. The symbol a_x is to be read as a definite description: "the acceleration of *x*" or, more fully, "the proportion in which the acceleration of *x* stands to the unit acceleration". The consequent of the law is thus a very complicated property; we prefer to call it the property of being a system in a certain *state*. It is the state of having an acceleration, and a distance from a suitable object, whose proportions to their respective units, in turn, stand to one another in a specific relation of proportion. The "antecedent" of the law is also a property of being a system in a certain state, the state of being in circumstances *C*. This state is also a complicated conjunction of properties such as "being near a body of mass *M*" and "having mass that is small compared to *M*" and so on. Thus the law is written as a relation between two states of a system.

This statement of the law is not, however, satisfactory as it stands. It is not a logical necessity that the proportion in question should hold in circumstances *C*: there is no self-contradiction in supposing there to be a world in which the circumstance obtains and the proportion fails to hold. The relation of proportion only holds between the relevant quantities provided that the laws of nature are obeyed, and that will not be so in all possible worlds; we think it need not even be so at all times and places in the actual world. To register this fact, we need to add a further qualification to the property of "being in circumstances *C*": we need to add something which ensures that the system we are considering is one which obeys the laws of nature. Thus, we need to feed into the law a condition we will label $N_{@}x$. The relevant proportion between physical quantities can be guaranteed to hold in circumstances *C* only on the supposition that some further condition $N_{@}x$ is met, and we call this condition the naturalness condition.

The relation of proportion identified in the law of gravity is, we maintain, one which holds only on condition that some condition is met that we have labelled $N_{@}x$. One way to read $N_{@}x$ is as asserting that "the system *x* obeys the laws of nature of our world". It would be hoped that this will not be the only possible reading of this condition, since it makes direct appeal to the very notion – laws of nature – that our theory is aiming to analyse. However, if our theory is to give a correct characterisation of laws of nature, then any adequate construal of the condition $N_{@}x$ will have to be extensionally equivalent to the reading "the system *x* obeys the laws of nature of our world". The account of the condition $N_{@}x$ we adopt in this paper is that "the states of x are natural states for the actual world". This account will be discussed further in subsequent sections, but under this account the meaning of the symbolism $N_{@}x$ can be explained as follows: the symbol @ refers to our world, the actual world, and the symbol $N_{@}$ refers to the property of instantiating only states that are, as we will say, "natural" for the actual world.

The statement of the law of gravitional acceleration, thus becomes

$$\hat{x}(Cx \;\&\; N_{@}x) \text{ entails } \hat{x}(a_x = k/r_x^{\,2}).$$

Incidentally, if we wish we can now express this law in terms of determinate quantities as follows:

$$\forall a' \, \forall r' \{\hat{x}(Cx \;\&\; N_{@}x \;\&\; a'(x) \;\&\; r'(x)) \text{ entails } \hat{x}(a_x' = k/r_x'^{2})\}.$$

For the sake of economy, we will only deal with the law expressed in terms of determinables rather than determinates.

We suggest that all laws can be expressed in a similar form, as relations between states of systems. The general form of a law can thus be written:

$$\hat{x}(Cx \;\&\; N_{@}x) \text{ entails } \hat{x}(Rx).$$

Many laws are similar in form to the Newtonian law of gravitational acceleration, constituted by the holding of a certain equation, a certain relation among various quantities, under certain circumstances. For these laws, the circumstances under which the relation holds are not usually explicitly included in the scientific representation of the laws. Other scientific laws explicitly state relations between different states of a system. For these laws, it is convenient to split the circumstances C into two parts, the explicitly mentioned state, and the presupposed and unmentioned background: A & C'. In this case the states explicitly mentioned when we describe the law in the science concerned will be just A and R. These foreground states may refer to the same time or to different times; and R is usually later in time although in some cases it may be earlier, as in the case of time-reversible laws. C' refers to the circumstances under which the law holds, within some degree of precision. Given this articulation of a law, what we have is the existence of the following entailment between states:

$$\hat{x}(Ax \ \& \ C'x \ \& \ N_{@}x) \text{ entails } \hat{x}(Rx).$$

Reference to background conditions and to the naturalness condition are generally left tacit, and so the scientific expression of the law thus often takes the abbreviated form "any A must be an R". This is the form focussed on by the Dretske, Tooley and Armstrong theory that laws are relations of necessitation holding among universals.

Consider an indeterministic law with two possible outcomes R_1x and R_2x. In this case the consequent of the law may be written $\hat{x}(R_1x$ or $R_2x)$. The indeterministic theory may also have laws which have consequents that give the probabilities of the two possible outcomes: $P(R_1x)$ and $P(R_2x)$. Thus the representation of indeterministic laws will be straightforward. It is also straightforward to accommodate the possibility that the laws change with time. In this case $N_{@}x$ means that x always instantiates only the states that are natural for our world at the appropriate time or over the appropriate period. A time index could be added ($N_{@(t)}x$) if one wished to refer to the natural states for the world at a particular time.

In conclusion, then, we urge that laws are relations between states. The scientific articulations of laws, often encapsulated in equations, are best construed as identifying such a relationship between states. This schema for lawhood is flexible and allows room for a variety of expressions, appropriate to the mathematical equations of theoretical physics, the less mathematical laws of other sciences, and laws with only local and approximate jurisdiction.

4. Universals and Natural Necessity

We have now constructed a schema for laws of nature; this schema reflects the fact that laws arise from relations among properties. Such relations can hold, as we have argued, irrespective of whether there are or even ever will be any actual systems instantiating these properties. This raises the question of the philosophical status of properties, and in particular the status of properties that are not instantiated and perhaps never will be.

In philosophical tradition, properties and relations are generally taken to be *universals.* There are other views on the nature of properties, but we will not consider these. Universals are generally thought to be those things that are or can be instantiated by more than one thing at once. While many universals are indeed of this sort, we do not assume that all properties or relations have to be of that kind. We allow that there may be properties which are necessarily unique, which can be instantiated at most once. Thus we do not take multiple instantiation to be an essential feature of universals. Rather, we take the defining characteristic of universals to be the fact that they *can* be instantiated, whether by individuals or by other universals. A universal can either instantiate or be instantiated: an individual can instantiate but cannot be instantiated. Bigelow and Pargetter (1990) defend this view. What is crucial here is the importance of the link between universals and the possibility of instantiation. For universals, to be is to be instantiable. There are no universals which nothing could possibly instantiate. This ontological exclusion of impossible properties will be crucial to our theory.

To say that a property exists, we maintain, is not to say that some entity exists in some kind of Platonic heaven. The claim that a property exists is logically equivalent to the claim that something could instantiate such a property. It might appear at first that "there are" some self-contradictory properties that nothing could instantiate. However, we maintain that a self-contradictory putative property, like that of "being less than two feet wide and more than five feet wide in the very same direction" does not exist. For any such putative property, there are predicates in English, but there are no properties that correspond to these predicates. We think that this sense of what it means for a property to exist corresponds well with ordinary scientific usage. We might say: "there exist certain discrete energies that are solutions to equation H, but there exist no energies between these that are solutions to this equation".

The account of the condition $N_{@}x$ we adopt in this paper is that "the states of x are natural states for the actual world". This idea of there being natural and non-natural states for our world can be enlarged upon by considering what is known as the semantic view of scientific theories. (See Giere [1988], Lloyd [1988], van Fraassen [1989].) On this view a theory can be identified with the set of possible states allowed by that theory. In the case of the approximate Newtonian theory we considered, the possible states of the theory are those that are either not in circumstances C or have their acceleration due to gravity lying within a given interval of the proportion k/r^2. These states share the general state $\hat{x}(\sim Cx \text{ or } Rx)$, where R is the state of having acceleration due to gravity lying in that interval. The impossible states according to the theory are those that are in circumstances C yet have acceleration that does not lie within that interval. These states share the general state $\hat{x}(Cx \,\&\, \sim Rx)$. We will express this by saying that the state $\hat{x}(\sim Cx \text{ or } Rx)$ is allowed by the theory and its contrary $\hat{x}(Cx \,\&\, \sim Rx)$ is disallowed by the theory.

On the semantic theory, some general state is allowed by the theory, its contrary is not; we cast a novel realist gloss on the distinction between the allowed and disallowed states. We claim that if the theory is true, then there is something in the world that corresponds to the distinction between the allowed state and the disallowed state; there is a reason why the theory is *right* to "allow" this state and "disallow" its contrary. We claim the allowed state has a property that the disallowed state lacks, and we claim that it shares this property with states allowed by other true theories. This common property of the states allowed by true theories is a property we are calling naturalness for our world. The states disallowed by true theories share the property of being non-natural for our world. It would be non-natural for things in our world to take on one of these states.

The general form of a law is:

$$\hat{x}(Cx \;\&\; N_{@}x) \text{ entails } \hat{x}(Rx),$$

that is, when this entailment holds between properties of these kinds, then it is a law that things in circumstances C manifest state R. Given the distinction between natural and non-natural states, it is not difficult to see that this entailment between properties will hold if we take the reading of the condition $N_{@}x$ to be "the states of x are natural" i.e. $N_{@}$ is the property of instantiating only those states that are natural for our world. First assume that for a system to be in circumstances C but not in state R would be non-natural. In other words the combined state $\hat{x}(Cx \;\&\; {\sim}Rx)$ is an non-natural state, for our world. We also assume that the state $\hat{x}(Cx \;\&\; Rx)$ is a natural state. Then if we make the assumption that the system x is one that takes on natural states only, it will follow that it cannot instantiate the state $\hat{x}(Cx \;\&\; {\sim}Rx)$, and it will follow from this that if it instantiates C then it must instantiate R.

What is the ontological ground of an entailment between properties? We will argue that an entailment between properties, such as that involved in a law of nature, will hold just in case a certain conjunction of properties does not exist. A law, we have said, takes the following form: if a system instantiates circumstances C, and given that the system meets the naturalness condition $N_{@}x$, then it cannot fail to instantiate state R. This means that the property of being a system in circumstances C, *and* being a system which instantiates only natural states, *and* having the property of not being in state R, would be a contradictory property. And so on our view, that conjunctive property does not exist. That is:

$$\hat{x}(Cx \;\&\; N_{@}x \;\&\; {\sim}Rx) \text{ does not exist.}$$

Given that something is in circumstances C and instantiates only natural states, therefore, it *must* be in state R, because there is no such property as that of not being in state R, when being in circumstances C and being in entirely natural states.

Thus it is the case that

$$\hat{x}(Cx \;\&\; N_{@}x) \text{ entails } \hat{x}(Rx)$$

just in case the property $\hat{x}(Cx \;\&\; N_{@}x \;\&\; {\sim}Rx)$ does not exist. Ontologically, what we can say is that an entailment rests on the nonexistence of a certain possibility. This is parallel to what we meet in elementary logic, where it is said that the premises of an argument entail its conclusion just in case there does not exist a possible world where the premises and the negation of the conclusion are true together. Similarly, one property entails another just in case it is not logically possible for something to instantiate the first property and not the second property, so the conjunctive property of instantiating the first and a contrary of the second is a putative property which simply does not exist.

There is a complication worth noting. When we have a law of nature, it should be impossible for a thing to be in a natural state in the circumstances C without satisfying R; but it should also be *possible* for a thing to be in a natural state and be in those circumstances while it does satisfy R. Otherwise the conditions of lawhood would be vacuously satisfied, since then $\hat{x}(Cx \;\&\; N_{@}x)$ would not exist, so one of the properties involved in the putative law would not exist. Thus, for there to be a law of nature, we should require it to be the case that $\hat{x}(Cx \;\&\; Rx)$ is itself a natural state, or in other words, that $\hat{x}(Cx \;\&\; N_{@}x \;\&\; Rx)$ must exist.

A second complication is also worth noting. It is a strength of our theory that laws of nature and laws of logic and mathematics are treated as subspecies of a single genus. But it is also important to distinguish laws of nature from laws of logic and mathematics. If we are dealing with a law of nature, as opposed to laws of logic or mathematics, then it is logically possible for something to fail to fit the law, that is, for something to be in conditions C without satisfying condition R. This shows us that, if we are to be focussing on laws of nature, then we need to add the requirement that $\hat{x}(Cx \mathbin{\&} \sim Rx)$ exists. For suppose that it did not: suppose that,

$\hat{x}(Cx \mathbin{\&} Rx)$ exists, but
$\hat{x}(Cx \mathbin{\&} \sim Rx)$ does not exist.

Then it would follow that

$\hat{x}(Cx)$ entails $\hat{x}(Rx)$,

and hence trivially that

$\hat{x}(Cx \mathbin{\&} N_{@}x)$ entails $\hat{x}(Rx)$.

Hence, if this were all that was required for something to be a law of nature, every law of logic or law of mathematics would count as a law of nature. The distinction between these species derives just from the fact that when what we have is a law of nature but not of logic or mathematics, then there is a logical possibility that a thing might have C without R. In a law of nature therefore, as distinguished from a law of logic, although there is no such property as $\hat{x}(Cx \mathbin{\&} N_{@}x \mathbin{\&} \sim Rx)$, there is such a property as $\hat{x}(Cx \mathbin{\&} \sim Rx)$.

Bringing these comments together, we propose the following conditions as being required for a law of nature:

$\hat{x}(Cx \mathbin{\&} N_{@}x \mathbin{\&} Rx)$ exists,
$\hat{x}(Cx \mathbin{\&} \sim Rx)$ exists,
$\hat{x}(Cx \mathbin{\&} N_{@}x \mathbin{\&} \sim Rx)$ does not exist.

When these conditions are met, then:

$\hat{x}(Cx \mathbin{\&} N_{@}x)$ entails $\hat{x}(Rx)$,

and it is a law of nature that R obtains in circumstances C.

We can now make sense of the idea that there is a kind of natural necessity that is not as strong as logical necessity. We can define a relation of natural entailment,

$\hat{x}(Cx)$ naturally entails $\hat{x}(Rx)$,

as logical entailment conditional on the supposition that the system has the property of instantiating only natural states for the actual world. And we can then say that it is a matter of natural necessity that *R* obtains given that *C* obtains.

Thus we have defined a notion of natural entailment that is weaker than logical entailment: it is logical entailment conditional on the states of *x* being natural states. This notion of natural entailment is not so weak as to reduce laws to mere regularities. A law can entail but is not entailed by, a regularity.

Consider the law

$$\hat{x}(Cx \mathbin{\&} N_{@}x) \text{ entails } \hat{x}(Rx).$$

The following regularity follows from this law:

$$\forall x(Cx \mathbin{\&} N_{@}x \supset Rx).$$

(This follows because the property $\hat{x}(Cx \mathbin{\&} {\sim}Rx \mathbin{\&} N_{@}x)$ does not exist, so it cannot be instantiated.) Then if we make the assumption that the states instantiated by all of the *x*'s are always natural:

$$\forall x\, N_{@}x,$$

then it follows that

$$\forall x(Cx \supset Rx).$$

This is the sort of thing which is taken by the regularity theory to be a candidate for a law of nature. Our theory allows that, when things always possess only natural states, and when the appropriate circumstances do come about, laws do result in such regularities. It is a mistake, however, to conflate a law with the regularity, if there is one, which it explains.

Consider the regularity "all the books now on this desk are hardbacks." If this is the case, then it is true that under the circumstances *C* of being a book now on this desk, it is the case *R* that this book is a hardback. However this is not a law of nature – it would be consistent with the laws of nature for there to be a paperback now on this desk. We would say that while the conjunction of state *C* and *R* is a natural state for our world, the state *C* and not-*R* is also a natural state for our world. Our theory gives a very natural account of what distinguishes laws from accidental generalisations such as this. There is a law relating circumstances *C* to state *R* if the conjunction of *C* and not-*R* is non-natural for our world. In the case of an accidental generalisation, the conjunction of *C* and not-*R* happens not to be actually instantiated, but it is a perfectly natural state.

It follows that laws and accidental generalisations differ in their treatment of counterfactual suppositions. Thus although in the case of the accidental generalisation it is true that all things in circumstances *C* do instantiate state *R*, it is possible that if something else were to be in circumstances *C* then it would not instantiate state *R*, given that it instantiates only natural states. In the case of the books on this desk, if counterfactually, there had been another book now on this desk, then it could have been a paperback, while still instantiating only natural states. In the case of the law, if anything else were to be in circumstances *C*, then it is not possible that it would not also be in circumstances *R*, given that it instantiates only natural states. Thus a law does not only tell us what does happen in circumstances *C*, it also tells us what must happen in those circumstances, on the assumption that things instantiate natural states. This is the sense in which laws tell us what must happen and what could not possibly happen, as well as what does and does not happen. In contrast, accidental generalisations only tell us what does and does not happen.

5. The Naturalness Condition

We have presented a recommendation about how the naturalness condition $N_{@}x$ should be read – as "the states of x are natural". However, it is possible to read this condition in other ways, and these rival readings will generate other necessitarian theories, some of which will resemble various pre-existing theories on the market. We will here discuss two of those theories, and the reasons we reject them. In conclusion we will say something more about our reading of the naturalness condition.

One option that we have explored in a conference paper, but now reject, is to suppose that the laws of nature are essential properties of our world. On this view, for something to be law-abiding, for it to meet the naturalness condition $N_{@}x$, all that is required is for this thing to be "natural"; and this requires only that it be part of "nature"; and this in turn requires only that it exist in this world we live in. This view identifies nature with the world. On this view, it is a sufficient condition for being law-abiding that a thing simply exist in this world: it could not exist in this world without obeying the laws of this world. This is a theory advanced by Bigelow [1990], taking up and adapting an idea which appeared in Bigelow, Ellis and Lierse [1992]. According to this theory, in formulating a law of nature, we can replace the condition $N_{@}x$ with the condition: "the world x is in @". In Lewis's counterpart theory, it would be clearer if we said that "the world x is in is a counterpart of @". What is important is that when something breaks a law of nature, on this theory, what we must say is that the world this happens in is another world there could have been instead of this one, and not another way that this actual world could have been. Under this replacement, the entailment which constitutes a law of nature becomes something of the form:

$$\hat{x}(\text{the world } x \text{ is in is @ \& } Cx) \text{ entails } \hat{x}(Rx).$$

On this view, the laws of nature are essential properties of the world: this world would not be the world it is if it did not have these fundamental features. Call this the world essence theory.

The idea developed in Bigelow, Ellis and Lierse [1992], and further elaborated by Ellis and Lierse in work in progress, harking back in some respects to Harré & Madden [1975], is slightly different from the world essence idea. Ellis's idea was that the world belongs to a natural kind, as do many of the things in it, and that laws of nature are essential properties of natural kinds. Something which obeys a law could not cease to do so without ceasing to belong to the natural kind that it does belong to. Laws become logical necessities relative not to the sheer existence of a thing, but to the fact that the thing belongs to some specific natural kind. On this view, what is required for a thing to obey a law, what is required in order for a thing to meet the naturalness condition $N_{@}x$, is just for that thing to belong to a specific natural kind. Replacing $N_{@}x$ by "x belongs to such-and-such a natural kind", then, the entailment which constitutes a law of nature:

$$\hat{x}(Cx \mathbin{\&} N_{@}x) \text{ entails } \hat{x}(Rx),$$

becomes something like this:

$$\hat{x}(Cx \mathbin{\&} x \text{ belongs to kind } K) \text{ entails } \hat{x}(Rx).$$

This is, according to Ellis and Lierse, the general form of a law.

This general form admits of one special case worthy of notice. Ellis and Lierse argue that some laws are grounded in the essential properties of the world as a whole: conservation laws, and laws of symmetry for example. In this case, the form of the entailment grounding a law will take the form:

$$\hat{x}(Cx \mathbin{\&} x \text{ is inhabiting a world of kind } K) \text{ entails } \hat{x}(Rx).$$

This gives an alternative to the world essence theory in Bigelow [1990].

There is, however, a heavy intuitive cost to all these theories. It seems easy to imagine little miracles and anomalies, small infringements of the laws of nature, occurring in our world. When we do this, we seem to be imagining things that could have happened in this very world, this very world of this very kind that we find ourselves in. When we imagine things in this world breaking the laws of nature – people walking on water for instance – we do not seem to be thinking of other things that there could have been instead of people and water, in worlds of a different kind from ours. We seem to be imagining people just like us walking on the same kind of water that we are familiar with, in a world which is of the same kind, indeed, which is numerically identical with the very world we are in. It is paradoxical to adopt a theory which would entail counterfactuals like this: "If Murphy had walked on water, as she said she did, then she would not have been a woman and it would not have been water she had walked on, and indeed this world, a world of this kind, would not have existed". We assert, in opposition to all these theories, that it is a logical possibility for a person to walk on water in this very world: if someone had done so then that would not have necessarily made this a different world, or a world of a different natural kind.

Miracles and anomalies may not actually happen in the actual world. Perhaps there are laws of nature that are truly exceptionless, who knows? Yet, we maintain, it is possible to be confident that there are laws of nature, without being confident that these laws are obeyed absolutely without exception, through all space and time. We urge a gentler, kinder conception of laws.

Return now to our theory of the naturalness condition $N_{@}x$. We have said that some states are natural for our world and others are non-natural for our world. We have used the term "non-natural" as the opposite to "natural". We have not used the term "unnatural", because that can be taken as a synonym for "artificial", which is not the meaning wanted. Another possible opposite to the natural is the "supernatural". One option would be to suppose that if there were any supernatural phenomena such as human psychic powers or the powers of other agents such as gods, then the exercise of these powers could contravene the laws of nature. States that result from supernatural intervention would then be classed as non-natural. However, it is not obvious that the supernatural must necessarily fail to be law-governed. If supernatural powers are exercised with any kind of pattern it may not be "contrary to the nature of the world" that they occur, and it might be possible to frame some law or laws to encompass them – even if it is only "the power is exercised when the agent wills it so". Thus on our view this would simply be to discover more laws of nature, and some former laws may need to have their circumstances of application limited to those where the so-called supernatural powers are not exercised. This means that the only truly non-natural states of affairs are ones where laws that previously held were broken by an "anomaly", something – whether "natural" or "supernatural" – that did not share the property, if there is one, which underlies all law-abiding phenomena.

There are, we maintain, natural states and non-natural states for our world. We have epistemic access to natural states because we observe, every day, things taking on natural states. Perhaps all the states we ever witness are perfectly natural. It is an ontologically serious commitment to suppose that there is some property which all these states have in common, and which we call the property of being natural. If there is such a property, then there really are laws of nature – there is a "nature" for there to be laws *of*. If we are right, then there is a real distinction between natural and non-natural states for our world, and the property of instantiating only natural states is shared by things which obey the laws of nature. If there were no such distinction among states, we maintain, then there would be no real distinction between laws and mere regularities. Hume would then have been right. But we think Hume is wrong; and that this means that there is a real property of naturalness of states, and it is in virtue of the fact that things take on only natural states that these things fall under the regularities they do.

Our theory requires that there must exist a property of naturalness, a real property common to law-abiding states, if there are to be laws of nature. For present purposes we can remain agnostic about what, in more detail, can be said about this property of naturalness. But one is free to speculate. The picture of laws we have given so far is consistent with the laws existing "before" the Big Bang – there could have been natural entailments among universals even when there were no individuals in our universe at all. If one adopts this view, one could take natural entailments among universals as primitive, or suppose they are grounded in other properties or propensities that may have been instantiated by other individuals that existed "before" the Big Bang – such as "ideas in the mind of God", for example. On the other hand one might take the view that natural entailments are grounded in the properties, or propensities, of things in the world, like the complex properties of the fields that first appeared with the Big Bang, and in the properties of other entities that have formed since. Our theory is neutral on the ultimate grounds of natural entailments among properties. All that our theory requires is that these natural entailments among properties exist.

Bibliography

Armstrong, D. [1978]: *Universals and Scientific Realism*, Vols. 1 and 2. Cambridge: Cambridge University Press.

Armstrong, D.M [1983]: *What is a Law of Nature?* Cambridge: Cambridge University Press.

Bigelow, J.C. [1988]: *The Reality of Numbers: A Physicalist's Philosophy of Mathematics*. Oxford: Clarendon Press.

Bigelow, J.C., Pargetter, R.J. [1990]: *Science and Necessity*. Cambridge: Cambridge University Press.

Bigelow, J.C. [1990]: "The World Essence", *Dialogue XXIX*, pp.205-217.

Bigelow, J.C., Ellis B.D., Lierse, C. [1992]: "The World as One of a Kind: Natural Necessity and Laws of Nature", *British Journal for the Philosophy of Science 43*, pp.371-388.

Butts, R.E. [1968]: *William Whewell's Theory of Scientific Method*. Pittsburgh: University of Pittsburgh Press.

Davies, P. [1992]: *The Mind of God*. Penguin Books.

Dretske, F.I. [1977]: "Laws of Nature", *Philosophy of Science, 44*, pp.248-68.

Forster, M.R. [1988]: "Unification, Explanation, and the Composition of Causes in Newtonian Mechanics", *Studies in the History and Philosophy of Science, 19*, pp.55-101.

Giere, R.N. [1988]: *Explaining Science: A Cognitive Approach*. Chicago: University of Chicago Press.

Gleick, J. [1992]: *Genius: Richard Feynman and Modern Physics*. London: Little, Brown and Company.

Halliwell, J. [1991]: "Quantum Cosmology and the Creation of the Universe", *Scientific American*, December.

Harré, R. and Madden, E.H. [1975]: *Causal Powers: A Theory of Natural Necessity.* Oxford: Basil Blackwell.

Leckey, M.J. [1991]: "Quantum Mechanics and the Consilience of Inductions", M.A. Thesis, The University of Melbourne.

Lloyd, E.A. [1988]: *The Structure and Confirmation of Evolutionary Theory*. New York: Greenwood Press.

Tooley, M. [1977]: "The Nature of Laws", *Canadian Journal of Philosophy*, *7*, pp.67-98.
Tooley, M. [1987]: *Causation: A Realist Approach*. Oxford: Clarendon Press.
Van Fraassen, B.C. [1989]: *Laws and Symmetry*. Oxford: Clarendon Press.

Ronald N. Giere

The Skeptical Perspective: Science without Laws of Nature

1. Interpreting the Practice of Science

It is a fact about humans that their practices are embedded in interpretive frameworks. This holds both for individuals and for groups engaged in a common enterprise. Of course any sharp distinction between practice and interpretation, whether drawn by participants or third-party observers, will be somewhat arbitrary. Nevertheless, drawing some such distinction is useful, perhaps even necessary, for those who, while not direct participants in a practice, seek to understand it from their own perspective.

Such is the situation of historians and philosophers of science regarding the practice of science and the concept of a 'law of nature.' The claim of some philosophers, for example, that scientists seek to discover laws of nature, cannot be taken as a simple description of scientific practice, but must be recognized as part of our interpretation of that practice. The situation is complicated, of course, by the fact that, since the seventeenth century, scientists have themselves used the expression 'law of nature' in characterizing their own practice. The concept is thus also part of the interpretative framework used by participants in the practice of science. That shows that the concept sometimes lives in close proximity to the practice, but not that it is divorced from all interpretive frameworks.

Insisting on the interpretive role of the concept of a law of nature is important for anyone like myself who questions the usefulness of the concept for understanding the practice of science as a human activity. I realize full well that many others do not share this skeptical stance. Being part of the characterization of the goals of science is but one interpretive role played by this ubiquitous concept. Laws played an essential role in Hempel's [1948, 1965] influential analysis of scientific explanation, and they continue to play a central role in more recent accounts [Salmon, 1984]. Nagel's [1961] classic analysis of theoretical reduction focuses on the derivation of the laws of one theory from those of another theory. Even critics of these analyses, including radical critics [Feyerabend, 1962], have generally focused on other features and left the role of laws unexamined. A concern with the status of laws has inspired many investigations into the confirmation or falsification of universal statements. Laws also figure in contemporary analyses of the concept of determinism [Earman, 1986]. And scientific realism is often characterized in terms of the truth, or confirmation, of laws referring to theoretical entities.

It is thus not surprising that, like Kant two centuries ago, many contemporary philosophers take it as given that science yields knowledge of claims that are at least universal, and perhaps necessary as well. Their problem, like Kant's, is to show how such knowledge is possible.[1] Doubting that such knowledge is actual, I have little interest in rebutting arguments that it is possible. More serious would be claims that knowledge of universal and/or necessary laws is not only actual, but necessary for understanding the practice of science. But I shall not here be concerned to rebut such arguments.

I will begin by advancing some general reasons for skepticism regarding the role of supposed laws of nature in science. Then I will outline an alternative interpretive framework which provides a way of understanding the practice of science without attributing to that practice the production or use of laws of nature as typically understood by contemporary philosophers of science. Finally, I will sketch explanations of how some expressions can play a fundamental role in science without being regarded as 'laws,' and how one can even find necessity in nature

1 Among recent philosophers, David Armstrong [1983] seems to me to come closest to the Kantian stance.

without there being 'laws of nature' behind those necessities. I shall thus be offering an interpretation of science even more radical than what David Armstrong once called the 'truly eccentric view ... that, although there are regularities in the world, there are no laws of nature' [1983, p.5]. On my interpretation there are both regularities and necessities, but no laws.

2. Historical Considerations

One way of understanding the role that a concept plays in an interpretation of a practice is to examine the *history* of how that concept came to play the role it now has. Through the history one can often see the contingencies that led to that concept's coming to play the role it later assumed and realize that it need not have done so.

Of course there is a standard answer to this sort of historical argumentation. The origins of a concept, it is often said, are one thing; its validity quite another. Philosophy is concerned with the validity of a concept, whatever its origins.[2] But this answer rings somewhat hollow in the present context. It is typically assumed that we need a philosophical analysis of the concept of a law of nature *because* that concept plays an essential role in our understanding of science.[3] Inquiring into how the concept came to play its current role may serve to undercut this presupposition.

---

2 This answer is an obvious generalization of Reichenbach's [1938, Ch. 1] famous distinction between the contexts of discovery and justification for scientific hypotheses.

3 Armstrong, for example, writes [1983, p. 4]: 'If the discovery of the laws of nature is one of the three great traditional tasks of natural science, then the nature of a law of nature must be a central ontological concern for the philosophy of science.' Similarly, John Earman describes the concept of laws of nature as 'a notion that is fundamental to the study not only of determinism but to the methodology and content of the sciences in general' [1986, p. 81].

Among the characteristics attributed to laws of nature by contemporary philosophers of science, several are especially prominent. Laws of nature, it is typically said, are *true* statements of *universal* form. Many would add that the truths expressed by laws are not merely contingent, but in some appropriate sense *necessary* as well. Finally, laws are typically held to be *objective* in the sense that their existence is independent of their being known, or even thought of, by human agents.[4]

These characteristics, I believe, came to be associated with some scientific claims not simply through reflection on the practice of science, but in large part because of particular circumstances obtaining in Europe in the seventeenth century when modern science began to take the form it now exhibits. Unfortunately, there seem to be few sources that focus directly on this question, and undertaking such a study is beyond both the purposes of this paper and my own expertise. So here I can offer only some suggestions and a few references.[5]

The main sources for the use of 'laws of nature' as a concept to interpret the practice of science are to be found, it seems, in the works of Descartes and then Newton. For both, the laws of nature are prescriptions laid down by God for the behavior of nature.[6] From this premise the predominant characteristics of laws of nature follow as a matter of course. If these laws are prescriptions issued by God the creator of the universe, then of course they are true, hold for the whole universe, are necessary in the sense of absolutely obligatory,[7] and independent of the

---

4 These are a subset of the assumed characteristics of laws that van Fraassen [1989, p. 38] picks out as pre-eminent.

5 Why the question of the origin of the notion of laws of nature has received so little attention from historians of science is itself a subject for still further speculation. My guess is that the correctness of the idea has been so taken for granted that few have felt the need to inquire into its origins.

6 That our modern use of the concept of 'laws of nature' is directly traceable back to Descartes and Newton, and flowed from their conceptions of the Deity, was argued both by Zilsel [1942] and Needham [1951].

7 Here it is important to observe the medieval distinction between what is necessary for God's creations from what is necessary for the deity itself.

wishes of humans, who are themselves subject not only to God's laws of nature, but to His moral laws as well.

There is at least one place in Newton's writings where this line of reasoning is explicit. In an unpublished draft of Query 31 of the *Optics*, dating from around 1705, Newton draws on his conception of the deity to support the universality of his laws of motion. 'If there be an universal life and all space be the sensorium of a thinking being who by immediate presence perceives all things in it,' he wrote, 'the laws of motion arising from life or will may be of universal extent.'[8] The modesty with which the connection is here asserted was appropriate. What *empirical* evidence did Newton have for the universality of his laws of motion? Only terrestrial motions, such as falling bodies, projectiles, and pendulums, and the motions of the Sun, Moon, planets, and, allowing the investigations of Edmund Halley, perhaps comets as well. The fixed stars posed a definite problem, for what prevented the force of gravity from pulling all the stars together into one place? Newton had need of his God.

Despite some arguments to the contrary, it seems pretty clear that the idea of laws of nature as emanating from the Diety did not originate with Descartes and Newton, or even in the seventeenth century at all.[9] Nor were all earlier uses of such notions

---

Both Descartes and Newton were 'voluntarists' in that they believed God could have chosen other laws for the world. Descartes notoriously even held that the laws of arithmetic and geometry could have been different if God had so willed.

8 Quoted by Westfall [1971, p. 397]. I owe this reference to Brooke [1991, p. 139]. Chapter VII of Westfall's book, particularly the last five pages, makes a strong case for the role of Newton's conception of the deity in his willingness to abandon direct mechanical interaction in favor of apparent action at a distance.

9 Both Zilsel [1942] and Needham [1951] claimed that the idea of God's laws for nature originated with the rise of powerful centralized governments in the early modern period. Thus Zilsel [1942, p. 258] argues unequivocally that 'the concept of physical law was not known before the seventeenth century' and suggests, more tentatively, that 'the doctrine of universal natural laws of divine origin is possible only in a state with rational statute law and fully developed central sovereignty' [p. 279]. Oak-

necessarily connected with that of a personified lawgiver. The distinction between divine laws for humans as opposed to laws for the rest of animate or inanimate nature can be traced back at least to Roman thinkers. On the other hand, by the thirteenth century, Roger Bacon seems to have thought of the laws of optics, reflection and refraction, in very much the secular way that became commonplace in the nineteenth. Galileo is famous for his employment of the 'two books' metaphor in which God is portrayed the author of both the Bible and the 'Book of Nature.' But the idea of 'laws of nature' in the sense of Descartes and Newton seems not to have been part of his understanding of the new science. Robert Boyle, who shared many of Newton's theological beliefs, nevertheless urged caution in using the notion of laws of nature on the grounds that, strictly speaking, only moral beings, and not inanimate matter, can appreciate the meaning of laws. One finds similar qualms in the writings of Aquinas.

There is another factor in the story which seems relatively distinct from theological influences, namely, mathematics. Would the concept of laws of nature have gained such currency in the absense of simple mathematical formulae which could be taken express such laws? And do not the qualities of universality and necessity also attach to mathematical relationships? These questions are as difficult as they are relevant. Galileo had the mathematical inspiration, but apparently did not think of the book of nature as containing 'laws.' Kepler, on the other hand, thought of laws in somewhat the same way as Descartes and Newton. Clearly the theological and mathematical influences both push in the same direction. In any case, the one does

ley [1961] objects that the idea existed long before in a theological tradition. Ruby argues that already in the thirteenth century Roger Bacon used the notion in a way that 'resembled that of modern science' [1986, p. 350] [p. 301 in this volume]. The comments which follow are based on my reading of all of the above mentioned authors.

not exclude the other. Perhaps both were necessary for the notion of a law of nature to have developed at all.[10]

In the end one may still ask why Descartes and Newton were so strongly inclined to interpret various mathematical formulae as expressions of God's laws for nature when thinkers a century earlier or a century later were far less inclined to do so. I would suggest the influence of the bloody religious conflicts exhibited in the Thirty Years War and the English Revolution respectively. These conflicts made it very difficult for anyone in France or England then to think about nature in significant ways without considering the possible role of God.[11] What matters, however, is not which ideas one can find when. At almost any period in history one can find a vast range of ideas existing simultaneously. The important question is which of the variety of ideas available at an earlier period got adopted and transmitted to later periods and shaped later interpretations. Here there can be no serious doubt that for Descartes and Newton the connection between laws of nature and God the creator and lawgiver was explicit. Nor can there be any doubt that it was Newton's conception of science that dominated reflection on the nature of science throughout the eighteenth century and most of the nineteenth as well.[12]

The secularization of the concept of nature's laws proceeded more slowly in England than on the continent of Europe. By the end of the eighteenth century, after the French Revolution, Laplace could boast that he had no need of the 'hypothesis' of God's existence, and Kant had sought to ground the universality

10 I own consideration of the importance of mathematics in this history to conversations with Rose-Mary Sargent. She also pointed out that Boyle's cautions regarding the use of the notion arose partly from his conviction that mathematical relationships abstracted too much from the complexities of nature.

11 Toulmin [1990] has recently emphasized the role of the Thirty Years War on Descartes' thinking.

12 The influence of Newton's conception of science on later British thought hardly needs documenting. For its influence on French Enlightenment thought, see Gay [1969, Book 3, Ch. 3].

and necessity of Newton's laws not in God or nature, but in the constitution of human reason. Comte's positivism found a large audience in France during the middle decades of the nineteenth century. But, in spite of the legacy of Hume, whether the laws of nature might be expressions of divine will was still much debated in the third quarter of the nineteenth century in Britain. Here the issue was whether Darwin's 'law of natural selection' might just be God's way of creating species. Not until Darwin's revolution had worked its way through British intellectual life did the laws of nature get effectively separated from God's will.[13]

It is the secularized version of Newton's interpretation of science that has dominated philosophical understanding of science in the twentieth century. Mill and Russell, and later the Logical Empiricists, employed a conception of scientific laws that was totally divorced from its origins in the theological climate of the seventeenth century. The main issue for most of this century and the last has been what to make of the supposed 'necessity' of laws. Is it merely an artifact of our psychological make-up, as Hume argued, an objective feature of all rational thought, as Kant argued, or embedded in reality itself?

My position, as outlined above, is that the whole notion of 'laws of nature' is very likely an artifact of circumstances obtaining in the seventeenth century. To understand modern science we need not a proper analysis of the concept of a law of nature, but a way of understanding the practice of science that does not simply presuppose that such a concept plays any important role whatsoever.

3. The Status of Purported Laws of Nature

What is the status of claims that are typically cited as 'laws of nature' – Newton's Laws of Motion, the Law of Universal Gravitation, Snell's Law, Ohm's Law, the Second Law of Thermody-

13 For an appreciation of the intensity of these debates see Brooke [1991, Ch. VIII] and Desmond and Moore [1991]

namics, the Law of Natural Selection? Close inspection, I think, reveals that they are neither universal nor necessary – they are not even true.[14]

For simplicity consider the combination of Newton's Laws of Motion plus the Law of Universal Gravitation around the year 1900, before the advent of relativity and the quantum theory. Could one find, for example, any two bodies, anywhere in the universe, whose motions exactly satisfied these laws? The most likely answer is 'no'. The only possibility of Newton's Laws being precisely exemplified by our two bodies would be either if they were alone in the universe with no other bodies whose gravitational force would effect their motions, or if they existed in a perfectly uniform gravitational field. The former possibility is ruled out by the obvious existence of numerous other bodies in the universe; the latter by inhomogeneities in the distribution of matter in the universe. But there are other reasons as well for doubting the precise applicability of the laws. The bodies would have to be perfectly spherical, otherwise they could wobble. They could have no net charge, else electrostatic forces would come into play. And they would of course have to be in 'free space' – no atmosphere of any kind which could produce friction. And so on and on.[15]

Many excuses have been given for not taking more seriously the lesson that, strictly speaking, most purported laws of nature seem clearly to be false. A recent one is that the laws actually discussed by scientists are not the 'real' laws of nature, but at

14 This thesis was argued thirty years ago by Michael Scriven [1961] and more recently by others including Nancy Cartwright [1983, 1989] and myself [1988a]. Even Armstrong [1983, pp. 6-7] and Earman [1986, pp. 80-81] admit the strict falsity of the traditional examples of laws.

15 For a more extended discussion of the strict falsity of the law of the pendulum see [Giere, 1988a, pp. 76-78]. That classical mechanics has be superseded by relativity theory and quantum mechanics does not materially change the argument. Cartwright [1983, 1989] provides examples from quantum theory. Similar examples could be developed for relativity theory as well.

best 'near' laws.[16] Here I wish only to examine a view that does take the lesson seriously, but remains still too close to the traditional view. This is the view, developed by Coffa [1973] and Hempel [1988], that laws are expressed not by simple universal statements, but by statements including an implicit 'proviso.' As I understand it, Coffa's and Hempel's account is that purported statements of laws of nature of the form 'All bodies, ..., etc.' are to be interpreted as really of the form 'All bodies, ..., etc., with the proviso that ...' My objection to this interpretation is that it is impossible to fill in the proviso so as to make the resulting statement true without rendering it vacuous.

This problem is particularly evident in cases where the implicit proviso must be understood to be expressed in concepts that are not even known at the time the law containing the implicit proviso is first formulated. Thus most of the laws of mechanics as understood by Newton would have to be understood as containing the proviso that none of the bodies in question is carrying a net charge while moving in a magnetic field. That is not a proviso that Newton himself could possibly have formulated, but it would have to be understood as being regularly invoked by physicists working a century or more later.[17] I take it to be a *prima facie* principle for interpreting human practices that we do not attribute to participants claims that they could not even have formulated, let alone believed.

It is important to realize that my objection is not just that the proviso account introduces indefiniteness into our interpretation of science. One of the major lessons of post-positivist philosophy of science is that no interpretation of science can make everything explicit. Important aspects of the practice of science must remain implicit. The issue is where, in our interpretation of science, we locate the unavoidable indefiniteness. The proviso account locates indefiniteness right in the formulation of what, on that account, are the most important carriers of the

16 For an elaboration of this view see Swartz [1985] and Swartz in this volume [Ch. II. pp. 67-91].

17 For a more extended development of this objection see [Giere, 1988b].

content of science, namely, its laws. I think a more faithful interpretation would locate the indefiniteness more within the practice of science and leave its products, including its public claims to knowledge, relatively more explicit.[18]

4. Models and Restricted Generalizations

Let us return to the example of Newton's equations of motion together with his equation for the force of gravity between two bodies. My reference here to Newton's *equations* of motion rather than his *laws* of motion is deliberate. Everyone agrees that Newton used these equations. The issue is how to interpret them, whether as 'laws,' which was Newton's interpretation, or as something else.

Interpreting the equations as laws assumes that the various terms have empirical meaning and that there is an implicit universal quantifier out front. Then the connection to the world is relatively direct. The resulting statement is assumed to be either true or false.

On my alternative interpretation, the relationship between the equations and the world is *indirect*. We need not initially presume either a universal quantifier or empirical meaning. Rather, the expressions need initially only be given a relatively abstract meaning, such as that m refers to something called the mass of a body and v to its velocity at a specified instant of time, t. The equations can then be used to construct a vast array of abstract mechanical systems, for example, a two-body sys-

18 Cartwright [1983] holds the superficially similar view that lower level laws, such as Snell's Law, are to be understood as *ceteris paribus* laws of the form: 'Everything else being equal, ..., etc.' But she does not claim that such laws are true, only that they are explanatory in a way not compatible with a covering law model of explanation. I would prefer a more radical interpretation that does away with law talk even though this departs from the way scientists themselves often present their science. I think this can provide us (philosophers) with a better understanding of what they (scientists) are doing.

tem subject only to mutual gravitational attraction. I call such an abstract system a *model*. By stipulation, the equations of motion describe the behavior of the model with perfect accuracy. We can say that the equations are exemplified by the model or, if we wish, that the equations are true, even *necessarily* true, for the model. For models, truth, even necessity, comes cheap.

The connection to the world is a provided by a complex relationship between a model and an identifiable system in the real world. For example, the earth and the moon may be identified as empirical bodies corresponding to the abstract bodies in the model. The mass of the body labeled m_1 in the model may be identified with the mass of the earth while the distance r in the model is identified with the distance between the center of the earth and the center of the moon. And so on. Then the behavior of the model provides a representation of the behavior of the real earth-moon system. For the purposes of understanding the relationship by which the model represents the real system, the concept of truth is of little value. A model, being an abstract object rather than something linguistic, cannot literally be true or false. We need another sort of relationship altogether.

Some friends of models invoke *isomorphism*, which is at least the right kind of relationship.[19] But isomorphism is too strong. The same considerations that show the strict falsity of presumed universal laws argue for the general failure of complete isomorphism between scientific models and real world systems. Rather, models need only be similar to particular real world systems in specified respects and to limited degrees of accuracy. The question for a model is how well it 'fits' various real world systems one is trying to represent. One can admit that no model fits the world perfectly in all respects while insisting that, for specified real world systems, some models clearly fit better than others. The better fitting models may represent more aspects

19 Van Fraassen [1980, 1989], for example, defines scientific realism as that one of a family of models is exactly isomorphic with the system it is intended to represent. I have objected [Giere, 1985; 1988a, Ch. 4] that this is too strong a requirement for a reasonable realism.

of the real world or fit some aspects more accurately, or both. In any case, 'fit' is not simply a relationship between a model and the world. It requires a specification of which aspects of the world are important to represent and, for those aspects, how close a fit is desirable.

In this picture of science the primary representational relationship is between individual models and particular real systems, e.g., between a Newtonian model of a two-body gravitational system and the Earth-Moon system. But similar models may be developed for the Earth-Sun system, the Jupiter-Io system, the Jupiter-Sun system, the Venus-Sun system, and so on. Here we have not a universal law, but the *restricted generalization* that various pairs of objects in the solar system may be represented by a Newtonian two-body gravitational model of a specified type. Restricted generalizations have not the form of a universal statement plus a proviso, but of a conjunction listing the systems, or kinds of systems, that may successfully be modeled using the theoretical resources in question, which, in our example, are Newton's equations of motion and the formula for gravitational attraction.

Other pairs of objects in the solar system cannot be well represented by the same sort of model, the Earth-Venus system, for example. Moreover, although one could in principle construct a single Newtonian model for all the planets together with the sun, the resulting equations of motion are not solvable by any known analytical methods. One cannot even solve the equations of motion for a three-body gravitational system, one intended to represent the Earth-Jupiter-Sun system, for example. Here one must approximate, for example, by treating the influence of the Earth as a perturbation on the motion of the two-body Jupiter-Sun system. Such approximative techniques have been part of Newtonian practice since Newton himself, but have been largely ignored by the tradition that interprets Newton's equations of motion as expressing universal laws of nature.

It is typically said to be a major part of Newton's success that he 'unified' the behavior of celestial and terrestrial bodies.

The equations of motion used to build models of the Jupiter-Sun system may also be used to construct models to represent the behavior of balls rolling down an inclined plane, pendulums, and cannon balls. This was a considerable achievement indeed, but it hardly elevates his equations of motion to universal laws. It had yet to be shown that similar models could capture the comings and goings of comets, and the fixed stars were beyond anyone's reach. What Newton had in 1687 were not God's all encompassing laws for nature, but a broad, though still restricted, generalization about some kinds of systems that could be modeled using the resources he had developed. That he had fathomed God's plan for the universe was an interpretation imported from theology.

5. Principles versus Laws

It may reasonably be objected that focusing simply on Newton's equations of motion does not do justice their role in the science of mechanics. They seem somehow to capture something fundamental about the structure of the world. One might express similar feelings about the Schrödinger equation in quantum mechanics. The problem is to capture this aspect of these fundamental equations without lapsing back into the language of universal laws.

An interpretative device that has considerable historical precedent would be to speak of Newton's *Principles of Motion* and the *Principle of Gravitational Attraction*. The title of his book, after all, translates as *The Mathematical Principles of Natural Philosophy*.[20] Whether or not thinkers in the seventeenth, or even eighteenth, century recognized any significant distinction between 'laws' and 'principles,' we can make use of the linguistic variation. Principles, I suggest, should be understood as rules devised by humans to be used in building models to

20 Recall also that Descartes' main work in natural philosophy was titled *The Principles of Philosophy*.

represent specific aspects of the natural world. Thus Newton's principles of mechanics are to be thought of as rules for the construction of models to represent mechanical systems, from comets to pendulums. The rules instruct one to locate the relevant masses and forces, and then to equate the product of the mass and acceleration of each body with the force impressed upon it. With luck one can solve the resulting equations of motion for the positions of the bodies as a function time elapsed from an arbitrarily designated initial time.

What one learns about the world is not general truths about the relationship between mass, force, and acceleration, but that the motions of a vast variety of real world systems can be successfully represented by models constructed according to Newton's principles of motion. And here 'successful representation' does not imply an exact fit, but at most a fit within the limits of what can be detected using existing experimental techniques. The fact that so many different kinds of physical systems can be so represented is enough to justify the high regard these principles have enjoyed for three hundred years. Interpreting them as universal laws laid down by God or Nature is not at all required.[21]

21 It is worth noting that in the twentieth century the expression 'principle of relativity' has had considerable currency, as in the title of the well-known collection of fundamental papers by Einstein and others [Einstein *et. al.* 1923]. Einstein himself [1934] distinguished between what he called 'constructive' theories and 'principle' theories. The special theory of relativity, he claimed, was of this latter type. One of its principles is the 'principle of the constancy of light in vacuo' [1934, p. 56]. I doubt, however, that Einstein's intent corresponds to my own, given that he describes the advantages of principle theories over constructive as being 'logical perfection and security of the foundations' [p. 54]. He seems to think of his principles' as expressing deep general truths about the world, and, like Newton, draws on religious, though not theological, inspiration.

6. Necessity Without Laws

Traditionally it has been the supposed universality of laws of nature that has seemed to require their necessity. For, as Kant argued, how could a universal association be just a regularity? For an association to be truly universal, he thought, there must be something making it be so. Thus, denying the existence of genuine universal laws in nature makes it possible to deny the existence of necessity as well. But such denial is not required. It is also possible to deny the existence of universal laws of nature while affirming the existence of causal necessities.[22]

Consider a model of a harmonically driven pendulum of the sort that one would use to represent the motion of a pendulum on a typical pendulum clock. Solving the classical equations of motion for the period as a function of length (assuming that the angle of swing, θ, is sufficiently small that $\cos\theta \approx 1$) yields the familiar result that the period is proportional to the square root of the length. Now this model provides us with a range of possible periods corresponding to various possible lengths. These possibilities are built into the model. But what of the real world?

Suppose the actual length of the pendulum on my grandfather clock is L. The model permits us to calculate the period, T. It also permits us to calculate a slightly greater period T' corresponding to a slightly greater length, L'. Suppose the clock is running slightly fast. I claim that turning the adjusting screw one turn counter-clockwise would increase the length of the pendulum to L', and this would increase the period to T', so that the clock would run slightly slower. This seems to be a claim not about the model but about the real life clock in my living room. Moreover, it seems that this claim could be true of the real life clock even if no one ever again touches the adjusting screw. These possibilities, it seems, are in the real physical system, and are not just features of our model.

22 These two positions are represented by van Fraassen [1980, 1989] and Cartwright [1983, 1989] respectively.

There are, of course, many arguments against such a realistic interpretation of modal (causal) claims. Here I will consider only the empiricist argument that there can be no evidence for the modal claim that is not just evidence for the regularity relating length and period for pendulums. The inference to possibilities, it is claimed, is an unwarranted metaphysical leap. Moreover, I will not try to argue that this empiricist interpretation is mistaken; only that it is no less metaphysical than the opposing view.

I claim that by experimenting with various changes in length and observing changes in period one can effectively *sample* the possibilities that the model suggests may exist in the real system. That provides a basis for the conclusion that these possibilities are real and have roughly the structure found in the model. The empiricist argument is that the most one can observe is the actual relationship between length and period for an actual series of trials with slightly different initial conditions. So the issue is whether experimentation can reveal real possibilities in a system or merely produce actual regularities in a series of trials. Whichever interpretation one favors, one cannot claim that the latter interpretation is somehow less metaphysical than the former. It is just a different metaphysical view. I think the modal realist interpretation provides a far better understanding of the practice of science, but that is not something one can demonstrate in a few lines, or even a whole paper.

References

Armstrong, D. [1983] *What is a Law of Nature?* (Cambridge: Cambridge University Press).

Brooke, J.H. [1991] *Science and Religion: Some Historical Perspectives.* (New York: Cambridge University Press).

Coffa, J.A. [1973] *Foundations of Inductive Explanation.* (Ann Arbor: University Microfilms).

Cartwright, N. [1983] *How the Laws of Physics Lie.* (Oxford: Clarendon Press).

Cartwright, N. [1989] *Nature's Capacities and Their Measurement.* (Oxford: Oxford University Press).

Desmond, A. and Moore, J. [1991] *Darwin.* (London: Penguin).

Earman, J. [1986] *A Primer on Determinism.* (Dordrecht: Reidel).

Einstein, A. *et. al.* [1923] *The Principle of Relativity*. (London: Methuen).
Einstein, A. [1934] 'What is the Theory of Relativity?' *Essays in Science*. (New York: Philosophical Library).
Feyerabend, P.K. [1962] 'Explanation, Reduction, and Empiricism.' *Minnesota Studies in the Philosophy of Science, Vol. 3, Scientific Explanation, Space, and Time*, ed. H. Feigl and G. Maxwell. (Minneapolis: University of Minnesota Press).
Gay, P. [1969] *The Enlightenment: An Interpretation*. (New York: Knopf).
Giere, R.N. [1985] 'Constructive Realism.' *Images of Science*, P.M. Churchland and C.A. Hooker, eds. (Chicago: University of Chicago Press).
Giere, R.N. [1988a] *Explaining Science: A Cognitive Approach*. (Chicago: University of Chicago Press).
Giere, R.N. [1988b] 'Laws, Theories, and Generalizations.' *The Limitations of Deductivism*, A. Grünbaum and W. Salmon, eds., (Berkeley: University of California Press) pp. 37-46.
Hempel, C.G., and Oppenheim, P. [1948] 'Studies in the Logic of Explanation.' *Philosophy of Science* 15, pp. 135-75.
Hempel, C.G. [1965] *Aspects of Scientific Explanation*. (New York: Free Press).
Hempel, C.G. [1988] 'Provisos: A Problem Concerning the Inferential Function of Scientific Theories.' *The Limitations of Deductivism*, A. Grünbaum and W. Salmon, eds. (Berkeley: University of California Press).
Nagel, E. [1961] *The Structure of Science*. (New York: Harcourt, Brace, and World).
Needham, J. [1951] *The Grand Titration: Science and Society in East and West*. (London: George Allen and Unwin Ltd).
Oakley, F. [1961] 'Christian Theology and the Newtonian Science: The Rise of the Concept of the Laws of Nature.' *Church History*, 30.
Reichenbach, H. [1938] *Experience and Prediction*. (Chicago: University of Chicago Press).
Ruby, J. [1986] 'The Origins of Scientific "Law".' *Journal of the History of Ideas*, pp. 341-59.
Salmon, W.C. [1984] *Scientific Explanation and the Causal Structure of the World*. (Princeton: Princeton University Press).
Scriven, M. [1961] 'The Key Property of Physical Laws – Inaccuracy.' *Current Issues in the Philosophy of Science*, H. Feigl and G. Maxwell, eds. (New York: Holt, Rinehart, and Winston) pp. 91-101.
Swartz, N. [1985] *The Concept of Physical Law. (Cambridge: Cambridge University Press).*
Toulmin, S. [1990] Cosmopolis: The Hidden Agenda of Modernity. (New York: Free Press).
van Fraassen, B.C. [1980] *The Scientific Image*. (Oxford: Oxford University Press).
van Fraassen, B.C. [1989] *Laws and Symmetry*. (Oxford: Oxford University Press).

Westfall, R.S. [1971] *Force in Newton's Physics.* (New York: American Elsevier).
Zilsel, E. [1942] 'The Genesis of the Concept of Physical Law.' *The Philosophical Review*, pp. 245-79.

Brian Skyrms and Karel Lambert[1]

The Middle Ground: Resiliency and Laws in the Web of Belief

1. Introduction

In the final section of "Two Dogmas of Empiricism" – entitled "Empiricism without the Dogmas" Quine paints a striking picture of the pragmatics of belief:

> The totality of our so-called knowledge or beliefs, from the most casual matters of geography and history to the profoundest laws of atomic physics or even of pure mathematics and logic, is a man-made fabric which impinges on experience only along the edges. Or, to change the figure, total science is like a field of force, whose boundary conditions are experience. A conflict with experience at the periphery occasions readjustments in the interior of the field. Truth values have to be redistributed over some of our statements. Reevaluation of some statements entails reevaluation of others, because of their logical interconnections – the logical laws being in turn simply further statements of the system, certain further elements of the field. Having reevaluated some statements we must reevaluate some others, which may be statements logically connected with the first or may be statements of logical connections themselves. But the total field is so underdetermined by its boundary conditions, experience, that there is much latitude of choice as to what statements to reevaluate in the light of any single contrary experience.

But however attractive this picture may be, Quine does not offer any methodology for modeling and mapping the networks of belief and representing the place of laws within them.

We believe that the best framework for a precise realization of these ideas is the theory of personal probability. The question then arises how to map a network of degrees of belief in a way which reveals the weak and strong resistances to disconfir-

1 Both authors are grateful for support from the University of California President's Research Fellowship in the Humanities.

mation and more generally how need for revision tends to be accommodated by the network. Here we need to pay attention not only to the probabilities of statements, but also their probabilities conditional on other statements. Conditional probabilities play a key role as rules of inference which guide belief revision, but which themselves are also modified in the process of belief revision. Based on the conditional probability structure, notions of invariance, independence, and conditional independence all play a role in determining the place of a statement on the web of belief.

The general framework we have in mind has room for another feature of Quine's account. That is that the treatment of laws of mathematics and even some laws of logic is of a piece with the treatment of laws of nature. Of course, some logical foundation is necessary if we are to use the theory of probability, but the amount that is needed is often overstated. Essentially we assume finite truth functional logic in order to get the fundamental Boolean structure on which to do probability theory. We certainly do not need to assume a probability value of one for every logical or mathematical truth, however those categories may be interpreted. As a result the truth-functional tautologies are given a special unrevisable status. Some may consider this a flaw in the model, but it is difficult to see how anything more than impressionism is possible unless some structure is assumed. Consequently, we are prepared to insist here that "departure from the law of excluded middle would count as evidence of revised usage of 'or' and 'not"' ["Carnap and Logical Truth" p.113]. Perhaps something interesting could be done on even more parsimonious assumptions, but we will not pursue that line of thought here.

The plan of the paper is as follows: In the next section, we introduce the framework and the definition of resiliency. In section 3 we claim that for a statement to achieve the status of a well-confirmed law of empirical science, it must have resiliently high degree-of-belief. Section 4 extends the discussion to *networks* of laws. We then indicate some ways that the account can be applied to mathematical and logical laws in section 5.

2. Probability and Resiliency

We assume that the set of statements for which the epistemic agent in question has degrees of belief is closed under finite truth functional operations. No further closure conditions are required but anything further is permitted. We assume that our agent's degrees of belief are probabilities on the resulting boolean algebra. That is, we assume that every statement has a probability, that probabilities are non-negative real numbers, that the probability of a finite truth-functional tautology is equal to one, and that if p & q is truth functionally inconsistent then pr(p or q) = pr(p) + pr(q).

Notice that there is nothing in the foregoing modest idealization that prohibits an agent from assigning probability .99 to the axiom of choice, .9 to Fermat's last theorem and .1 to the 100th digit of pi being 7. For that matter, even the logic of quantifiers is up for grabs. An agent may be in doubt about: "If No horses fly and Pegasus is a horse then Pegasus does not fly." This is not to deny the magnitude of the idealization that our agent assigns probability one to all truth functional tautologies of whatever complexity, but only to contrast it with far more sweeping idealizations that have not been made.

We assume that conditional probabilities are of central importance for belief change. We have in mind either belief change which takes place either by straightforward conditioning or the generalized form of conditioning introduced by Richard Jeffrey as probability kinematics. Then it is plausible to characterize resistance to belief change in terms of *invariance* under conditioning. This is just probabilistic independence with respect to the statements which may serve as conditions. The central idea in *Causal Necessity* was to use independence, conditional independence, and approximate independence to elucidate interrelationships in the web of belief. High probability with a large domain of invariance is *robust belief* and it is the robust beliefs which form the central core of the network of beliefs.

Robust beliefs have a certain kind of necessity. They are highly probable not only in the current belief state, but also in all belief states accessible from the current belief state by conditioning on statements in the domain of invariance. From the pragmatic point of view, it is this sort of necessity that constitutes the felt necessity of laws. The greater the domain of invariance, the more compelling the sense of necessity.

One type of invariance was given a special name there: *resiliency*. Suppose we have a set of statements, P, and we are interested in invariance of the probability another statement, q, with respect to all possible situations that we can describe by truth-functional combinations of members of P. We consider only truth-functional combinations of members of P which are consistent with q and its denial and which have positive probability. If the probability of q is invariant with respect to conditioning on all such propositions, then we say that it is *absolutely resilient* with respect to the set P. If conditioning causes the probability of q to vary by at most ε, we say that the probability of q is *ε-resilient* with respect to the set, P.

One can define *resilient* and *ε-resilient* for conditional probabilities along similar lines. The probability of q conditional on r is *resilient* with respect to a set of statements if it is invariant under conditioning on any truth functional compound of the set which is consistent both with r & q and with r & not-q. A probability (or a conditional probability) might not be resilient with respect to a set, S, but might become so if the agent conditioned on some other statement, k. There might even be some finite partition, k1, ... kn, such that upon conditioning on any member of this partition, the probability (or conditional probability) in question would become resilient over S. In this case we say it is resilient over S conditional on the partition. A few examples may help to introduce these ideas:

Example 1: Paradoxes of Material Implication

The truth functional conditional is notorious for the fact that it is perfectly legitimate – in fact compulsory – to believe it just because one disbelieves the antecedent or just because one believes the consequent. That is to say that the probability of the material conditional must be at least a great as that of the denial of its antecedent and at least as great as that of its consequent. Thus, if one believes that Oswald killed Kennedy one must also believe at least as strongly that if Oswald didn't kill Kennedy, then Bertrand Russell did – where "If-then" is the material conditional. Likewise, one must also believe that if Kennedy was killed by the Mafia, then Oswald killed Kennedy. If one believes the stated conditionals only for the stated reasons, and one's other beliefs about the matter are those endorsed by the Warren commission, then one's probabilities for these conditionals, although high, will not be resilient over the set of antecedent and consequent. Thus, if one conditions on the statement that Oswald didn't kill Kennedy one will give up one's belief in each of these conditionals. If belief in the material conditional, If p then q, is resiliently high over {p,q} then the conditional will support dynamic modus ponens and modus tollens. That is to say that conditioning on p will lead to a high probability of the denial of q and conditioning on the denial of q will lead to a high probability of the denial of p.[2]

Example 2: Probabilistic Causation

A probabilistic cause is prior to and positively probabilistically relevant to its effect, but the positive relevance must not be *spurious* - that is to say it must not disappear when one conditions on confounding factors. One variety of standard example arises when the putative cause and the putative effect are correlated not because the first brings about the second but because they are effects of a common cause. The probabilistic dependence disappears when one conditions on the presence or ab-

2 The example and the key points of the analysis are due to Ernest Adams. See also the collections edited by Harper, Stalnaker and Pearce and by Eells and Skyrms.

sence of the common cause.[3] Thus if the *prima facie* correlation is spurious, one or both of the conditional probabilities $p(E|C)$; $p(E|-C)$ will fail to be resilient over a confounding factor.

Example 3: Causal Decision Theory

We want to evaluate decisions in terms of what they are likely to bring about. We don't want to base decisions on spurious correlations. The infamous Newcomb problem[4] is an invitation to commit the fallacy of basing one's decisions on spurious correlation, as is the case of Prisoner's Dilemma with a clone. The fallacy is avoided by causal decision theory.[5] Causal decision theory can be seen as a straightforward application of the methdology for avoiding spurious correlations to the analysis of rational decisions. This is most evident in the version of causal decision theory due to Skyrms where conditioning on a background partition plays a key role.[6]

Example 4: Ergodic Probabilities

Consider a physical system whose state gets mapped onto a new state by a transformation, T. We think of T as modeling the repetition of an experiment. Repetition of an experiment should leave unchanged the physical probabilities which characterize the experiment, these probabilities should be *invariant* under the transformation. We also think that the physical probabilities should take into account all the relevant experimental background factors. One should not be able to change it by conditioning on some previously unspecified experimental factor.

3 See Suppes (1970) and Skyrms (1980)(1988).

4 See Nozick(1969).

5 See the contributions by Gibbard and Harper and by Stalnaker in Harper, Stalnaker and Pearce (1981), Lewis (1981) and Skyrms(1980),(1984).

6 See Skyrms (1980)(1982)(1984).

Physical probabilities must be *resilient* with respect to the experimental factors. Experimental factors are modeled as invariant sets in the probability space. If the experimental parameters are set one way, then repetition of the experiment does not alter them. Then the physical probability measures must be ones which are *ergodic* with respect to the transformation, T. Invariant probabilities are always mixtures of ergodic probabilities. Non-trivial mixtures represent uncertainty about settings of some experimental parameters; ergodic probabilities represent certainty about all such settings. The ergodic theorem can then be seen as a general characterization of conditions for successful induction.[7]

3. Laws and Resiliency

To count something as well – confirmed *qua* law, one must give it not only high degree-of-belief, but *resiliently* high degree-of-belief. The necessity of laws like the necessity of causes is resiliency. This is the leading idea of Skyrms (1980).

There are two main approaches to degrees-of-belief associated with laws, and the resiliency idea could be applied to either. One school holds that to believe a law is just to believe a universal statement. According to this view, to believe the law that ravens are black is give high probability to the universally quantified statement:

$$(\forall x)[Rx \supset Bx]$$

Ramsey advocated a different, pragmatic, view according to which the probabilities of interest are always probabilities of one or a finite number of instances. There are two versions of this view, one which concentrates on the unconditional probability of a instance regarding a new individual:

$$Pr[Ra \supset Ba]$$

7 See Skyrms 1984 Ch.3 and Skyrms forthcoming.

(or a finite number of instances), and one which concentrates of the natural "predictive" conditional probability:

$$Pr[Ba|Ra]$$

Carnap, who followed Ramsey in emphasizing the importance of these concepts, called the first the *instance confirmation* of a law and the second the *qualified instance confirmation* of a law.

We will use Hempel's paradox of confirmation to illustrate resiliency at the level of instance confirmation and qualified instance confirmation. Suppose someone had seen and indeed knew only of instances which were non-Ravens. Such a person might give very low probability to the proposition that some new unexamined individual, *a*, is a Raven. If so, she must give high probability to the material conditional, "If *a* is a Raven then *a* is black", and high instance confirmation to the law. But if the high probability for the material conditional is based solely on the belief that *a* is not a Raven, the *qualified* instance confirmation of the law will not be high. Conditional on a being a Raven one would be loath to predict that *a* is Black. What good is a law, if the hypothesis that it applies disqualifies its use for prediction?

Why not, then, just take qualified instance confirmation as the appropriate measure of confirmation? Hempel replies that laws are used many ways in inferences, and a reasonable account must take all such ways into account. For instance, the law might be used to predict that a non-Black thing is a non-Raven. So there is a whole bundle of conditional probabilities that should be high in order for the web of belief to be such that the law is applicable in normal ways. Qualified instance confirmation pays attention to only one of these conditional probabilities.

Notice, however, that in the example just considered the *Resiliency* of the probability material conditional is defective. It is not resilient over the antecedent of the conditional. If we require that the probability of the material conditional be resiliently high, we guarantee not only high qualified instance confirmation, but also high probability conditional on the denial of the consequent. If we require that it be high relative to a set of statements not only including *Ra* and *Ba* but also statements specifying the habitat of *a*, etc. one guarantees that instantiation of these conditions do not invalidate the application of the law.

If one were a Carnapian robot, starting off with a clean slate and forming one's beliefs by simple induction, then the question as to how one got such resilient confirmation would have a simple answer: *variety of evidence.* You must look at non-Black things as well as Ravens; at Arctic Ravens as well as temperate Ravens, etc. Of course, as was pointed out early on in the discussion[8], background knowledge makes a difference. If we already know that there are lots of non-black things, we don't have to worry about vacuous confirmation coming in the back door.

These points could restated in the context of confirmation as high probability of the universal generalization. High probability that there are no Ravens at all gives high, but perhaps not resiliently high probability for the universally quantified material conditional. A resiliently high probability for the universally quantified material conditional buys the instantial virtues at a somewhat higher price.

Sometimes laws are used without a precise specification of scope. Thus, "All Ravens are Black" might mean something like "All normal Ravens are black" without any exact explication of normalcy. Such imprecise specification of scope leaves lots of room for Duhemian maneuvers. In this regard, there is a kind of trade-off between resiliency and construal of scope. If Raven really means "Raven from a temperate climate" then instantial resiliency over "it's from the arctic" is no problem; things from the arctic are automatically things either not Ravens from a temperate climate or black. Likewise, resiliency of the probability of the universal generalization over "There are arctic Ravens" is not problematic since they would not be potential counterexamples. Restriction of the scope of a law gets resiliency on the cheap. But this is not the kind of resiliency that we value most

8 See Hosiasson-Lindenbaum (1940).

highly. The scientific ideal calls for both wide scope and resilient high probability.

4. Networks of Laws

Laws, however, do not stand in isolation. They operate as part of a network containing both other laws and auxiliary statements which are not laws. To understand the pragmatic structure of the network, we want to know the resiliency of beliefs relevant to the network as a whole, and of beliefs relevant to subnetworks. An analysis of this kind of a real network of scientific laws would be a major undertaking – as indeed it should be if the idea is worth its salt.

Here, we illustrate some of the principles involved with some very simple models. We will deal with a finite set of laws of the form "All Fs are Gs", where the Fs and Gs are just Boolean combinations of simple monadic predicates. We will apply the idea of resiliency here at the level of instances. That is, we will take a name, '*a*', for some new, unexamined individual and for each law:

$$(\forall x)[Rx \supset Bx]$$

consider its instance:

$$Ra \supset Ba$$

For a single law in isolation, we considered the probability of this instance and asked whether it was resilient. For a network of laws, we consider the conjunction of these instantiations, the probability of this conjunction and the resiliency or robustness of this probability.

The idea is straightforward enough, but it contains within it an interesting twist. The probability of q is resilient over the set P, if it is invariant with respect to conditioning on all truth functional combinations, s, of members of P *which are consistent with q and its denial and which have positive probability.* For some networks and some sets, P, the class of truth functional combinations which are exempted from serving as test conditions by the italicized clause may be strictly larger for the network than for one of its subnetworks. For instance, this can happen if some such s is a counterexample to some member of the network, but not to any member of the subnetwork. In such a case, it is possible that the whole network has resilient high probability, but that the high probability of the subnetwork is not resilient, with respect to P. Resiliency is not, in general, monotonic with respect to the formation of subnetworks. This phenomenon can indicate that belief in the subnetwork is robust only relative to the maintenance of assumptions stated elsewhere in the full network. On the other hand, resilient high probability for the subnetwork over a wide variety of relevant sets, P, can indicate that belief in the subnetwork is relatively independent of assumptions stated elsewhere in the network.

At a more general level, we can say that the internal pragmatic structure of the web of belief can be mapped by looking at the invariance (or approximate invariance) classes for probabilities attaching to its subnetworks. These show us the consequences of giving up a member of the network for the robustness of belief in the subnetwork.

5. Logical Laws

Earlier we asserted it as an ideal of science that its laws be both resilient and have wide scope. The wider the scope of the law and the greater its resiliency, the closer it is to the center of the web of belief. In natural science this means that the goal is to develop resilient laws holding unrestrictedly of all physical objects, in formal science it means ultimately developing resilient laws which hold unrestrictedly of all "mathematical" (abstract) objects, and in logic it means developing resilient laws holding unrestrictedly of all objects – period! Indeed, the rationale for the system of predicate logic (called 'Ontology') developed by the Polish logician S. Leszniewski – a logic whose laws purport to be about all objects, whether concrete and abstract – seems to accord nicely with the spirit of this scientific ideal.

However, the notion of scope is tricky. Scope can be restricted either in the language or in the rules for interpreting the language. Consider again the putative law of earlier sections, 'All ravens are black', where the topic of concern was really ravens in temperate regions. So construed, Arctic white ravens do not really constitute a counterexample. What is intended by 'scope' in this case is a restriction put on the antecedent of the generalization which, if made explicit, would read 'All ravens in temperate regions are black'. As we said earlier, this is a way of getting resiliency on the cheap, and it is done at the expense of wider scope in the sense just explained.

But there is another way of interpreting 'scope' (which also will be germane to our concerns later), namely, that the universe of discourse over which the quantifier 'All' ranges is restricted to ravens in temperate regions. This understanding of scope requires no more precise rephrasal of 'All ravens are black', but it also makes irrelevant Arctic white ravens as a class of objects with respect to which the generalization in question is resilient. So when we say that science aims at laws of the widest possible scope we mean that it aims at unrestricted laws ranging over all objects of a generous universe of discourse. Usually the context makes clear which use of the word 'scope' is at issue.

The scientific ideal that laws be both resilient and have the widest possible scope can be used to explain developments not only in the empirical sciences but also in mathematics and logic. We shall consider some simple but well known examples.

First mathematics. Prior to and during the period when Leonhard Euler flourished, numbers were understood as quantities. So negative "numbers," -1, for instance, were deemed "impossible" and thus not genuine. All genuine numbers were positive. In that environment, consider the plausible looking generalization

$$\forall x \forall y (x - y = \iota z(z + y) = x) \qquad (1)$$

where the bound variables range over the positive numbers and '*i*' is the definite description operator (read: 'the'). This generalization was not resilient with respect to the class of number pairs as number was then understood because if the pair $<x, y>$ is such that x is not greater than y, then the requisite number did not exist. Resiliency, on the cheap, however, is obtained by restricting (1) as follows:

$$\forall x \forall y\ (\mathit{If}\ x>y\ \mathit{then}\ x-y = iz(z+y)=x) \qquad (2)$$

(2) *is* resilient with respect to the class of number pairs x,y such that x is not greater than y. Just as an occurrence of an Arctic white raven does not reduce the resiliency of the generalization that ravens in temperate regions are black, so the number pair $<2,3>$ no longer reduces the resiliency of the new law (2). But it is obtained at a price, namely, scope; (2) is less fundamental than the overturned (1) because the former generalization no longer applies to *all* classes of number pairs.

With the advent of the conceptual change which rejected the identification of numbers with quantities, and admitted negative numbers as genuine numbers, (1) was revived. A change in the scope of (1), in the sense that the universe of discourse was enlarged, and, hence, also the set of entities over which the quantifiers ranged, now made (1) not only resilient with respect to all pairs of positive numbers, but also with respect to all pairs of numbers of *any kind.* This is an example where the initial plausibility of (1) and the desire to make it resilient by altering its scope in the sense of enlarging the things over which the quantifiers range may have initiated or at least encouraged the conceptual change of identifying numbers with something other than quantities; for example, sets of a certain kind.

Turning to the special case of the laws of (deductive) logic, which is the ultimate topic of concern in this section, they are unique in the sense that in contrast to the laws of the other sciences, empirical and formal, they are supposed to be resilient with respect to all sets of objects in the universe of discourse; 'All ravens are ravens', for example, is resilient with respect to ravens living in regions, temperate or otherwise, nonravens of any sort, etc. Indeed, they are supposed to be resilient with respect to parameters of any kind, or at least that is the ideal. But in the case of the logic of general terms (or predicates) this has not always been the case, and realization of that fact led to abandonment of some candidates for the status of logical law.

In the Port Royal theory of immediate inference, the inference from an A-statement to an I-statement was counted valid. So the generalization

Some *S* are *P* if all *S* are *P*, (3)

for all general terms (predicates) *S* and *P*, could be counted as a truth of logic. But, as is well known, (3) is not resilient with respect to the class of empty general terms if the antecedent of (3) is construed as a universal generalization and the consequent as an existential conjunction. For if *S* is the general term unicorns' and *P* is the general term 'horned animals', (3) comes to grief; in short, (3) is not resilient with respect to class of empty general terms, of which ' unicorns' is a member.

A popular and cheap way to restore the resiliency of (3) in the late middle ages was to restrict the class of general terms to nonempty general terms. This is the famous doctrine of existential import. But such a move restricts the scope of (3) in the sense that the range of entities over which the (bound) variables *S*, *P* range is narrowed, and hence restricts the applicability of formal logic. (3) gets replaced (in effect) by:

Some *S* are *P* if both all *S* are *P* and some *S* exists. (4)

For all general terms *S* and *P*. (4) is resilient with respect to the class of all general terms including those contradicting (3). In modern terminology, (4) is rendered as

$$\forall x(S(x) \supset P(x)) \wedge \exists x S(x)) \supset \exists x(S(x) \wedge P(x)) \qquad (5)$$

for all general terms *S* and *P*, and, indeed, apparently satisfies the ideal of deductive logic that its laws be resilient with respect to *all* parameters. Again one sees how the desire to have generalizations which are highly resilient and which have the widest possible scope encouraged, and explained, the move from (3) to (4) (or (5)).

The explanation is portable; it can be used to explain the development of free logic. A law of modern predicate logic is the generalization

$$\forall x P(x) \supset P(t) \tag{6}$$

for all general terms *P*, simple or complex, and all singular terms t. But, as is well known, where t is the expression Vulcan' (the "name" of the putative planet), an expression which refers to no existing object, and *P* is, for example, the predicate, E! (exists), (6) is false because, on the conventional interpretation of the quantifiers, the antecedent is true but the consequent is false. From the free logician's point of view, acquiring resiliency on the cheap for (6) by restricting the class of singular terms over which the bound variable t ranges – its scope, in one sense of the word[9] – to singular terms that refer to existents has all of the unpalatable consequences of restricting the bound variables ranging over general terms to restore high resiliency in the case of (3). In free logic, (6) gets replaced by

$$\forall x P(x) \supset (E!(t) \supset P(t)) \tag{7}$$

for all general terms *P* and all singular terms *t*, in accordance with the ideal of combined high resiliency and widest possible scope. For (6) does not require any restriction on the class of singular terms.

Could the continuous pressure toward wider and wider scope while maintaining high resiliency lead to a conceptual revolution similar to the introduction of negative and imaginary numbers? There are those – Dana Scott, Kit Fine and Terence Parsons, to name three – who find the notion of object as coextensive with existent individual too confining both technically and philosophically. There are theories of "arbitrary objects" and "non-existent objects". For such putative objects to gain the

9 Note that "scope" here does not refer to the domain of the quantifiers but rather to the singular terms which serve as parameters.

status we give to negative and imaginary numbers, they must show that they can pay their way in terms of increased utility of the resulting conceptual structure.

We will not attempt to judge the utility of these theories here. We only want to maintain that the issue is entirely pragmatic. As our ontological beliefs change, our judgements about what it means to say that the laws of logic should be resilient with respect to "everything" will also change, and hence so will the candidates for the status of logical laws -except at the truth-functional center of the web of belief. The laws in this truth-functional center, and they alone, have the status of what Carnap used to call "meaning postulates".

6. Conclusion

In summary, we believe that the structure of the web of belief should be analyzed within the theory of subjective probability. In this endeavor, the concept of conditional probability is of central importance. Relations of probabilistic dependence, independence and conditional independence – and of invariance and resiliency all have important parts to play in the analysis. The account treats empirical science, mathematics and most of logic as part of a seamless cloth of knowledge. Only the truth-functional boolean structure required as a foundation for probability has a special status. Laws – both laws of nature and laws of logic -have a kind of necessity grounded not in occult properties or entities, but rather in invariance within the belief structure. Their necessity is pragmatic, not metaphysical.

References

Adams, E. (1975) *The Logic of Conditionals: An Application of Probability Theory to Deductive Logic* Dordrecht: Reidel.

Carnap, R. (1962) *Logical Foundations of Probability* 2nd ed. Chicago: University of Chicago Press.

Diaconis, P. and Freedman, D. (1981) "Partial Exchangeability and Sufficiency" In *Statistics: Applications and New Directions* Calcuta: Indian Statistical Institute. 205-236.
Diaconis, P. and Zabell, S. (1982) "Updating Subjective Probability" *Journal of the Americal Statistical Association* 77:822-830.
Eells, E. and Skyrms, B. eds. (forthcoming) *Probability and Conditionals* Cambridge: Cambridge University Press.
Fine, K. (1985) *Reasoning with Arbitrary Objects* Oxford University Press: Oxford.
Harper, W. , Stalnaker, R. and Pearce, G. eds. (1981) *Ifs* Dordrecht: Reidel.
Hempel, C. G. (1965) *Aspects of Scientific Explanation and Other Essays in the Philosophy of Science* New York: Free Press.
Hosiasson-Lindenbaum, J. (1940) "On Confirmation" *Journal of Symbolic Logic* 133-148.
Jeffrey, R. (1965) *The Logic of Decision* New York: MacMillan. Rev. ed. (1983) Chicago:University of Chicago Press.
Lambert, K. (1969) "Logical truth and Microphysics" in *The Logical Way of Doing Things* ed. K. Lambert New Haven: Yale University Press.
Lambert, K. (1971) "Explanation and Intention" in *The Psychology of Private Events* ed. A. Jacobs and L. Sacks New York: Academic Press.
Lambert, K. (1980) "Explanation and Understanding: An Open Question?" in *Rationality in Science* ed. R. Hilpinen Dordrecht: Reidel.
Lambert, K. and Brittan, G.(1987) *An Introduction to the Philosophy of Science* 3rd. ed. Atascadero, Ca.: Ridgeview.
Lambert, K. (1988) "Prolegomenon zu einer Theorie des Wissenschaftlichen Verstehens" in *Erklären und Verstehen in der Wissenschaft* ed. G. Schurtz Munich: Olderburg.
Lambert, K. (1990) "On whether an answer to a why-question is an explanation if and only if it yields scientific understanding" in *Causality, Method and Modality: Essays in Honor of Jules Vuillemin* ed. G. Brittan Dordrecht: Kluwer.
Lambert, K. ed. (1991) *Philosophical Applications of Free Logic* Oxford: Oxford University Press.
Lambert, K. and Schurz, G. (1994) "An outline of a formal theory of scientific understanding" forthcoming in *Synthese*.
Lambert, K. and Simons, P. (1994) "Characterizing and Classifying: A Study in the Philosophy of Biology" forthcoming in *The Monist*.
Lewis,D. (1981) "Causal Decision Theory" in *Australasian Journal of Philosophy* 59, 5-30.
Nozick, R. (1969) "Newcomb's Problem and Two Principles of Choice" in *Essays in Honor of C. G. Hempel* ed. N. Rescher Dordrecht: Reidel.
Parsons, T. (1980) *Nonexistent Objects* New Haven: Yale University Press.
Pearl, J. (198) *Probabilistic Reasoning in Intelligent Systems:Networks of Plausible Inference* San Mateo: Morgan Kaufmann.

Quine, W. V. O. (1951) "Two Dogmas of Empiricism" *Philosophical Review* 60, 20-43 [reprinted in *From a Logical Point of View* Harvard University Press: Cambridge, 1953.

Quine, W. V. O. (1963) "Carnap and Logical Truth" in *The Philosophy of Rudolf Carnap* La Salle, Illinois: Open Court.

Ramsey, F. P. (1931) *The Foundations of Mathematics and Other Logical Essays* London: Routledge and Kegan Paul.

Skyrms, B. (1980) *Causal Necessity* (Yale University Press: New Haven).

Skyrms, B. (1982) "Causal Decision Theory" *The Journal of Philosophy* LXXIX 695-711.

Skyrms, B. (1984) *Pragmatics and Empiricism* (Yale University Press: New Haven).

Skyrms, B. (1986) "EPR and the Metaphysics of Causation" in *Symposium on the Foundations of Modern Physics:50 Years of the EPR Gadanken Experiment* ed. P. Lakki and P. Mittlestadt (World Scientific Publishing Co.:Singapore) 609-623.

Skyrms, B. (1988) "Probability and Causation" *Journal of Econometrics* 39: 53-68.

Skyrms, B. (1988) "Conditional Chance" in *Probability and Causality: Essays in Honor of Wesley Salmon* ed. J.Fetzer (Reidel:Dordrecht,Holland) 161-178.

Skyrms, B. (forthcoming) "Bayesian Projectibility" in *Grue* ed. D. Stalker.

Skyrms, B. (forthcoming) "Adams Conditionals" in *Probability and Conditionals* ed. Eells and Skyrms (Cambridge University Press:Cambridge).

Spohn, W. (1993) "Causal Laws are Objectifications of Inductive Schemes" in *Philosophy of Probability* ed. J. P. Dubucs (Kluwer: Dordrecht).

Suppes, P. (1970) *A Probabilistic Theory of Causality* Amsterdam: North Holland.

van Fraassen, B. (1989) *Laws and Symmetry* (Oxford University Press: Oxford).

Zabell, S. L. (1988) "Symmetry and Its Discontents" In *Causation, Chance and Credence* ed. W. Harper and B. Skyrms. Dordrecht: Kluwer 155-190.

Margaret Morrison

The New Aspect: Symmetries as Meta-Laws – Structural Metaphysics

1. Introduction

When we talk at an intuitive level about symmetries of nature we usually have in mind objects that have perfectly symmetrical shapes, for example, the geometric symmetry of crystals (snowflakes) or the spherical shapes and motions of the planets. These symmetries supposedly reflect the inner simplicity and harmony of the universe. One of the great examples of the power of symmetry arguments comes from Kepler's *Mysterium Cosmographicum*. Believing that God created the solar system according to a mathematical pattern, Kepler attempted to correlate the distances of the planets from the sun with the radii of spherical shells that are inscribed within and circumscribed around a nest of solids. The goal was to find an agreement between the observed ratios of the radii of the planets and the ratios calculated from the geometry of the nested solids. Although unsuccessful, Kepler remained convinced that there were underlying mathematical harmonies that could explain the discrepancies between his theory and ratios calculated from observations.

Kepler's story is interesting for two reasons. First, belief in the mathematical harmony of the universe still holds a prominent place in scientific theorizing, however, the manifestation of that belief has changed dramatically since Kepler's investigations. To quote Hermann Weyl "we no longer seek this harmony in static forms like the regular solids, but in dynamic laws".[1] Modern definitions of symmetry usually emphasise laws as the

1 Weyl, [1952] p.77.

important component in establishing symmetries.[2] Very simply, a symmetry of nature exists whenever the laws of physics are unaffected by a change in the frame of reference. The other important aspect of Kepler's approach is his unfaltering reliance on deeper symmetries that would somehow explain the apparent lack of symmetry at the level of observable objects. This role for symmetries is also mirrored in modern physics, especially in high energy physics. For example, even though the electroweak theory incorporates so-called broken symmetries, the theory itself is ultimately governed by a symmetrical Lagrangian that reflects an underlying symmetry that is hidden from us. As a result the actual symmetry of the physical system is less than the symmetry of its underlying laws.

The idea that there could be hidden symmetries of physical laws that are not manifest in the system itself brings me to the main distinction I want to explore in the paper, namely, the differences and relationship between the symmetries of laws and symmetries of objects. Initially this distinction seems to be reflected in two different kinds of symmetries; the global symmetries of rotation and translation whose transformations do not depend on position in spacetime and affect each point in the same way, and the local symmetries whose transformations involve changes that can vary freely from point to point in spacetime. In the former case even though we speak about laws being invariant under such transformations we usually also think of these symmetries as having a physical component like the rotation of molecules. In the case of local symmetries we are dealing with more abstract properties like gauge invariance and hence the symmetries themselves are higher order abstractions involving inferences about structural constraints on the laws of nature.

2 This emphasis on laws as the manifestation of symmetries can be found in Wigner [1979] and Feynmann [1967]. For a more popular account of the relation between laws and symmetries see Barrow [1990] and Weinberg [1993]. Van Fraassen [1989] also discusses the relationship between laws and symmetries although his view is quite different; he claims that symmetries can obviate the need for laws and that science can proceed without requiring any laws at all.

These local symmetries function as a system of laws from which we can derive and explain other laws like the conservation of electric charge and laws that govern individual physical systems. But, unlike most laws they are not generalizations from empirical data, but rather mathematical structures that impose constraints on what can happen in a particular physical system. Part of the argument in this paper is an attempt to show that in some cases the empirical component in these kinds of symmetries is all but absent and where it does exist the symmetries provide no information over and above that which is empirically discoverable. I claim that the use of symmetries to generate empirical theories does not legitimate symmetries as empirical phenomena. Particularly, the use of hidden symmetries to ground laws that govern non-symmetrical phenomena involves a decidedly non-empirical process. It is important to stress that my intention is not simply to argue for a traditional kind of instrumentalist versus realist view about the status of symmetries in general. Rather, I want to show how the use of symmetries sometimes results in a blurring of the line between science and metaphysics. While sometimes seen as part of the empirical dimension of theories and methodology, the use of symmetries can also go beyond mere scientific conjecture creating a domain of a priori principles that govern the process of theory construction.

Part of the argument regarding the empirical versus abstract role of symmetries concerns their connection with conservation laws. This connection, first formalized by Emmy Noether, is important for answering questions about what symmetries can tell us about laws of nature. I will begin by looking at this connection in the context of global symmetries and then go on to explore the case of local symmetries in an effort to show why the latter, and perhaps even the former, should not be viewed straightforwardly as concrete claims about the world. Implicit in my argument is the idea that one cannot separate the notions of symmetry and law. Conservation laws provide the empirical component or manifestation of symmetries and, as structural principles, symmetries are laws about what the laws of nature must be like. And, as we shall see below, local symmetries function as further constraints insofar as they determine fundamental aspects of particular kinds of physical systems and theories.

2. Symmetries, Invariances and Conservation Laws

There are a variety of symmetries used in physics; global and local, continuous and discrete as well as geometrical and internal.[3] Global symmetries deal with transformations that are not affected by position in space and time. They can be either geometrical, reflecting the homogeneity of space and time, or they can be internal symmetries that apply to the intrinsic nature of particles rather than their position or motion.[4] Geometrical symmetries (sometimes called simply space-time symmetries) involve the operations of translation through space, rotation about an axis and translation through time. Such symmetries entail that there are no fixed reference points in space or preferred instants in time, hence the spatial displacement of a system as a whole or a shift in time will not change the mechanical properties of a closed system. Similarly, rotation does not change a system's properties since space is isotropic (i.e. all directions are equivalent). The basic idea of a global internal symmetry is that it involves transformations of several field components into one another in such a way that the physical situation remains unchanged. For example, if we have a three component field with each component corresponding to an axis in a 3-dimensional internal space, and if we mathematically rotate the axes in this internal space then each field component in real space is rotated to the same degree.[5] If the total field energy remains constant under these rotations then we have a global internal symmetry since each component is rotated to the same degree over all

3 Some local symmetries are also referred to as dynamical symmetries.

4 By intrinsic nature I mean properties like the quantum number of particles and the conservation of various charges.

5 An internal space is simply an imaginary space where these mathematical operations are performed. It bears no relation to real space.

of space. In the case of local internal symmetries the rotation of field components varies from point to point so that it is no longer the same over all of space.[6]

Symmetries can also be discrete or continuous. An example of a continuous symmetry is the rotation of a circle; it is a continuous operation describable by groups that possess an infinite number of elements (Lie groups) as opposed to discrete symmetry groups which contain a finite number of elements. Rotations of a square or triangle are instances of discrete symmetries as are spatial reflections where the associated transformation group contains only two elements, reflection of spatial coordinates and identity. Unlike continuous symmetries the changes involved in discrete symmetries are not arbitrarily small. They imply constants of motion in quantum mechanics and are multiplicative rather than additive. A discrete symmetry like parity (i.e. reflection or space inversion) is observed in strong and electromagnetic interactions but is violated in weak interactions.

There are also discrete internal symmetries such as invariance under charge conjugation where there is an exchange of particles with their anti-particles. Continuous internal symmetries involve specific properties of particles and continuous transformations of quantized fields. These symmetries are described in the language of group theory where the symmetry is defined as an invariance under a set of transformations that have a particular mathematical structure.[7] Here a specific symmetry group is associated with particular transformations. This is an extension of the ordinary geometrical concept which involves invariance

6 This destroys the symmetry of the field which can then be recovered by the introduction of another field – a non-abelian gauge field. The discovery of non-abelian gauge field theory was made by Yang and Mills in 1954. Non-abelian symmetry operations fail to obey the commutation rule unlike their abelian counterparts. These groups were named after the Norwegian mathematician Neils Abel.

7 A group is simply a collection of elements with specific interrelationships described by group transformations. One of the interesting features of a group is that repeated transformations between elements are equivalent to a transformation from the initial to the final element.

of particular figures under transformations of reflection, rotation and translation. For example, the U(1) group governs the continuous rotations of a circle and describes the symmetry of a single field. The SU(2) group is non-abelian (cf. n.6) and can be represented by two states made to coincide with each other. One of its most important applications is the construction of gauge theories like the electroweak theory.

Associated with each symmetry is a conserved quantity or a conservation law. Invariance under time translation, which means that the laws of physics predict the same evolution of identical processes regardless of when they occur, implies that the conservation of energy is built into the laws describing the process. Invariance under a translation in space (meaning that the laws of physics are the same in Toronto and London) implies conservation of momentum and invariance under spatial rotations implies conservation of angular momentum. This latter invariance is a consequence of the isotropy of space-time while the former two involve homogeneity. All of these invariances are assured by requiring that the equations describing physical systems are covariant under the Lorentz transformations.[8]

---

8 When the equations or laws of a theory are unchanged under specific transformations we say that they are covariant. The technically precise use of the term invariance involves reference to specific objects (geometrical) that remain unchanged under certain transformations. Most often the term invariance is used in discussing the relationships between symmetries and laws. However, this carries with it specific ontological implications since it also implies that there is a corresponding object which is also left unchanged. Hence, to say that symmetries entail this kind of invariance is somewhat of a misnomer since it certainly goes beyond what can be demonstrated regarding the relationship between laws and symmetries. In the case of conservation laws that which remains invariant is, of course, the conserved quantity; so, in that sense once could speak of the laws as being invariant. Since invariance seems to be the term of choice in the symmetry literature I will continue to use it, keeping in mind that I understand it to carry no ontological commitments over and above the empirically established conservation laws and associated quantities. All of the symmetries described above are continuous; finite changes are made

The connection between conserved quantities was first made explicit by Emmy Noether in 1918.[9] When a symmetry group governs a particular phenomenon and the appropriate Lagrangian does not change under the group transformations then a particular quantity is conserved. Noether's theorem says that for every continuous symmetry of a Lagrangian there is a quantity which is conserved by its dynamics. And, conversely, for every conservation law there is a symmetry. Essentially, in the Lagrangian formulation of a physical theory the symmetry in question is actually a symmetry of the Lagrangian. Because the form of the Lagrangian determines the equations of motion of the system, the symmetry is a symmetry of the equations of motions, i.e. of the theory itself. What makes Noether's theorem particularly interesting is its most important assumption, namely, that the equations of motion of a system are derivable from Hamilton's principle. Although there is no reason why the equations should have this form, they always do, so much so that when physicists are constructing a new theory and looking for the correct Lagrangian they inevitably assume that the new theory will have to obey Hamilton's principle.[10] Robert Mills [1989] (who, along with Yang Chen Ning discovered the massless gauge fields for the SU(2) group) has suggested the possibility that it is not Hamilton's principle that is fundamental to theory construction but rather Noether's theorem. And, if so, then the theories we devise have the strucutre they do because of the fundamental relationship between symmetries and conservation laws. As a result Noether's theorem becomes a prin-

---

in incremental amounts with the resulting constants of motion being additive classical quantities.

9 See Noether [1918].

10 Hamilton's principle states that for the true trajectory of a system – its history as a function of time – the time integral of the Lagrangian is stationary with respect to small changes of that trajectory away from its true shape, $\delta \int_{t_1}^{t_2} (T - V) = dt = 0$. It is sometimes stated simply as the fact the nature tends to equalize the mean kinetic and potential energies during a motion. Noether's theorem is true also in quantum mechanics. See Hill [1951] and Mills [1989].

ciple of theory construction like the principle of equivalence in special and general relativiy. We would then say that classical physical laws take the Lagrangian form and the quantum theoy takes its characteristic Hamiltonian form as a consequence of Noether's principle.[11]

This way of thinking about the role of symmetries raises an interesting issue; specifically what the implications of Noether's theorem are for a substantive view about symmetries. First, recall that the symmetries referred to in the theorem are really requirements that there be a symmetry of the Lagrangian, i.e. that the equations of motion are symmetrical; they don't involve symmetries of objects per se. Moreover, one cannot simply think of Noether's 'principle' in isolation from Hamilton's principle, since the latter is an assumption crucial to facilitating the proof of the theorem. Mills's suggestion that Noether's theorem is better thought of as a 'principle' analogous to the principle of equivalence in relativity seems to accord it a kind of independence, which implies that the connection between symmetries and conservation laws is just a brute fact about the world. However, even in relativity the principle of equivalence is not independent of the constancy of the ratio of inertial to gravitational masses of material bodies; it was this fact that led Einstein to realize (or postulate) that it would be impossible to distinguish locally between inertial and gravitational effects on material bodies, what he referred to as the principle of equivalence in general relativity. The close connection between Hamilton's principle and Noether's theorem is important and needs to be maintained because it shows that the relation between symmetries and laws is really a consequence of basic assumptions that we make in formulating physical theories; specifically the assumption about least action. So, the idea that Noether's theorem could, in some way, show that the so-called symmetries of nature are fundamental results from interpreting the theorem independently of the context that makes possible the connection between laws and symmetries in the first place.

---

11 See Mills [1989].

If we look at an example of how Noether's theorem works in explicating the relationship between conservation laws and symmetries, we see that her result can be construed as a special application of the properties of ignorable variables.[12] Her basic concern was variational problems where the action integral remains invariant with respect to a group of transformations. She showed that every parameter associated with such transformations leads to a corresponding conservation law. If we consider a system whose Lagrangian does not depend explicitly on time we can transform the independent variable t by $t=t'+\alpha$ where α is an infinitesimal constant. The $q_i(t)$ of the variational problem change as a result of the transformation but they can now be considered functions of t′. Hence the Lagrangian has the form $L=L(q_i, c_i', t'+\alpha)$. Because L does not depend explicitly on time the new variational integral becomes $A=\int_{t_1-\alpha}^{t_2-\alpha} L(q_i, q_i')dt'$ (note that α only appears in the limits). If we assume α as a function of t' where $\alpha(t_1)=\alpha(t_2)$ then the action integral will be a function of α but only $\alpha'(t')$ will appear explicitly in the integral. Hence $\frac{dq_i}{dt} = \frac{dq_i}{dt'}(1-\alpha')$ and the Lagrangian becomes

$$L(q_i\dot{q}_i) = L(q_i, \dot{q}_i(1-\alpha')) = Lq_i\dot{q}_i - (\sum_{i=1}^{n} \frac{\partial L}{\partial q_i} q_i)\alpha'$$

In addition $dt = (1+\alpha')dt'$. Hence the new variational integral is

$$A = \int_{t_1}^{t_2} L(q_i q_i')dt' - \int_{t_1}^{t_2} (\sum_{i-1}^{n} \frac{\partial L}{\partial q_i} q_i - L)\alpha' dt'$$

---

12 The Langrangian function L is usually a function of all the position coordinates q_i and the velocities $\dot{q}_i$. It is possible that a certain variable q_k does not appear in the Lagrangian function although $\dot{q}_k$ does. If we introduce the momenta p_k according to the definition $p_i = \frac{\partial L}{\partial \dot{q}_i}$ where p_i are the components of generalized momentum we can write the Lagrangian equations of motion in the following form $\dot{p}_k = \frac{\partial L}{\partial q_k}$. If q_k is an ignorable variable then $\frac{\partial L}{\partial q_k} = 0$ and $\dot{p}_k = \frac{\partial L}{\partial q_k}$ can be integrated to give p_k = constant = c.

Hence the momentum connected with an ignorable variable is a constant of motion.

In this form an additional degree of freedom $\alpha(t')$ has been added to the previous $q_i(t')$. The Lagrangian associated with $\alpha(t')$ yields $\frac{d}{dt}(\sum_{i=1}^{n} \frac{\partial L}{\partial q_i} q_i - L) = O$ which is $\sum_{i=1}^{n} p_i q_i - L = E$ which corresponds to the conservation of energy. This result is a consequence of the Lagrangian equations, deduced from Noether's requirement regarding transformations that leave the variational integral unchanged.[13]

The interesting point in all of this relates to the issue I mentioned at the beginning of the discussion about Noether, specifically, the relationship between the requirement of a formal symmetry with respect to the Lagrangian and the supposed symmetries of nature or objects in nature. Although we can deduce conservation laws from symmetry constraints this is not crucial for establishing the laws themselves. But what about the other component of the theorem, the conservation laws? The ontological status of conserved quantities is relatively straightforward since most conservation laws are simply the result of generalizations based on empirical data. For example, the conservation of energy has a history based on empirical discoveries which gradually led to the formulation used today.[14] What is more problematic, however, is the inference from symmetries of the equations to actual physical symmetries of the world. An argument could possibly be made that in the case of the geometric space-time symmetries the connection between the formal

13 For more on this see Lanczos [1970]. We can also think of symmetry with respect to a displacement of a system in space, which means that the Lagrangian L is unchanged, i.e.$\delta L = 0$ if every particle is displaced by the same infinitesimal amount δq. Hence we get $\delta L = \sum_{i=1}^{n} \frac{\partial L}{\partial q_i} \cdot \delta q = \delta q \frac{d}{dt} \cdot (\sum_{i=1}^{n} \frac{\partial L}{\partial \dot{q}_i}) = 0$

So, throughout the motion the vector $P = \sum_i \frac{\partial L}{\partial q_i} = \sum P_i$ remains constant. As a result, conservation of momentum P can be seen as a consequence of invariance under spatial transformation. A similar proof can be given for other conservation laws, see Gitterman and Halpern [1981] and Longair [1984].

14 Kuhn [1972] has documented part of this history.

and the physical is more easily forged since the invariances are usually described in terms of spatial or temporal characteristics, which in turn are seen as consequences of the homogeneity or isotropy of space and time themselves. But, when we say that the laws governing a physical system are invariant with respect to the origin of the time coordinate, what exactly does this mean for a physical account of symmetries? We can think of the conservation of energy and momentum as associated with the impossibility of measuring an absolute position in time and space respectively. This, in turn, is explained by the homogeneity of space and time, an inference we make to explain the symmetries of the equations of motion. Because we are dealing with familiar kinds of geometrical symmetries there is a temptation to simply assume that a symmetrical Lagrangian implies a straightforward correspondence with a physical symmetry. Although this is by no means obvious even with space-time symmetries the situation becomes increasingly complex in the case of local symmetries generated by group theory. And, as we shall see below some of the inferential practices used in dealing with the symmetries of modern physics bear a remarkable similarity to Kepler's arguments designed to save his account of nested solids.

3. Symmetries Without Objects

The symmetries I have considered thus far have all been global; that is, there is an invariance (or covariance) under a set of transformations described by a single parameter that applies regardless of location in space or time. When it comes to finding the appropriate symmetry for conservation of electric charge we need to move beyond the geometric spacetime symmetries to a local symmetry where physical laws are invariant under a local transformation involving a large number of separate transformations that are different at every point in space and time. In general a theory that is globally invariant will not be invariant under locally varying transformations. But, by introducing new force fields that transform in certain ways and interact with the original particles in the theory, a local invariance can be restored. In other words, these force fields are postulated in order for certain local invariances to hold. The global form of charge conservation would necessitate the instantaneous propagation of signals (i.e. charge cannot be created at one point and destroyed at another) which would conflict with special relativity. However, the local conservation requirement stems from Maxwell's own modification of Ampere's law and the introduction of the displacement current.[15] In the electromagnetic case the local symmetry is realizable only by introducing a new field called a gauge field. The new theory which incorporates the gauge field specifies interactions between the field and the conserved quantity electric charge. So, by requiring local gauge invariance one can predict the existence of the photon and its properties. In order to understand the origins of the symmetry let me first describe some background to the notion of gauge invariance associated with the gauge field.

Gauge invariance had its beginnings in 1918 in the work of Hermann Weyl who wanted to draw attention to the requirement that the laws of physics should remain the same if the scale of all length measurements is changed by a constant factor. Weyl wanted to extend this idea to a local invariance where scale (or gauge) changes could vary at different points in space and time in a manner analogous to the curvilinear coordinate transformations of general relativity. Weyl thought he could show that from this gauge symmetry one could, using Noether's theorem, derive the conservation law for electric charge. The attempt was unsuccessful but the idea was resurrected in 1927 by Fritz London who showed that the proper symmetry for electric charge was phase invariance rather than scale invariance; that the electromagnetic field in fact allowed for an arbitrary variation in the phase factor from point to point in spacetime. Because Weyl was

15 The continuity equation states that the rate of decrease of charge in a volume is due only to the flux of current out of its surface; no net charge can be created or destroyed in the volume. Because the volume can be made arbitrarily small charge must be locally conserved.

also involved in this reformulation it continued to be labelled gauge invariance even though it had nothing to do with the notion of a gauge which refers to a choice of length scale.

The original idea involved the introduction of a scaling factor S(x). If we think of a vector at *x+dx* the scale factor is given by $S(x+dx) = 1 + \partial_\mu S dx^\mu$.[16] The derivative $\partial_\mu S$ is the connection associated with the gauge change which Weyl identified with the electromagnetic potential A_μ because it transformed like a potential.[17] The reinterpretation in light of quantum mechanics involved a change in phase of the wave function $\Psi \rightarrow \Psi^{-ie\lambda}$ with the gauge transformation for the potential A_μ becoming $A_\mu \rightarrow A_\mu - \partial_\mu \lambda$. If we write the non-relativistic Schrödinger equation

$$[\frac{1}{2}m(i\hbar\nabla - eA)^2 + e\phi + V]\Psi = i\hbar\frac{\partial\Psi}{\partial t}$$

(where the canonical momentum operator $P_\mu - eA_\mu$ is replaced by $-i\hbar\nabla - eA$) then after the phase change an additional gradient term proportional to $e\nabla\lambda$ emerges, the result of the operator $i\hbar\nabla$ acting on the transformation wave function. This additional term spoils the local phase invariance which can then be restored by introducing the new gauge field A_μ. The gauge transformation $A_\mu \rightarrow A_\mu - \partial_\mu\lambda$ cancels out the new term. This new gauge field is simply the vector potential defining the electrogmagnetic field.[18] A different choice of phase at each point can be accommodated by interpreting A_μ as the connection relating phases at different points. In other words, the choice of a phase function $\lambda(x)$ will not affect any observable quantitiy as long as the gauge transformation for A_μ has a form which allows the phase change and the change in potential to cancel each other.

16 The norm of the vector is *f(x+dx)* and expanded to first order in *dx* it is $f(x+dx) = f(x) + \partial_\mu f dx^\mu$ where ∂_μ is just $\frac{\partial}{\partial x^\mu}$.

17 A second gauge change with a scale factor Λ will transform the connection in the following way $\partial_\mu S \rightarrow \partial_\mu S + \partial_\mu \Lambda$ which is similar to the transformation of the potential in electromagnetism $A_\mu \rightarrow A_\mu + \partial_\mu \Lambda$ which leaves the electric and magnetic fields unchanged.

18 See Moriyasu [1983].

What this means is that we cannot distinguish between the effects of a local phase change and the effects of a new vector field. Combining the additional gradient term with the vector field A_μ prescribes the form of the interaction between matter and the field since A_μ is what provides the connections between phase values at nearby points. The phase of a particle's wave function can be identified as a new physical degree of freedom which is dependent on spacetime position. Although it is too complicated to go through the derivation here it is possible to show in specific detail that from the conservation of electric charge we can, on the basis of Noether's theorem, assume the existence of a symmetry, and the requirement that it be local forces us to introduce a gauge field, which turns out to be the electromagnetic field. The structure of this field which is dictated by the requirement of local symmetry in turn dictates, almost uniquely, the form of the interaction; that is, the precise form of the forces on the charged particle and the way in which the electric charge current density serves as the source for the gauge field.[19]

The phase transformations of quantum electrodynamics are one parameter transformations and form what is called a one dimensional abelian group (meaning that any two transformations commute). This group is called the U(1) group of a U(1) gauge symmetry. The mathematical description of all symmetry operations are dealt with in group theory which uses algebra to describe operations like rotation.[20] These symmetry

19 For a detailed discussion see Mills [1980]. Again, we can see this very simply by thinking in terms of the invariance of the Lagrangian. Because of the space dependence of the phase of the electron wave the Lagrangian is changed by this gauge transformation. The presence of the photon interacting with the electron and the changes in the photon wave function cancel out the changes in the Lagrangian, thereby resotring its invariance. So, the invariance of the Lagrangian under local gauge transformations requires the existence of the photon, i.e. the electrogmagnetic field. For discussion of the derivation of charge conservation in quantum field theory see Gitterman and Halpern [1893] and for a general treatment of gauge theory see Moriyasu [1983].

20 The axioms of group theory include the associative law, the identity law I which involves a simple rotation and corresponds essentially to the act

groups, however, are more than simply mathematizations of certain kinds of tranformations since in the non-abelian case it is the mathematical structure of the symmetry group that determines the structure of the gauge field and the form of the interaction. In these more complicated situations there are several wave functions or fields transforming together as in the case of SU(2) and SU(3) transformations which involve unitary matrices acting on multiplets.[21] These symmetries are internal symmetries and are typically associated with families of identical particles. In each case the conserved quantities are simply the quantum numbers that label the members of the multiplets (such as isospin and colour) together with operators that induce transitions from one member of a multiplet to another. Hence the operators correspond on the one hand to the conserved dynamical variables (isospin etc.) and to the group of transformations or the symmetry group of the multiplets on the other.[22]

The extension of gauge invariance beyond electromagnetism began with the work of Yang and Mills in the 1950s (cf. n.5). The connection between charge conservation and local gauge invariance, together with the fact that the latter could determine

---

of leaving the object unchanged; i.e. $I \times R_1 = R_1 \times I = R_1$, and an inverse operation which corresponds to rotating an object backward. This has the effect of undoing the original rotation, i.e. $R_1 \times R_1^{-1} = I = R_1^{-1} \times R_1$. These axioms apply to several kinds of symmetry operations in multidimensional spaces – reflections as well as the interchange of objects etc.

21 These can also be thought of as phase transformations where the phase is considered a matrix quantity. See Aitchinson [1982]. See note 23 below for a definition of multiplets.

22 Isospin actually refers to similar kinds of particles considered as two states of the same particle in particular types of interaction. For example, the strong interaction between two protons and two neutrons is the same which suggests that for strong interactions they may be thought of as two states of the same particle. So, hadrons with similar masses but differing only in their charge can be combined into groups called multiplets and regarded as different states of the same object. The mathematical treatment of this characteristic is identical to the one used for spin (angular momentum). Hence the name isospin, even though it has nothing to do with angular momentum. The SU(2) group is the isospin group and is also the symmetry group of spatial rotations that give rise to angular momentum.

the structural features of electromagnetism, suggested that perhaps the argument could be generalized to other kinds of conservation laws. If so, gauge invariance would become a powerful tool for generating a theory of strong interactions. They began with the conserved quantity isospin, (which is violated in electromagnetic and weak interactions). Because isospin conservation entails that the proton and neutron can be considered as two states of the same particle, a local gauge invariance means that although we can label the proton as the up state of isospin 1/2 and the neutron as the down state, the up state need not be the same at another location. Because this labelling of protons and neutrons at each point is arbitrary some form of connection or rule is required to coordinate the various choices at different positions. In electromagnetism phase values are connected by the potential A_μ and so an analogous isospin potential field was introduced by Yang and Mills. But, because the SU(2) symmetry group which governs isospin is also the group that governs rotations in a three dimensional space the potential field associated with this group is very different from the U(1) phase group of electromagnetism. In the SU(2) case the "phase" is replaced by a local variable that specifies the direction of the isospin.[23]

The interesting aspect of the Yang Mills theory is that the isospin connection and hence the potential behaves like the SU(2) symmetry group. The components of isospin are transformed into one another by the elements of the group, consequenly the connection (potential) is capable of perfoming the same isospin transformations as the transformations of the

23 To do this one first specifies the up component of an isotropic spin vector at a certain position x in an abstract isotropic space. One then determines how much that up state at x needs to be rotated in order to correspond to an up state at position y. In other words, the connection between the isospin states at different points acts like an isospin rotation. The particle in the up state is moved through the potential field to position y in such a way that the spin direction is rotated by the field so that it is pointing up at y. The set of these rotations forms a symmetry group and hence the connection between internal space directions at each different point must also act like a symmetry group. My discussion follows Moriyasu [1983].

SU(2) group. In order to see how the potential generates a rotation in the isospin space we need to first consider how the potential relates to a rotation.[24] A three dimensional rotation is a wavefunction $R(\theta)\Psi = e^{-i\theta L}\Psi$ where θ is the angle of rotation and L is the angular momentum operator. It is important to note that the potential is not a rotation operator. In the electrodynamics case the amount of phase change has to be proportional to the potential to ensure that the Schrödinger equation remains invariant. In the Yang Mills case the potential is considered as a "generator" of a rotation and must be proportional to the angular momentum operator to preserve gauge invariance.[25] Consequently the potential must include three charge components corresponding to three angular momentum operators L_+, L_-, L_3. So, $A_\mu = \sum_i A^i_\mu(x)L_i$ is a linear combination of the angular momentum operators where the coefficients A^i_μ depend on space time positon. So, the potential acts like a field in spacetime and an operator in the abstract isospin space. Each of the potential components behaves like an operator, as in the case of the raising operator L_+ which transforms a down state into an up state. This formal operation corresponds to the case of a neutron absorbing a unit of isospin form the gauge field and becoming a proton.[26]

In the Yang Mills case gauge invariance can dictate the form of the potential, the fact that the gauge field must, unlike the electromagnetic potential, carry electric charge. In addition, the requirement of local invariance dictates that the mass of the potential field be zero. However, unlike the photon which is

---

24 Isospin space is the imaginary internal space mentioned in n. 4.

25 The generators of a group are a fixed subset of independent elements of the group such that every member of the group can be expressed as the product of powers of the generators and their inverses. In general the number of generators equals the number of independent parameters in the group.

26 The component L_+ would have to carry one unit of isospin charge. The number of components of the internal part of the potential field in isospin space is equal to the dimension of the symmetry group.

massless, the particles involved in weak interactions must have a very large mass. Unfortunately in the Yang and Mills case the standard mass term that is usually introduced into the Lagrangian is not invariant under gauge transformation; hence, it could not be added and their theory failed to provide an account of strong interactions. It did, however, furnish an important framework for dealing with the relations between conserved quantities, especially quantum numbers and symmetries. Yang and Mills showed that gauge invariance was more than just an "accidental" symmetry of electromagnetism or merely a property of phase changes. Gauge symmetry had become powerful tool that could determine the form of the matter-field interaction and could define much of the dynamical content of a theory of strong interactions.[27] Not only did it provide a methodology for theory construction, a way to go about building a theory of strong (and eventually weak) interactions; but, as a requirement that dictated specific physical laws, gauge invariance functioned as a structural law that in turn determined certain constraints on physical systems.

Before going on, let me sum up what the discussion has revealed thus far about the relationship between laws and symmetries, and what the latter can tell us about the former. Earlier on I mentioned Noether's theorem which states that for every conservation law there is an accompanying symmetry and vice versa. The implications of this theorem for modern physics have, for the most part, focused on the search for symmetries

27 Although I have discussed the qualitative ideas relating symmetry groups and the gauge potential field it is possible to show in quantitative detail exactly how a symmetry group transformation gives rise to the connection between the states that we identify as the potential field. The connection can be calculated from the symmetry transformations by moving the charged particle through the potential field. It isn't necessary to go into the details here since my purpose is to show the relation and order of progression from an empirical phenomenon (a conservation law or conserved quantity) to a mathematical representation of a symmetry transformation to a dynamical account of a potential field. For details see Moriyasu [1983], chapter 3 and Mills [1989].

corresponding to empirically established conservation laws like electric charge. In other words the correspondences between conservation laws and symmetries have been established using a "bottom-up" method rather than a "top-down" approach where symmetries are employed as structural principles that enable one could go on to derive new conservation laws or discover conserved quantities. But, regardless of whether one views symmetries as factual claims about the structural organization of the world or as methodological constraints on theory construction, there is little doubt that symmetries do function in a structural capacity by providing a framework of laws that govern the laws of physical systems. To the extent that all laws are subject to conservation laws which in turn yield symmetries, symmetries that regulate our understanding of how the world functions. However, there are several senses in which one can understand or give content to this relationship.

The most straightforward expression of the relationship is simply that the existence of a particular symmetry entails that the laws of physics will remain the same regardless of a change in the frame of reference. For example, the geometric symmetries of space and time (i.e. translation and rotation) guarantee that the laws of nature, regardless of what they are, will be the same at all points in space and time. This requirement is an important constraint on what constitutes a law, (that it be covariant) but it provides no substantive information about the kinds of laws that we have or must have. Although these symmetries are commonly referred to as expressions of the homogeneity and isotropy of space-time, the physical manifestation of these features is the invariance of physical laws. And in that sense the notions of laws and symmetries are inseparable in that they furnish necessary structural constraints on the practice of science, i.e. that experiments performed in Toronto will not differ from experiments performed in London under the same consitions. Here the symmetries are simply laws about what the laws of nature must be like.

Even in the case of electromagnetism where one can move from the requirement of gauge invariance to a field with a uniquely determined structure and specified form for the forces on the charged particles, there is no additional dynamical contribution to the theory over and above what was already present in Maxwell's equations. It was commonly known that the field equations were invariant under a gauge transformation, but prior to the work of Weyl, London and Fock in the 1920s gauge invariance was considered a merely accidental symmetry of electromagnetism, resulting from the arbitrariness of the potential in classical physics. It was retained because it facilitated calculations but played no role in defining the dynamical content of electromagnetism.

The second issue concerns connection between conservation laws and symmetries. Embedded in this issue is a question about the ontological status of symmetries themselves. Although Neother's theorem tells us that there must be a symmetry for each conservation law, we saw above that this requirement is dependent on the assumption that the equations of motion are derivable from Hamilton's principle. The symmetries from which we are able to derive a particular conservation law are themselves exact symmetries of the equations of motion, the Lagrangian, rather than exact symmetries of nature itself. This is particularly evident in the case of dynamical symmetries like gauge invariance but is also applicable to space-time symmetries, each of which is represented in terms of the invariance of laws under certain kinds of transformations. Although symmetries enable us to derive conservation laws, the symmetry itself provides no additional information about the conservation laws over and above what can be established empirically. Both the conservation of energy and momentum have a rich historical background, and even the conservation of isospin, although formulated as a symmetry principle under the SU(2) rotation group, had its beginnings with the discovery of the charge independence of the nuclear force.[28] Hence, the physical manifestation of symmetries in this case is the conservation laws.

28 The strength of the interaction between a proton or a neutron or between two neutrons or two protons was measured and found to be the same. The

Finally, there is a more abstract way of understanding the relationship between symmetries and laws, specifically laws that govern the interaction of forces. Above I said that nothing about symmetries commits us to a positive thesis about laws of nature over and above the requirement that if there are any, then their form must be invariant under transformations. Although symmetries do commit us to the existence of conservation laws such laws are established empirically, with their stability coming *not* from their connection with symmetries but rather from empirical fact that we know of no cases where they are violated. Hence, where conservation laws are known it is they that are prior insofar as they provide the structural constraints on the physical aspects of our theories.[29] But, as the Yang-Mills case indicated symmetries can also be used to generate physical theories that include laws about how particles and fields interact. The philosophical question that emerges concerns the status of these symmetries once the theories themselves have been empirically confirmed. The final example I want to discuss is the electroweak theory, a theory that is well confirmed and whose foundation is based on symmetry principles. The development of the electroweak theory presents an interesting case insofar as symmetry constraints proved to be at odds with well developed empirical laws governing the characterisitics of particles. The problem was solved and the symmetry of the Lagrangian retained by invoking the idea of a hidden symmetry that is spontaneously broken at certain energy levels. However, as we shall see, this appeal to hidden symmetries gives the theory an a priori quality more appropriate to metaphysics than empirical science.

---

isospin parameter was first introduced by Heisenberg to describe the two charge states of a nucleon (the proton and neutron). The idea that they could both correspond to two states of the same particle was suggested by the fact that their masses are approximately equal and that light, stable, even nuclei contain even numbers of them. See Yang and Mills [1954].

29 For example, even though Maxwell's later formulations of electrodynamics were derived from Lagrange's equations the conservation of energy played an important substantive role in Maxwell's thinking. See M. Morrison [1992].

4. Symmetries and Metaphysics

In the discussion of the Yang Mills gauge theory I said that as a theory of strong interactions it failed because it required a massive interaction quanta, yet the requirement of gauge invariance dictates that the mass of the potential field be zero. Hence if one introduces the usual mass term into the Lagrangian gauge invariance is destroyed.[30] A few years later Sheldon Glashow and Julian Schwinger attempted to develop a theory of weak interactions based on the suggestion that leptons must also carry isospin like the strongly interacting particles.[31] In order to use gauge theory it is necessary to identify a conservation law that would imply a symmetry for the Lagrangian. Because the lepton's wavefunctions could be grouped into families of the same lepton number (i.e. electron number and muon number) where $l_e = \begin{pmatrix}\nu_e \\ e^-\end{pmatrix} l_\mu^2 \begin{pmatrix}\nu_\mu \\ \mu^-\end{pmatrix}$, and because this bore a distinct similarity to the isospin doublet containing a proton and neutron it was conceivable that weak interactions were independent of the electrical charge of the particles on which they acted. In other words, the lepton cannot distinguish between a neutrino and an electron. Hence, it was thought that leptons carried a "weak isospin" that was exactly analogous to the isospin of nucleons for strong interactions. Consequently weak interactions could be considered invariant under rotations in the isospin space and the Lagrangian would be invariant under the group of isospin rotations characterized by SU(2).

---

30 This also makes it almost impossible for the theory to be renormalizable. Although space does not permit me to discuss the relationship between gauge invariant theories and renormalizability, the procedure is a way of getting rid of infinities that arise when dealing with any kind of relativistic quantum field theory. The requirement of renormalizibility is a constraint placed on the acceptability of quantum field theories and as a result it limits the kinds of theories that can be considered. Those with an infinite number of parameters are non-renormalizible. For more see Weinberg [1980].

31 See Schwinger [1957] and Glashow [1961].

But, the situation became more complicated than this. Recalling the electromagnetic case we know that the electromagnetic force is mediated by photon exchange and in analogy with electrodynamics it was thought that weak decay was mediated by the exchange of a spin-1 quantum call the W boson (i.e. the bosons were required to communicate between the interacting leptons whatever the local convention that governed the mixtures of electrons and neutrons).[32] Hence, this suggested that weak and electromagnetic processes could be described using the same kind of theory. The most serious problem whith the W particle was that its mass prevented the theory form being renormalizable.[33] Moreover, local symmetry required the introduction of such gauge particles to ensure invariance of the Lagrangian, but, gauge invariance allows only massless particles, so the W would effectively break the symmetry. Hence, one obtained an asymmetric solution (massive bosons) from a symmetric theory (gauge theory). In order for the theory to work it had to be possible for the gauge particles to acquire a mass in a way that would preserve gauge invariance. Moreover, if the electroweak and the electromagnetic theory were truly unified and mediated by the same kind of gauge particles then how could such a difference in the masses of the bosons and the photons exist? The answer to these questions was provided by a mechanism konwn as spontaneous symmetry breaking.

---

32 There were however significant differences between the photon and the W boson. The boson had to exist in at least two charged states W+ and W- since all weak interactions involved the exchange of electric charge between the currents J_μ and J'_μ. In addition the boson had to be extremely massive in order to explain the short range of the weak force. Also if neutral current phenomena are to be explained there must be a neutral boson as well, a particle commonly known as the Z^0.

33 In perturbation theory the probability of an event is given by the sum of contributions representing increasingly complex Feynman diagrams. Each contribution should, for all intents and purposes, be a dimensionless number. The mass factor in the W boson propagator leads to an ever-increasing number of infinite contributions to the perturbation series which cannot be reabsorbed into redefinition of the masses and couplings. Hence it cannot provide physically meaningful answers.

Since 1964 it was known that when a local symmetry is spontaneously broken the vector particles acquire a mass through a phenomenon known as the Higgs mechanism.[34] This principle of spontaneous symmetry breaking implies that the actual symmetry of a system can be less than the symmetry of its underlying physical laws; in other words, the Hamiltonian and commutation relations of a quantum theory would possess an exact symmetry while physically the system (in this case the particle physics vacuum) would be non-symmetrical. In order for the idea to have any merit one must assume that the vacuum is a degenerate state (i.e. it is not unique) such that for each unsymmetrical vacuum state there are others of the same minimal energy that are related to the first by various symmetry transformations that preserve the invariance of physical laws.[35] The phenomena observed within the framework of this unsymmetrical vacuum state will exhibit the broken symmetry even in the way that the physical laws appear to operate. Although there is no evidence that the vacuum state for the electroweak theory is degenerate it can be made so by the introduction of the Higgs mechanism which is an additional field with a definite but arbitrary orientation in the isospin vector space.[36] The orientation breaks the symmetry of the vacuum.

---

34 This phenomenon was an interesting counterexample to a theorem published by Goldstone, Salam and Weinberg in 1962 which showed that broken symmetries must always be accompanied by massless bosons known as the Goldstone bosons. See Goldstone, et. al. [1962]. The Higgs mechanism was due to the work of Peter Higgs [1964].

35 See Mills [1989]. The most famous example of the phenomenon is the ferromagnet which when brought to its state of lowest energy (cooled to absolute zero) must exhibit a magnetized state with a definite orientation, even though any orientation is allowed. The latter represents perfect symmetry while a definite orientation breaks the symmetry. This ground state of the ferromagnet is the analog of the vacuum state for our universe. It is perhaps important to point out that the vacuum state referred to in a quantum mechanical theory is not what one normally thinks of when we see the term "vacuum". Instead of a state of nothingness or emptiness the partical physics vacuum is the level at which physics actually takes place.

36 For a clear account see Collins, et. al. [1989] esp. sec 3.2, pp. 70-74

The Higgs field (or its associated particle the Higgs boson) is really a complex SU(2) doublet $\begin{pmatrix}\phi^+\\ \phi^-\end{pmatrix}$ consisting of four real fields which are needed to transform the massless gauge fields into massive ones. A massless gauge boson like the photon has two orthogonal spin components transverse to the directon of motion while massive gauge bosons have three including a longitudinal component in the direction of motion. In the electroweak theory the $W^{\pm}$ and the Z^0, which are the carriers of the weak force, absorb three of the four Higgs fields, thereby forming their longitudinal spin components and acquiring a mass. The remaining neutral Higgs field is not affected and should therefore be observable as a particle in its own right. The Higgs field breaks the symmetry of the vacuum by having a preferred direction in space (like the ferrogmagnet) – but the symmetry of the Lagrangian remains invariant. When the W and Z particles acquire masses the charges I_3 and Y (weak isospin and hypercharge) associated with the SU(2) × U(1) symmetry are not conserved in weak interactions because the vaccuum absorbs these quantum numbers.[37] The electric charge Q associated with the U(1) symmetry of electromagnetism remains conserved. The electroweak theory predicts the existence of four gauge quanta, a neutral photon-like object, sometimes referred to as the X^{o} and associated with the U(1) summetry as well as a weak isospin triplet $W^{\pm}$ and $W^{\pm}$ associated with the SU(2) symmetry. As a result of the Higgs symmetry breaking mechanisms the $W^{\pm}$ particles acquire a mass and the X^{o} and W^{o} are mixed so that the neutral particles one sees in nature are really two different linear combinations of these two. One of these neutral particles, the Z^{o} has a mass, while the other, the photon, is massless. Since the masses of the $W^{\pm}$ and the Z^{o} are governed by the structure of the Higgs field they do not affect the basic gauge invariance of the theory. The so-called "weakness" of the weak

37 Hypercharge is a quantum number associated with an elementary particle and characterizes an isotopic multiplet. It is equal to twice the mean electric charge (in e units) of the particles that form the multiplet.

interaction which is mediated by the $W^{\pm}$ and the Z^{o} is understood as a consequence of the masses of these particles.[38]

In summary then the Higgs phenomenon plays two related roles in the theory. It explains the discrepancy between the photon and the intermediate vector boson masses – the photon remains massless because it corresponds to the unbroken symmetry subgroup associated with the conservation of charge, while the bosons have masses because they correspond to symmetries that are broken. Secondly, the avoidance of an explicit mass term in the Lagrangian allows the theory to be renormalizable.

Returning to our original question, what does this episode tell us about the relationship between laws and symmetries? What makes the electroweak case different from the other examples is the role played by symmetries in generating empirical laws and explaining phenomena. The discovery that electromagnetism was gauge symmetric added nothing to the dynamical structure of the theory as formulated by Maxwell; however, from the beginning the development of the electroweak theory depended on symmetry principles. Although lepton isospin was the empirical starting point for constructing the theory, it was soon disovered that symmetry could not be preserved at the level of weak interactions; the phenomena simply did not obey the symmetry. Then, in a manner reminiscent of Kepler's appeal to the underlying harmony of the universe, it was found that the overall symmetry of the system could be retained by invoking a hidden or deeper level of symmetry that is not manifest in nature, thereby retaining the symmetry of the Lagrangian in the face of nonsymmetrical solutions. Whether or not it is the Higgs mechanism that is responsible for "breaking" the symmetry has not been confirmed; but that is a scientific problem that can supposedly be resolved using accelerators that are powerful enough to produce interactions at very high energies. But, the idea that there is a symmetry that is hidden from us, a reality represented by the symmetrical equations, seems more like a metaphysical assumption than an empirical hypothesis; at least

38 See Mills [1989].

at this stage in the development of physics. My reasons for suggesting that the Higgs hypothesis is different from claims about hidden symmetries stem from the fact that even if the Higgs particle is observed, resulting in an understanding of the physics of spontaneous symmetry breaking, even smaller distances would be required to observe the SU(2)× U(1) symmetry. Why should this matter? Well, simply because the particle physics vacuum, the fundamental level of physical interactions, embodies a broken symmetry causing the physical states in the theory (the solutions to the Lagrangian) to be nonsymmetrical. Given that the particle physics vacuum is where all of physics supposedly takes place, why should we believe that there really is a hidden symmetry that is reflected in the Lagrangian?

One of the reasons symmetry is so important is the fact that it is required for renormalizability; in other words, the theory will not give physically meaningful predictions and results unless the symmetry of the Lagrangian is observed. What 'tHooft and others showed was that a theory with a spontaneously broken symmetry could also be renormalizable. This was done by showing how renormalizability needn't be affected by what occurs at the scale where the symmetry is broken. As Georgi [1989] puts it, "'tHooft was able to disentangle the symmetry from the dynamics". A new renormalization scheme was introduced that discarded short distance effects through a procedure that makes no explicit reference to the physical parameters at that level. As a result there was no longer any dependence of physical results on the physics that operates at short distances. Intially, then, it would seem that symmetry is a crucial requirement for the theory to be realistic, so instead of adding metaphysics it has exactly the opposite effect. Moreover, since the symmetry constraints on the theory lead to some testable predictions (like the coupling constants that give the probability that W's and Z's will be emitted and reabsorbed in particle interactions) the hypothesis of an underlying symmetry seems to fit the classical hypothetico-deductive theory structure. But, to my mind the assumption of an underlying symmetry is not so straightforward. First of all the hypothetico-deductive pattern is not one that takes you directly from the hypothesis to all observable consequences since the symmetry constraints on the Lagrangian require massless particles. In that sense the structure deviates from the one displayed in the relatively unproblematic case of electrodynamics, where one can go from gauge symmetry to the dynamics of the theory. Secondly, although the hidden symmetry is required to perform the renormalizations it isn't immediately clear that that presents unquestionable justification for the assumption of a hidden symmetry; especially since the philosophical status of renormalizability is somewhat questionable and, moreover, the effective field theory programme no longer sees renormalization as a fundamental condition.[39] An effective quantum field theory refers only to particles that are important at the energy being studied. As a result the infinities that were once renormalized away now become limiting conditions indicating that, to quote Georgi [1989], "we do not know what happens at distances much smaller than those we can look at directly. Whatever happens at short distances, it doesn't affect what we actually do to study the theory at the distances we can probe". The advantage of effective field theories is that they do not assume that we undertand what takes place at very short distances. For instance, it is reasonable to suppose that an effective field theory might be able to tell us about the causes of the broken symmetry in the SU(2)× U(1) model but not necessarily about a hidden symmetry that becomes manifest at much smaller distances. I am not trying to suggest here that this is an open and shut case; the status of renormalization as an independent problem and its role in EFT are complicated issues requiring another paper. However, it is exactly because they are complicated

39 This is not to say that renormalizability is no longer important. If the gap between two neighbouring mass scales is large the effective field theory near the lower scale is approximately renormalizable since the nonrenormalizable interactions have an extremely small effect. This occurs in QED near the electron mass scale. See Georgi [1989]. There has been growing interest among philosophers of science in the status of renormalization, see for example papers by Paul Teller [1989a&b] and Cao and Schweber [1993].

that one should not uncritically accept the existence of hidden symmetries simply because they are necessary for renormalization. For now the best argument for hidden symmetry is that we need it in order for the electroweak theory to work, but that in no way legitimizes it as an *empirical* assumption.

To sum up: There is no direct empirical test for verifying hidden symmetries, or any symmetries for that matter. But in most cases symmetries allow for either the derivation of conservation laws or the covariance of physical laws in general; both of which are empirically confirmed by the phenomena. In that sense symmetries function as the hypothesis in a hypothetico-deductive style argument. The difference in the electroweak case is that the symmetry of the laws is not found in the corresponding phenomena. In that sense the theory is driven by a metaphysics that falsely presents itself as, at best, empirical methodology or at worst, a substantive claim about the world. My emphasis on hidden symmetry as metaphysics involves an attempt to recognize the status of certain hypotheses within the overall theory. Although the mass of the Higgs particle is not derivable from within the electroweak theory, its existence is, for practical purposes, an empirical claim that is directly verifiable. Not only are hidden symmetries not verifiable but the phenomena observed in experiment exhibit nonsymmetrical behaviour. So, although the symmetrical Lagrangian functions as a structural law in the derivation of the theory, that function does not legitimate an accompanying empirical foundation for the symmetry itself. I recognize of course that decisions about what counts as metaphysics is largely a matter of temperament; nevertheless I still maintain that there are good arguments to be made regarding when something should be considered metaphysics and why one should be sceptical about metaphysical hypotheses. Platonists, for example, will find nothing disturbing about the reality of a symmetry that is found only in equations. Steven Weinberg [1992] describes the issue in the following way: "Broken symmetry is a very Platonic notion: the reality we observe in our laboratories is only an imperfect reflection of a deeper and more beautiful reality, the reality of the equations that display all the symmetries of the theory". My claim is that there is only one reality, the reality of the laboratory.

The claim that symmetries are part of the empirical framework of physics is based on the idea that the invariance of the Lagrangian for a particular system reflects a conservation law whose effects or instantiations are observable in nature. However, the empirical connection in these cases, namely, the existence of conserved quantities, is rooted in experimental investigation and discovery. One could then infer the existence of a symmetry based on Noether's theorem, and in the case of electromagnetism, go on to show how that symmetry (gauge invariance) could provide us with the necessary structural features already present in electrodynamics. A crucial feature of this process is absent from the electroweak case. When symmetries are used as structural laws for generating theories by determining the form of interactions, etc. the link to empirical phenomena becomes even more important. But it is in exactly this case where the direct connection between symmetries and conserved quantities like isospin breaks down. The move to deeper hidden symmetries to explain this breakdown simply adds more metaphysics to the already dubious status of symmetries as a substantive property of the physical world.[40]

40 I would like to thank Nancy Cartwright for helpful conversations and for her encouragement and support. The paper was greatly improved as a result of comments and suggestions by Paul Forster. Versions of the paper were delivered at the University of Western Ontario and Rutgers, I would like to thank the audiences for their comments. Support of research by the Social Sciences and Humanities Research Council of Canada is gratefully acknowledged.

Bibliography

Aitchinson, I.J.R. [1982]: *An Informal Introduction to Gauge Field Theories.* Cambridge: CUP

Barrow, J. [1990]: *Theories of Everything.* New York: Oxford University Press.

Cao, T.Y. and Schweber S. [1993]: "The Conceptual Foundations and Philosophical Aspects of Renormalization Theory." *Synthese 97*, pp.32-108.

Collins, P.D., Martin A.D. and Squires, E.J. [1989]: *Particle Physics and Cosmology.* New York: Wiley.

Feynmann, R. [1967]: *The Character of Physical Law.* Cambridge MA: MIT Press, p.84.

Georgi, H. [1989]: "Effective Quantum Field Theories." In Paul Davies (ed.): *The New Physics.* Cambridge: Cambridge University Press, pp.446-457.

Gitterman, M. and Halpern, V. [1981]: *Qualitative Analysis of Physical Problems.* New York: Academic Press.

Glashow, S. [1961], *Nucl. Phys. 22*, p.579.

Goldstone, J., Salam, A., Weinberg, S., [1962], *Phys. Rev. 127*, p.965.

Higgs, P. [1964], *Phys. Rev. Lett., 12*, p.132.

Hill, E.L. [1951]: "Hamilton's Principle and the Conservation Theorems of Mathematical Physics", *Reviews of Modern Physics, 23*, pp.253-60.

Kuhn, T. [1972]: "Conservation of Energy: A Case of Simultaneous Discovery." In *The Essential Tension.* Chicago: University of Chicago Press.

Lanczos, C. [1970]: *The Variational Principles of Mechanics.* New York: Dover

Longair, M.S. [1984]: *Theoretical Concepts in Physics.* Cambridge: Cambridge University Press.

Mills, R. [1989]: "Gauge Fields." *American Journal of Physics, 57*, pp. 493-507.

Morrison, M. [1992]: "A Study in Theory Unification." *Studies in History and Philosophy of Science, 23*, pp.103-145.

Moriyasu, K. [1983]: *An Elementary Primer for Gauge Theory.* Singapore: World Scientific.

Noether, E. [1918], *Nachr. kgl. Ges. Wiss..* Göttingen 235.

Schwinger, J. [1957], *J. Ann. Phys., 2*, p.407.

Teller, P. [1989a]: "Three Problems of Renormalization." In H. Brown and R. Harré (eds.): *The Philosophical Foundations of Quantum Field Theory.* Oxford: Clarendon Press.

Teller, P. [1989b]: "Infinite Renormalization." *Philosophy of Science, 56*, pp.238-57.

van Fraassen, B.[1989]: *Laws and Symmetries.* Oxford: Clarendon Press.

Weinberg, S. [1980]: "Conceptual Foundations of the Unified Theory of Weak and Electromagnetic Interactions." *Reviews of Modern Physics, 52.*

Weinberg,S.[1993]: *Dreams of a Final Theory.* New York: Pantheon.

Weyl, H. [1952]: *Symmetries.* Princeton: Princeton University Press.

Wigner,E. [1979]: *Symmetries and Reflections.* New York: Woodbridge.

Yang. C.N. and Mills, R.L. [1954]: "Conservation of Isotropic Spin and Isotropic Gauge Invariance." *Physical Review*, *96*.

III. Scientific Views

Allan Franklin

Laws and Experiment

Traditionally, the role of experiment has been limited to the testing of laws or theories.[1] During the past decade we have recognized that the relation between laws and experiment is more complex, and that experiment may play varied roles. In addition to the confirmation or refutation of theory, experimental results may also call for a new theory, give us hints as to the structure of that theory, help to decide the mathematical form of the theory, or provide evidence for some future theory to explain. In addition, experiment may also have a life of its own, independent of theory.[2] In this essay I will discuss some of the problems associated with the role of experiment in the evaluation of laws.

1 In this paper I will use the terms "laws" and "theory" interchangably. In general, I think of laws as generalizations of empirical data, (I expect that other contributors to this volume will discuss the nature of laws) such as Kepler's Laws of Planetary Motion: 1) the planets move in ellipses with the sun at one focus; 2) the line joining the planet to the sun sweeps out equal areas in equal times; and 3) the ratio of the square of the period to the cube of the semi-major axis of its elliptical orbit is the same for all planets. I tend to think of theories as being somewhat deeper and having explanatory power. Nevertheless some laws clearly go beyond experimental generalizations. We may justly regard the conjunction of Newton's three Laws of Motion with his Law of Universal Gravitation as a theory. They certainly entail Kepler's laws, and indeed Kepler's laws are regarded as a confirmation of Newtonian mechanics.

2 For a discussion of these roles see Franklin [1990, Ch. 7].

1. Enabling Laws

I will begin, however, with a less-discussed interaction of experiment and theory. This is the use of laws in the design of experiment, what Peter Galison has called an "enabling theory." He illustrates this in his discussion of the experiments to measure the gyromagnetic ratio of the electron. The experiments were begun by Maxwell and failed because his apparatus was not sensitive enough to detect the effect. Maxwell had no model of the phenomenon which might have given him an estimate of the size of the effect he was looking for and indicated to him that his experiment would not work. It was only when Einstein and de Haas used the model of an electron orbiting an atomic nucleus that anyone had an idea of the size of the effect expected and experiments could be designed to measure them. (For details see Galison 1987, Ch.2).

I would like to further illustrate this use of laws in experimental design by considering an experiment to measure the K^{+}_{e2} branching ratio, the fraction of all K^{+} mesons that decay, as the subscript indicates, into two particles, namely a positron, e^{+}, and an electron neutrino, ν_e. The $K^{+}_{\mu 2}$ branching ratio signifies the decay of K^{+} mesons into a positive muon, μ^{+}, and a muon neutrino, ν_{μ}. [Bowen et al. 1967. See also Franklin 1990, Ch.6 for a discussion of this experiment]. In addition, the experiment was a test of the V-A theory, the accepted theory of weak interactions at the time the experiment was done in the 1960s. The V-A theory predicted that the ratio of the decay rates for $(K^{+} \rightarrow e^{+} + \nu_e) / (K^{+} \rightarrow \mu^{+} + \nu_{\mu})$, that is the ratio $(K^{+}_{e2}/K^{+}_{\mu 2})$, would be 2.6×10^{-5} for a pure axial vector interaction, A, (without radiative corrections), while a pure pseudoscalar interaction, P, gave a value of 1.02. Thus, a measurement of the K^{+}_{e2} branching ratio provides a stringent test of the theory, and on the presence of the P interaction.

Because $K^{+}_{\mu 2}$ decays had a branching ratio of approximately 63 percent, this gave a predicted K^{+}_{e2} branching ratio of 1.44×10^{-5}. (This includes the effect of radiative corrections). This sets the scale for the experiment. If one wanted to detect say 10

K^+_{e2} decays, one would need to have a sample of approximately 700,000 K^+ decays.[3]

The experimental apparatus is shown in Figure 1. The kaons were separated from pions and protons in the beam and identified by their range in matter and by time of flight. They were then stopped in counter C_3.

Particles from decays at approximately 90° to the incident beam passed through a set of thin plate spark chambers, the momentum chambers, located in a magnetic field. This permitted the measurement of the particle momentum with an experimentally measured resolution of 1.9%. The decay particles were detected by two scintillation counters, C_5 and C_6, and then passed through a gas Cerenkov counter, which was set to detect high energy positrons. This Cerenkov counter had been independently checked, and had a measured efficiency for 250 MeV positrons of between 95%and 99%, depending on the trajectory of the positron through the counter. For particles other than electrons, the measured efficiency was approximately 0.38%. Thus, the counter served to identify decay positrons.

A thick plate range spark chamber was placed behind the Cerenkov counter. Its total thickness of 80 g/cm^2 was enough to stop all particles resulting from the decay of the K^+ meson, so that one could use the range in matter, along with momentum and detection by the Cerenkov counter, to help identify decay particles.

Let us consider what signal the experimenters expected to find, and what the possible sources of background, processes that might simulate K^+_{e2} decay, were. This sort of calculation is often required in order to estimate whether or not an experi-

---

3 This is an oversimplification. The detector had a small solid angle, and thus detected only a small fraction of the decays of the stopped K^+ mesons. In addition, the solid angle was a function of the decay particle momentum. The result of this was that one needed a considerably larger number of stopped kaons to observe the K^+_{e2} decay. For the entire experimental run a total of approximately 1.3 x 10^9 kaons were stopped and a total of 6 K^+_{e2} events observed.

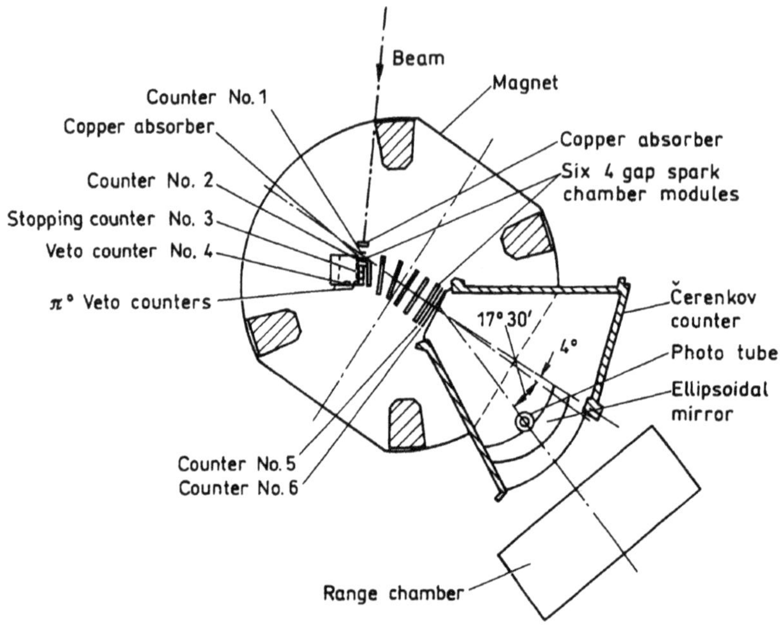

Figure 1. Details of the experimental apparatus for the K^+_{e2} experiment of Bowen et al. [1967].

ment is feasible. If the backgrounds are too large the experiment cannot be done.

The positron from K^+_{e2} decay has a momentum of 246.9 MeV/c in the K^+ center of mass. This is higher than the momentum of any other direct decay product of K^+ decay. The closest competitor is the muon from $K^+_{\mu 2}$ decay, which has a momentum of 235.6 MeV/c. The principal sources of high-momentum positrons that might mimic a real K^+_{e2} decay are

1) $K^+ \rightarrow e^+ + \Pi^o + \nu_e$, K^+_{e3} decay, (where Π^o is an additional particle, namely the uncharged Π meson) with a maximum positron momentum of 228 MeV/c and a branching ratio of about 5%.
2) $K^+ \rightarrow \mu^+ + \nu_\mu$, followed by $\mu^+ \rightarrow e^+ + \nu_e + \bar{\nu}_\mu$, (where $\bar{\nu}_\mu$ is the muon antineutrino), with a maximum momentum of

> 246.9 MeV/c and a branching ratio of approximately 1.2 x 10^{-4} per foot of muon path. I note that this decay rate per foot is about a factor of 5 larger than the total expected K^+_{e2} decay rate. If this source of background could not be eliminated or greatly reduced the experiment could not be done.

Using the measured momentum resolution of 1.9%, the accuracy with which the momentum of a particle could be determined in the apparatus, and the known K^+_{e3} decay rate and momentum spectrum one could calculate that the number of K^+_{e3} events expected in the K^+_{e2} decay region, 242 MeV/c – 252 MeV/c, was less than 5% of the expected K^+_{e2} rate. If the K^+ decayed in flight then the momentum of the positron from K^+_{e3} decay might be higher than 228 MeV/c. This possible source of background was completely eliminated when "prompt" events were removed from the event sample.

The background due to K^+ decay into a muon, followed by muon decay into a positron was also calculated. This involved a detailed calculation which included the decay rate, the momentum and angular distribution of the decay positrons relative to the muons, the momentum and angular resolution in the thin plate chambers, and the ability to extrapolate the decay particle trajectory from the thin plate chambers into the range chamber. The result of this calculation was an expected background of about 15% of the expected K^+_{e2} decay rate. The experimenters expected a total background of approximately 20% of the expected K^+_{e2} rate. There were other sources of background, due to the operation of the experimental apparatus, but these were not calculable in advance. Thus, calculations based on the law the experimenters were proposing to test were vital in determining the feasibility of the experiment. Without such calculations the experimenters would not have known the relative sizes of the predicted effect and the sources of background.

An interesting point is that a theory does not have to be correct in order to serve as a useful enabling theory. In the case of the Fifth Force hypothesis, to be discussed later, the law is now generally considered to be incorrect, yet it was extremely important in the design of the experiments that ultimately refuted it. The suggestion was of a small addition to the normal gravitational force, which had a strength of approximately one percent that of the normal gravitational coupling and a range of about 100 meters. The strength allowed experimenters to estimate the size of the effects expected, whereas the range suggested improvements in the experimental sensitivity by placing the apparatus in an environment with a large matter asymmetry, such as a cliff or a hillside, or near a large laboratory mass.[4] I also note that the model of an orbiting electron to explain magnetism is incorrect. The current explanation is that magnetism is due to electron spin, a quantity unknown to either Einstein or de Haas. The order of magnitude of the effect is, however, the same.

2. Do Experiments Test Laws?

a) The Theory-Ladenness of Observation

One might well ask whether or not the importance of the theory of the phenomenon in the experimental design precludes using the experiment as a test of that theory. Recall that the K^+_{e2} experiment was designed to be a rather stringent test of the V-A theory. In this case it seems clear that there is no real problem. The theory was involved only in determining the feasibility of the experiment, and did not, in any way, influence the results of the experiment, nor did it guarantee results in agreement with that theory. The experiment would have produced valid data and results even if the theory was wrong. The experimental apparatus and analysis procedure were shown to operate properly, independent of the theory. If the actual K^+_{e2} branching ratio had been smaller than that predicted, then the experimenters would not have seen any excess events. If it had been larger, the observed number of events would have been far larger than the es-

4 For details see Franklin [1993].

timated background. In the actual case the experimenters found $6^{+5.2}_{-3.7}$ events. This gave a branching ratio of $(2.1^{+1.8}_{-1.3}) \times 10^{-5}$ in good agreement with the theoretical prediction of 1.44×10^{-5}.

This discussion is related to the more general problem of the theory-ladenness of observation.[5] This is the fact that terms used in observation statements or measurement reports are laden with a particular theory. For example, although terms like "force"and "mass" appear in observation statements, they take their meanings only within the context of Newtonian dynamics. The question then arises whether or not such theory-laden terms can be used in experiments that test the theories. Kuhn [1970], Feyerabend [1975], and Barnes [1982], have all argued that there can be no comparison of competing paradigms or theories based solely on experimental evidence. "There is no appropriate scale available with which to weigh the merits of alternative paradigms: they are incommensurable [Barnes 1982. p. 65]."[6]

Briefly stated the argument is as follows. There can be no theory-neutral observation language. All observation terms are theory-laden, and thus we cannot compare experimental results, because in different paradigms, terms describing experimental results have different meanings, even when the words used are the same. An example of this would be the term "mass," which is a constant in Newtonian mechanics, whereas in special relativity it is a function of velocity.

---

5 Hacking [1983, pp. 171-2] and Hanson [1969].

6 In fairness to Kuhn, it should be noted that he has modified his views somewhat in later work [Kuhn 1983]. He argues there for "local holism," that individual terms can be learned only as part of a system of terms, and for "local incommensurability," which implies that vocabularies of two rival successor theories preserve meanings across theory change, and that only "core terms" remain incommensurable. Here, incommensurable means simply untranslatable. Although Kuhn has not specified clearly what he means by core terms, I believe that under any reasonable view, mass must be such a term. Thus, my argument still holds. I am grateful to Friedel Weinert for suggesting this point.

Let us consider the following thought experiment. I shall demonstrate that an experiment described in procedural, theory-neutral (between the two competing theories or paradigms) terms, gives different results when interpreted within the two alternative paradigms. Thus, a measurement of the quantity derived will unambiguously distinguish between the two. I shall take as an example one of Kuhn's own exemplars: the difference between Newtonian and Einsteinian mechanics. The case in point can be loosely described as the scattering of equal-"mass" objects.

The experimental procedure is as follows: Consider a class of objects, let us say billiard balls. The objects are examined pairwise by placing a compressed spring between them. The spring is allowed to expand freely, and the velocities of the two objects measured. Because we restrict ourselves to a single frame of reference in the laboratory, the measurement is theory-neutral. We then select two balls whose velocities were equal. A Newtonian would say that they have the same mass, M_N, whereas an Einsteinian would say that the relativistic masses $M_R = M_{OR} (1 - v^2/c^2)^{-1/2}$ are equal and thus that the rest masses M_{OR} are equal. This is agreed, but the point is that the procedure itself is theory-neutral. One of the objects is placed at rest in the laboratory and the other is given a velocity V_1 (again, theory-neutral), and the objects are allowed to scatter off each other. Care is taken to make the collision elastic (no energy given off). The angle between the two objects after the collision is measured. Although the assignments of energies and momenta will be different, the measurement of the angle does not depend on these assignments. As is well known, the predicted value of this angle differs in Newtonian and relativistic mechanics (for details see Franklin 1986 pp. 109-13). For a Newtonian $\theta_N = 90°$, whereas for an Einsteinian, $\theta_R < 90°$. A measurement of the angle between the velocities of the two outgoing objects will clearly distinguish between the two paradigms. They are commensurable.

This is not, of course, a general proof that such experiments can always be constructed. It should be pointed out, too, that the experiments of Lummer and Pringsheim and of Rubens and

Kurlbaum on the spectrum of black-body radiation, which led to Planck's suggestion of quantization, also seem to be neutral with respect to classical and quantum physics. If this type of theory-neutral experiment can be found for what are two of the most revolutionary changes in physics, then I suspect that they can be found for almost all cases. However plausible the notion of the theory-ladenness of observation is, it has not been demonstrated that it prohibits theory testing.

b) The Duhem-Quine Problem

One might ask, however, whether or not one should trust the results of any particular experiment. I have previously suggested an epistemology of experiment, a set of strategies that provides grounds for reasonable belief in the validity of results. These include: 1) experimental checks and calibration, in which the experimental apparatus reproduces known phenomena; 2) reproducing artifacts that are known in advance to be present; 3) intervention, in which the experimenter manipulates the object under observation; 4) independent confirmation using different experiments; 5) elimination of plausible sources of error and alternative explanations of the result (the Sherlock Holmes strategy); 6) using the results themselves to argue for their validity; 7) using an independently well-corroborated theory of the phenomena to explain the results; 8) using an apparatus based on a well-corroborated theory; and 9) using statistical arguments.[See Franklin 1986, Ch. 6 and 1990, Ch. 6 for details]. I believe that such results may then be legitimately used in testing theory.

Philosophers of science have raised another question concerning the use of experiment in the testing of theory. This is known as the Duhem-Quine problem [See Harding, 1976]. In the usual *modus tollens* if a hypothesis h entails an experimental result e then $\neg e$ (not e) entails $\neg h$. As Duhem and Quine pointed out it is not just h that entails e but rather h and b, where b includes background knowledge and auxiliary hypotheses.

Thus, $\neg e$ entails $\neg h$ or $\neg b$ and we don't know where to place the blame. As Quine put it, "Any statement can be held true come what may, if we make drastic enough adjustments elsewhere in the system [Quine, 1953, p. 43]." One might, of course, also question the experimental result $\neg e$. This is where one applies the epistemology of experiment, to justify our belief in the experimental result.

In practice, however, scientists never confront all the logically possible explanations of a given result. There are usually only a reasonable number of plausible or physically interesting alternatives on offer. Scientists evaluate the cost of accepting one of these alternatives in the light of all existing evidence. One of these alternatives may be far better supported by the evidence than any of the others. The alternatives themselves may also be tested and we may be left with only one explanation.

Are there any hints that one might offer as to how one goes about solving the problem of where to place the blame? Noretta Koertge [1978] has made some very useful suggestions.[7] She suggests two strategies.

1) Check the most accessible source of trouble first. In other words, check those alternative explanations that are most easily tested.
2) Check on the most probable source of trouble early on.[8]

These are, in fact, the strategies scientists tend to use. She goes on to discuss the appraisals that go into a solution, X, of the Duhem-Quine problem.

1) "How interesting or informative or explanatory would X be if it were true.
2) What is the probability that X is true?" [Koertge, 1978, p. 263].

7 Koertge's discussion is influenced, in part, by *Zen and the Art of Motorcycle Maintenance* [Pirsig, 1974]. In case of ignition failure test the spark plugs before dismantling the carburetor.

8 If your carburetor has a history of trouble, check it early on.

This will, of course, involve appraisals of the plausibility or probability of the alternatives. It will also involve estimates of the scientific interest of the alternatives. As Koertge points out, it is here that the most serious differences of opinion will occur within the scientific community. Even if one had agreed upon and clear measures of content, simplicity, depth, heuristic power, etc. different scientists might well give these criteria different weights. These criteria may also be applied in different ways in different situations.

Should the fact that there is no prescriptive algorithm for the solution of the problem worry us excessively. I don't think so. It is precisely because of these differing judgments that more of the alternatives are likely to be explored. Sometimes, even a very implausible hypothesis turns out to be correct. The history of science shows us that in practice the Duhem-Quine problem is solved in the way Koertge and I suggest, using valid experimental evidence. (See Franklin 1990, Ch, 7 for details).

c) Discordant Results

Although the epistemology of experiment, discussed earlier, provides us with reasonable grounds for belief in experimental results, it does not guarantee that those results are correct, or even that two different experiments will give the same result. If we are to use experimental results to evaluate theory then we must solve this problem. How does the scientific community deal with this question of the fallibility of results or decide between two conflicting results? Fallibility can be dealt with. One uses the best available evidence, with the knowledge that it is fallible and that decisions made may turn out to be incorrect. Reasonable belief does not guarantee certainty.[9]

9 See discussions of the fallibility of experimental results, of theoretical calculation, and of the comparison between theory and experiment in Franklin [1990].

Discordant results are not unusual occurrences in the history of physics, particularly when a new experimental technology is used to investigate previously unobserved phenomena. Thus, in the early days of experimental investigation of the Weinberg-Salam unified theory of electroweak interactions, several groups reported such discordant results. In 1976 and 1977 groups at the University of Washington and Oxford University [Baird et al. 1976, 1977 Lewis et al. 1977] reported results on atomic parity violation that disagreed with the predictions of W-S theory. In 1978 and 1979 groups at Berkeley [Conti et al. 1979] and Novosibirsk [Barkov and Zolotorev 1978a,b, 1979] reported results in agreement with the theory. (I note here that only two of these experiments were replications in the sense that they measured the same quantity. Oxford and Novosibirsk measured parity violation in a particular transition in atomic bismuth, while Washington and Berkeley looked for similar effects in a different transition in bismuth and in thallium, respectively. All of the experiments did, however, investigate parity violation in atomic systems.)[10] How was the decision made between the conflicting atomic-parity violation results? The early Washington and Oxford results were internally inconsistent and also contained systematic uncertainties that were approximately as large as the predicted effects. In addition, both experiments used a new type of experimental apparatus. Both the Washington and Oxford groups continued their investigations into the 1980s, making improvements in their apparatus. Other atomic-parity violation experiments were also performed. All of the subsequent experiments gave results in agreement with

10 The interesting and important question of whether or not the evidence supported the W-S theory was made more difficult when another experiment on the scattering of polarized electrons from deuterium, the SLAC E122 experiment, confirmed W-S theory [Prescott et al. 1978, 1979]. This decision was based on the greater evidential weight of the carefully done and checked E122 experiment in comparison to the uncertain atomic parity results. For details of this see Franklin [1990, Ch. 8]. For differing views of the episode see Ackermann [1991], Lynch [1991], and Pickering [1991].

the theoretical predictions.[11] Repetition of the experiments and improvements in the apparatuses provided convincing support for one set of early results, those that agreed with W-S theory. In this case there were also theoretical attempts to reconcile the discordant early results. This, so-called, hybrid model was experimentally refuted.

A similar case of discordant results occurred recently in the investigation of the "Fifth Force," a proposed modification of the law of gravity.[12] The two initial experimental results, those of Thieberger [1987] and of the Eöt-Wash group, a group at the University of Washington,[13] [Stubbs et al. 1987], gave conflicting results, one favoring the existence of the Fifth Force and one opposed. These were followed shortly thereafter by Boynton's [1987] "marginally observed" (his words) positive result.

The subsequent history seems to be an illustration of one way in which the scientific community deals with conflicting experimental evidence. Rather than making an immediate decision as to which were the valid results, this seemed extremely difficult to do on methodological or epistemological grounds, the community chose to await further measurements and analysis before coming to any conclusion about the evidence. The torsion-balance experiments of Eöt-Wash and Boynton were repeated by others including Fitch [1988], Cowsik [1988, 1990], Bennett [1989], and Newman [1989, and Nelson 1990] and by Eöt-Wash [Adelberger et al. 1989, Stubbs et al. 1989a,b] and Boynton [1990] themselves. These repetitions, in different locations and using different substances, gave consistently negative results. In addition, Bizzeti [1989a,b], using a float apparatus similar to that of Thieberger, also obtained results showing no evidence of a Fifth Force. There is, in fact, no explanation of either Thieberger's or of Boynton's original, presumably in-

11 By this time the theoretical calculations had also changed.

12 For details see Franklin [1993].

13 The Eöt-Wash group took its nickname from the fact that some of the original evidence for the Fifth Force came from a reanalysis of an experiment done early in the twentieth century by Eötvös and his collaborators.

correct, results. The scientific community has chosen, I believe quite reasonably, to regard the preponderance of negative results as conclusive.[14] In this case too, there were theoretical attempts to reconcile the discordant results. Once the decision as made that the results arguing against the existence of the Fifth Force were correct those theoretical attempts became academic. There was nothing left to explain.

3. Conclusion

In this essay I have discussed several aspects of the complex interaction between laws and experiment. We have seen that laws may often be important in determining the feasibility of experiment and in experimental design. I have also argued that this important role of laws does not preclude the testing of those same laws by the experiments they were used to design. We have seen a case, that of the Fifth Force, where the hypothesis used in the experimental design was, in fact, refuted by those experiments. We have also examined several other problems concerning the role of experimental evidence in the evaluation of laws. These include the theory-ladenness of observation, the Duhem-Quine problem, the fallibility of experimental results, and the question of discordant experimental results. I have argued that all of these problems can be reasonably solved, so that we may use experimental evidence to evaluate laws. There is an evidential basis for scientific laws.

14 It is a fact of experimental life that experiments rarely work when they are initially turned on and that experimental results can be wrong, even if there is no apparent error. It is not necessary to know the exact source of an error in order to discount or to distrust a particular experimental result. Its disagreement with numerous other results can, I believe, be sufficient.

References

Ackermann, R. [1991]. "Allan Franklin, Right or Wrong," in Fine [1991], 451-7.

Adelberger, E.G. [1989]: "High-Sensitivity Hillside Results from the Eöt-Wash Experiment," in Fackler [1989], 485-99.

Baird, P.E.G., et al. [1976]: "Search for Parity Nonconserving Optical Rotation in Atomic Bismuth," *Nature*, *264*, 528-9.

Baird, P.E.G., et al. [1977]: "Search for Parity-Nonconserving Optical Rotation in Atomic Bismuth," *Physical Review Letters*, *39*, 798-801.

Barkov, L.M. and Zolotorev, M.S. [1978a]: "Observations of Parity Nonconservation in Atomic Transitions," *JETP Letters*, *27*, 357-61.

Barkov, L.M. and Zolotorev, M.S. [1978b]: "Measurement of Optical Activity of Bismuth Vapor," *JETP Letters*, *28*, 503-6.

Barkov, L.M. and Zolotorev, M.S. [1979]: "Parity Violation in Atomic Bismuth," *Physics Letters*, *85B*, 308-13.

Barnes, Barry [1982]: *T.S. Kuhn and Social Science*. New York: Macmillan.

Bennett, W.R. [1989]: "Modulated-Source Eötvös Experiment at Little Goose Lock," *Physical Review Letters*, *62*, 365-8.

Bizzeti, P.G. et al. [1989a]: "Search for a Composition Dependent Fifth Force: Results of the Vallombrosa Experiment," in Fackler [1989], 511-24.

Bizzeti, P.G. et al. [1989b]: "Search for a Composition-Dependent Fifth Force," *Physical Review Letters*, *62*, 2901-4.

Bowen, D.R., et al. [1967]: Measurement of the K^{+}_{e2} Branching Ratio," *Physical Review*, *154*, 1314-22.

Boynton, P.E. et al. [1987]: "Search for an Intermediate-Range Composition-Dependent Force," *Physical Review Letters*, *59*, 1385-9.

Boynton, P. [1990]: "New Limits on the Detection of a Composition-Dependent Macroscopic Force," in Fackler [1990], 207-24.

Conti, R., et al. [1979]: "Preliminary Observation of Parity Nonconservation in Atomic Thallium," *Physical Review Letters*, *42*, 343-6.

Cowsik, R. et al. [1988]: "Limit on the Strength of Intermediate-Range Forces Coupling to Isospin," *Physical Review Letters*, *61*, 2179-81.

Cowsik, R. et al. [1990: "Strength of Intermediate-Range Forces Coupling to Isospin." *Physical Review Letters*, *64*, 336-9.

Fackler, O. and Tran Thanh Van, J (eds.) [1989]: *Tests of Fundamental Laws in Physics: Ninth Moriond Workshop* (Editions Frontieres, Gif sur Yvette).

Fackler, O. and Tran Thanh Van, J. (eds.) [1990]: *New and Exotic Phenomena '90: Tenth Moriond Workshop* (Editions Frontieres, Gif sur Yvette).

Feyerabend, Paul [1975]: *Against Method*. New York: New Left Books.

Fine, A., Forbes, M., and Wessels, L. (eds.) [1991]: *PSA 1990*, Volume Two(Philosophy of Science Association, East Lansing, MI).

Fitch, V.L., Isaila, M.V., and Palmer, M.A. [1988]: "Limits on the Existence of a Material-Dependent Intermediate-Range Force," *Physical Review Letters*, *60*, 1801-4.

Franklin, Allan [1986]: *The Neglect of Experiment*. Cambridge: Cambridge University Press.

Franklin, Allan. [1990]: *Experiment, Right or Wrong*. Cambridge: Cambridge UniversityPress.

Franklin, A. [1993]: *The Rise and Fall of the Fifth Force: Discovery, Pursuit, and Justification in Modern Physics* (American Institute of Physics, New York).

Galison, Peter. [1987]: *How Experiments End*. Chicago: University of Chicago Press.

Hacking, Ian [1983]: *Representing and Intervening*. Cambridge: Cambridge University Press.

Hanson, Norwood R. [1969]: *Patterns of Discovery*. Cambridge: Cambridge University Press.

Harding, S. (ed.) [1976]: *Can Theories Be Refuted?* Dordrecht: D. Reidel.

Koertge, N. [1978]: "Towards a New Theory of Scientific Inquiry," in *Progress and Rationality in Science*, G. Radnitzky and G. Andersson (eds.), 253-78. Dordrecht: D. Reidel.

Kuhn, Thomas [1970]: *The Structure of Scientific Revolutions*, second edition. Chicago: University of Chicago Press.

Kuhn, Thomas [1983]: "Commensurability, Comparibility, and Communicability," in *PSA 1982*, Volume 2, P.D. Asquith and T. Nickles (eds.) 669-88. East Lansing, MI: Philosophy of Science Association.

Lewis, L.L., et al. [1977]: "Upper Limit on Parity-Nonconserving Optical Rotation in Atomic Bismuth," *Physical Review Letters*, *39*, 795-8.

Lynch, M. [1991]: "Allan Franklin's Transcendental Physics," in Fine [1991], 471-85.

Nelson, P.G., Graham, D.M., and Newman, R.D. [1990]: "Search for an Intermediate-Range Composition-Dependent Force Coupling to N – Z," *Physical Review D*, *42*, 963-76.

Newman, R., Graham, D., and Nelson, P. [1989]: "A 'Fifth Force' Search for Differential Acceleration of Lead and Copper Toward Lead," in Fackler [1989], 459-72.

Pickering, A. [1991]: "Reason Enough? More on Parity-Violation Experiments and Electroweak Gauge Theory," in Fine [1991], 459-69.

Pirsig, R.M. [1974]: *Zen and the Art of Motorcycle Maintenance:* An Inquiry into Values. New York: William Morrow.

Popper, Karl. 1959: *The Logic of Scientific Discovery*. New York: Basic Books.

Prescott, C.Y., et al. [1978]: "Parity Non-Conservation in Inelastic Electron Scattering," *Physics Letters*, *77B*, 347- 52.

Prescott, C.Y., et al. [1979]: "Further Measurements of Parity Non-Conservation in Inelastic Electron Scattering," *Physics Letters*, *84B*, 524-8.

Quine, W. [1953]: *From a Logical Point of View*. Cambridge: Harvard University Press.

Stubbs, C.W. et al. [1987]: "Search for an Intermediate-Range Interaction," *Physical Review Letters*, *58*, 1070-3.
Stubbs, C.W. et al. [1989a]: "Eöt-Wash Constraints on Multiple Yukawa Interactions and on a Coupling to 'Isospin,"' in Fackler [1989], 473-84.
Stubbs, C.W. et al. [1989b]: "Limits on Composition-Dependent Interactions Using a Laboratory Source: Is There a 'Fifth Force' Coupled to Isospin?" *Physical Review Letters*, *62*, 609-12.
Thieberger, P. [1987]: "Search for a Substance-Dependent Force with a New Differential Accelerometer," *Physical Review Letters*, *58*, 1066-9.

Erhard Scheibe

Laws and Theories: Generality vs. Coherence

I. Introduction

Natural laws are omnipresent in modern natural science. But it was reserved for physics to make them the proper subject of research and to advance towards the fundamental laws of nature. Kepler's laws of planetary motion and Galileo's law of free fall mark the beginning of the new epoch, and the present endeavours for the theoretical unification of the elementary interactions could, if successful, mean its end. 'The age in which we live is the age in which we discover the fundamental laws of nature' (Feynman 1965, p.72). A totally different question is whether we know what a law of nature is. Inspite of repeated and strenuous efforts to define the concept of natural law there is today no generally accepted definition and presumably never will (cf. Hempel 1965, Ch.10:Part III, and Ch.12:2.1). It is wiser, therefore, to focus one's efforts on some interesting aspects that we can read off from the physical laws already at hand.

Physical laws are not viable in isolation. We find them in the midth of more or less strictly axiomatically ordered *physical theories*, like mechanics, some as central axioms, like Newton's lex secunda, others as derived and more peripheral, like the laws of the pendulum. According to the axiomatic method laws are propositions (or: propositional functions, see below) within a theory that is endowed with a logical apparatus allowing the derivation of laws from other laws. Besides there are the physical concepts in which the laws are formulated, such as mass, force, energy etc. They, too, come in an hierarchical order, some being basic concepts and others defined by them. Just as classical mechanics is a literal extension of euclidean geometry, it took over its axiomatic method, and from mechanics axiomatics has entered other fields of physics as, for instance, thermodynamics and quantum mechanics.

Do we know, then, what theories are better than what laws are? Certainly not. The concept of physical theory has departed from its euclidean ideal corresponding to the advancement of physics. It has become more complicated and has incorporated new elements (Suppe 1974). A field that has left especially deep traces is the problem of the interplay between *theory and observation*. Observations serve – this is their most important function – to test theories. Inductive logic and theories of empirical evidence take care of this side of the matter. There is, however, also the inverse relation: observations are made with the help of theories. All actually performed measurements of physical quantities were made more or less indirectly: To measure a quantity occurring in theory T either theories other than T are used or even T itself. The external use originates in the necessity to use observational equipment of all kinds and to consider various perturbing influences. The internal use of a theory T for measuring quantity G occurs whenever G is irreducibly introduced by a law of T as, for instance, force is by Newton's lex secunda. In such a case every measurement of G presupposes the validity of the law in question. Consequently, further complications intrude into the theory concept in order to save a theory's empirical content.

Theory and observation – this certainly is an important and widely recognized aspect concerning physical theories and their laws. It is, however, not the aspect to be discussed in this paper. In this paper I am going to discuss a problem remarkably unnoticed that may be called the *polarity of coherence and generality* in a physical law. Put into common jargon a physical law says that all physical systems of a given kind behave in a certain manner. All planets move in elliptical paths, with the sun at one focus of the ellipse. All gases have pressure, volume and temperature according to van der Waals' equation. All hydrogen atoms behave according to Schrödinger's equation with Coulomb potential etc. In a law reconstructed in this way two things come together: 1) the contents of the law, and 2) its universal form. The contents are what physicists have to find out by investigating nature. It concerns exclusively the *individual system* and is different from case to case according to the different kinds of systems investigated. That besides this *all* systems of a given kind satisfy the same condition (according to the law) adds nothing to the law's contents. Yet the second part – the universal form – does something immensely important: according to the common view it is only its *generality* - *this* kind of generality – that gives a law its dignity *as* a universal law and, therefore, as a law at all.

So far, so good. There is, however, the problem that the two parts in which we have dissected a law *work against each other*: In a natural law two adverse tendencies can be identified. The first one concerns its contents. It is the task of theoretical physics to analyse the *interconnectedness* in nature. Thus by Kepler's laws a connection between the sun and a planet is established. In a gas law pressure, volume and temperature of a gas are connected by some equation etc. These connections are the contents of the corresponding laws, and – as I said – they are completely expressed in that part of the law that deals with the individual system. By contrast, the other tendency concerns the generality of a law and consists in the demand that the single instances of a general law be *independent* of each other – indeed as completely independent as different worlds. In the quantum mechanical example already the single H-atoms making up an ensemble have to be independent of each other, and the same has to hold for the different ensembles that realize different solutions of the Schrödinger equation: There should be no connection between them – neither real nor imaginary.

So here we have a difficulty: It looks as if a natural law, by its very universality and the *independence* of its instances implied by it, sets a limit to the proper task of the law, namely the establishing of a lawful *connection* between physical entities. For what we accomplish by performing this task – it seems – we have to pay the prize that in a different place we have to provide for the exact opposite. Moreover, it could even happen, and indeed does happen in certain cases, that the contents of a law produce such an integral connection that simply *disallows* (in principle) any independence and universality in one and the same world. In such cases the whole conception in question might lead to an internal contradiction, and at any rate an extreme coherentism according to which everything is related to everything else in some important way would not admit any natural law. I cannot help finding this situation a curiosity worthy of closer inspection – so much the more since it did not yet receive the attention that is deserves (cf. Scheibe 1989 and 1991).

II. The Generality of Laws 1

As I said, physical laws are embedded in theories consisting of a logical apparatus, non-logical symbols, rules of language formation, correspondence rules etc. It is really a complicated matter (cf. Ludwig 1978). In the following only those features of this equipment will be mentioned that are absolutely necessary for an understanding of the main argument.

Let me begin by recalling that many presentations of the subject offer the aristotelean

B belongs to all A (1a)

or its Fregean version

for all x: if Ax then Bx (1b)

or even the counterfactual conditionals

for all x: if Ax were (or: had been) the case,
then Bx would be (or: have been) the case (1c)

as the proper and typical form of a lawlike statement (cf. Stegmüller 1983, Ch.V). Obviously it is the universal form of a law that shall catch the eye. But already this would not be the right kind of emphasis. The generality of natural laws shares the fate of causality as another alledged ingredient of laws. On the latter Russell said already in his 1912, p.1:

'All philosophers...imagine that causation is one of the fundamental axioms...of science, yet, oddly enough, in advanced sciences such as gravitational astronomy, the word "cause" never occurs.' Likewise, if we look into the papers of physicists, especially the systematically composed text books on physics, we do not by any means find formulations like (1). Neither the actual presentations of fundamental laws like Newton's lex secunda, Maxwell's equations etc. nor those of derived and empirical laws like a gas law, Ohm's law etc. remind us even remotely of anything like (1). I do not say that physicists would not occasionally emphasize that physics is about to find the most general laws of nature or that it attempts to give its statements a universal form in some sense. But such emphasis does not appear in the professional papers and often is given a more concrete wording, quite different from the standard formulations preferred by philosophers of science. Very common is the idea of reproducibility of events and the delimitation of physics from cosmology. 'One may characterize physics as the doctrine of the repeatable, be it a succesion in time or the co-existence in space. The validity of physical theorems is founded on this repeatability...By contrast 'cosmology' is the doctrine of the unique universe, of its special, perhaps historically, developed features.' (Hund 1972, p.274) Such statements, then, can be found in spare time publications of physicists, and also in this paper the generality of laws is not dismissed. But formulations like (1) are certainly not the right starting point.

What else is? The key concept for an answer is the concept of a *physical system* and its *contingent descriptions*. As we shall see, laws are primarily statements about individual systems. The same is true for contingent descriptions. But whereas the latter can be formulated without referring to laws, in the formulation of laws we *must* have recourse to contingent descriptions. There is an obvious but important asymmetry here. Take the example of a charged particle, moving in an electromagnetic field under the Lorentz force. The equation of motion – the law of this kind of motion – simply cannot be formulated without knowing how the particle and the field are to be decribed by elementary statements. We have to develop the idea of the path of the particle in Minkowski spacetime, including its characterization by (relativistic) velocity and acceleration. We have further to fix the kind of description of the field as a covariant tensor field of rank 2 on the same spacetime. It is only after these preparations that we are in a position to write down the equation of motion. And as in electrodynamics it is in general: Contingent descriptions of systems in terms of elementary propositions bundled up in parameters are the proper starting point of theory formation in physics.

In the next step of our analysis we observe that the series of parameters making up a contingent description is divided into two parts: the constants and the variables. In the usual presentation of theories it is not always made explicit which parameters are the constants and which ones the variables. In our example of the charged particle moving in an electromagnetic field it is clear that the path of the particle is a variable of the theory: here, if anywhere, the generality of the theory is localized. On the other hand, spacetime and its Minkowski metric are described by constants: in classical electrodynamics we do not intend to have a theory about various spacetimes. In the case of the field one may waver whether it should be counted as a constant or a variable. As seen from outside various fields can be and have to be considered. From within the field appears to be a constant. The reaction of the particle on the field is neglected so that the particle moves in a fixed surrounding. For us the important thing to recognize is the *existence* of variables among the parameters of a contingent description: In this stadium of our development *it is these variables that represent the generality of the theory.*

Having the contingent descriptions at hand it can now be said that a *physical law in the primary sense* is an essential property Σ of a physical system formulated in terms of its contingent descriptions. Let us symbolize this by the formula

$$\Sigma(a_1 \ldots a_m; x_1 \ldots x_n) \qquad (2a)$$

where the a are the constants and the x the variables. Thus the conspicuous thing about (2a) is that free variables occur in it, and it is for this reason that I speak of the *free axioms* of a theory when referring to (2a). To avoid misunderstanding it must immediately be added that (2a) is not an axiom in the usual sense. If it were it would be equivalent to the statement resulting from (2a) by an *unrestricted* quantification over the variables x. This, however, never is the meaning of (2a). In our example we do not want to assert that every a priori possible motion of a particle satisfies the Lorentz equation. On the contrary, this equation and (2a) in general are meant to *select* a usually rather small (finite-dimensional) subset of solutions of the equation or models in the general case as representing those physical systems that are declared to be *physically possible* by the theory. Thus (2a) so far is not a statement at all. It is rather a concept, and it is called an axiom, if only a free one, to remind us that we can, of course, draw conclusions from it as long as we maintain the status of the x as free variables.

Besides the *external* generality expressed by the free variables x there is an *internal* generality hidden in the theory property Σ. The complexity of the propositional form (2a) as regards the composition of Σ is not restricted a priori in any way. In our example from electrodynamics all concepts and theorems of Minkowski geometry of spacetime are internal generalities. The same is true of all kinematical concepts and theorems such as the concept of 3-velocity and the relativistic addition theorem for these velocities. Indeed the main achievement of a physical theory, namely the establishment of connections between physical quantities, subsystems, contingent properties, probabilities etc., is completed on the level of the individual system, and it is founded exclusively on the internal generalities available already on that level. In this respect the usual explanations of the concept of law is particularly misleading because it suggests that we enter the temple of physical importance and universality only with the universal quantification over the variables indicating whole systems whereas before this we have to do only with questions of the type whether this swan is white or that raven is black. Against this it has to be emphasized that, for instance, the whole empirical material that is provided by the astronomy of the solar system since the Greeks refers, strictly speaking, to *one* individual system. If we view this material as a triumph of Newtonian celestial mechanics we testify that at least in certain cases the individual system is of considerable interest.

III. Coherence

The considerations of the previous section will be supported when we now pay some attention to the *coherence* that is achieved by a physical theory between its statements. For it will turn out that all coherence is obtained between statements concerning one individual physical system only. (For a deviating view see Sneed 1971, Ch. IV) In questions of coherence physics has paved its way somewhere between two extreme philosophical positions: the rationalistic coherence theory of truth and reality and, on the other hand, logical atomism. An outstanding advocate of philosophical rationalism in our days – Brand Blanshard – describes the ideal of fully coherent knowledge, the final goal of all thinking, as

> knowledge in which every judgement entails, and is entailed by, the rest of the system...The integration would be so complete that no part could be seen for what it was without seeing its relation to the whole, and the whole itself could be understood only through the contribution of every part.

(1939, vol.II,p.264 and 266) These somewhat exaggerated statements seem to lie far away from what we know about physics and its goals. And yet these goals are closer to them than to their philosophical opposite. A goal frequently mentioned by physicists is – in a wording by Einstein (1929,p.126f): 'We want to know not only *how* nature is like...but also...why nature is *such and not otherwise.*'

Let us now contrast this with Wittgenstein's formulation of logical atomism (in his 1921):

> 1.2 The world divides into facts

1.21 Each item can be the case or not the case while everything else remains the same

6.3 The exploration of logic means the exploration of *everything that is subject to law* and outside logic everything is accidental

Obviously this is a position of total contingency of the world. One is not surprised to hear that Blanshard has called logical atomism 'the most formidable attack ever made on reason as an independent source of knowledge.' (1962,p.92) And even if the matter is seen from the viewpoint of modern science this other extreme could not be accepted either. You cannot meet the efforts of physics, all this struggle for law and order, with simply pointing out that before the throne of logic all are equal – from Newton's laws of mechanics down to the most trivial statements on your present sense impressions.

The idea of coherence or interconnectedness as it is described in Blanshard's statement can easily be illustrated by physical examples. A case in point is Newton's theory of gravitation. What this theory has to say about one body as being a gravitating body cannot be said other than by relating it to every other body in the universe. Moreover, if we were to find a system of bodies moving exactly according to Newton's theory this very same theory would permit us to conclude that the system is all-inclusive. In other words, the part can only be understood by referring to the whole, and a system of bodies moving exactly according to Newton's equations must be the whole. Physical laws of the type of Newton's gravitational equations fulfill the rationalistic idea of coherence with an accuracy of which other intellectual circles that like to boast themselves of a holistic way of thinking could only dream.

However, when it is asked what precisely the coherence of a physical theory is like we are at a loss for an answer. In view of Einstein's question we certainly cannot satisfy ourselves by introducing particular irreducible modal concepts (cf. Armstrong 1983, Ch.6). Nor should we take refuge to the coherence theory of truth. For it rests on quite uncertain ground, especially with its view that truth is a matter of degree. There is something to the view insofar as it implies the idea of the *growth* of our understanding of nature. One could even entertain the view that this growth is proportional to the growth of the degree, not of truth, but of the coherence of our theories. The unification of hitherto unrelated theories by reducing them to a higher level theory is an important step in the process of increasing coherence. Since the unification of Kepler's and Galileo's laws in Newton's (gravitational) theory the idea of the unity of physics has accompanied and fertilized its development up to our times (cf. Weizsäcker 1971, Part III).

Apart from this diachronic assessment coherence must, of course, also be recognized in a single theory. This is a wide field, but there is one aspect particularly attached to the occurrence of a physical law in the free axiom of a theory. The contingent desciptions of the theory are made *redundant* by the law and are reduced to *initial and boundary conditions.* A contingent description of the motion of a particle or of the spacetime distribution of a field or of whatever can be analyzed as being composed of (infinitely many) elementary propositions, totally independent of each other in the sense of logical atomism: no subset of the set of these propositions allows to infer an elementary proposition that is not already a member of that subset. Now a law typically changes this situation by inducing dependences between the otherwise independent elementary propositions: Depending on the law at hand various reductions to characteristic subsets become possible, the most famous ones being sets of initial and boundary conditions. The classical case is the *determinism* of a dynamical law, often appearing mathematically as a differential equation. With celestial mechanics as paradigm determinism says (for systems with finitely many degrees of freedom) that the description of the course of all relevant parameters of a physical system can be reduced to a (finite) set of data referring to one time point only. A similar determinism exists for classical electrodynamics and other field theories, and even the most abstract and formal theories of present elementary particle physics induce correlations between single contingent propositions and allow the calculation of cross sections. Most remarkable in view of our main argument is that in all these cases we infer statements from other statements about the *same system* to which the premises refer. Indeed it is inferences of the form

$$\Sigma(a;x) \wedge \alpha\,(a;x) \rightarrow \beta\,(a;x) \qquad (3)$$

with no reference to more than one system under which all important calculations in physics can be subsumed.

IV. The Generality of Laws 2

The considerations of the two preceding sections were based on version (2a) of a physical law: the free axiom of a theory. When we now pass to the *bound axiom* corresponding to (2a) we are in for a difficulty. We know in advance that the proposition now to be formed is a universal conditional

$$\text{for all y: if Ky then } \Sigma(a_1 \ldots a_m; x_1{}^y \ldots x_n{}^y) \qquad (2b)$$

Here the bound variable *y* indicates physical systems, and

$$<a_1 \ldots an; x_1{}^y \ldots x_n{}^y)$$

is the description of system *y* in our theory. The constants *a* having fixed interpretations and *K* pointing out an *intended range of application* (2b) really is a *statement* made on behalf of the theory. But what statement is made in *Ky* about *y*? In II it has already been argued that the unrestricted universal quantification of (2a) does not come into question. In this respect Σ is a predicate like 'house', and we do not want to assert that all things are houses. On the other hand, if *K* were a predicate like Σ, i.e. if it were formulated in the language of the theory, the answer to the question 'which systems obey Kepler's laws?' could hardly be other than 'those that obey Kepler's laws'. We, therefore, conclude that by *Ky* we actually want to say something different *in kind* from what we say in the object language of a physical theory: We want to say, not how the systems in question, as described in physical terms, are *like*, but how they are *given*, are presented to us. This may be done by ostension, by enumeration, by pointing out paradigms, by predescriptions, as they are obtained by words like 'bodies', 'atoms', 'stars', 'gases', 'crystals' etc., and by qualifications of predescriptions (dilute gases, incompressible fluids, slowly moving particles). The common characteristic of these methods of pointing out an intended range of application is that they always contain a rather *vague* component which then incriminates the fundamental axiom (2b) (cf. Stegmüller 1973, Ch.IX.4).

I think that the difficulty pointed out is due to the nature of the matter and not produced by the reconstruction. In particular, one should not be mislead by the following consideration. If a gas is assumed to satisfy van der Waals' equation and if p is its pressure and v its volume then its temperature has the value T that can uniquely be calculated from the given data and the equation. Similarly, Kepler's laws allow to calculate the position x' at time t' if position x and velocity v are given at an earlier time t. It seems, therefore, that here we have, as conclusions from laws, universal implications whose premises are quite harmless propositions about pressure and volume of a gas or position and velocity of a planet. But it would be extremely misleading to present such implications as typical illustrations of definite lawlike statements, *unless* one emphasizes that the premises in question include a conjunctive member of the K-type, e.g. 'gas' or 'planet', that is only qualified by the other member, e.g. 'of pressure p and volume v' or 'with position x and velocity v'. Indeed this qualifying, Σ-type member, but not the K-type one, can meaningfully be switched into the conclusion which, however, would bring us back to (2b).

A further peculiarity of the generality displayed by (2b) is the relation of the instances being quantified over to each other. It is felt immediately when one compares the generality in question with those universal quantifications participating in the formation of the free axiom (2a). The essential statement defining an harmonic oscillator, for instance, is that for all times the acceleration is proportional to the elongation of the particle. This, in a self-explanatory sense, is a *system-constitutive* generality: The pieces here put together – these momentary connections between acceleration and elongation – make up a reasonable oscillator only *collectively*, for a whole time interval. By contrast, the various solutions of the equation of motion defining an oscillator are *competitors* of each other. A given oscillator can be described only by one of them, and other solutions cannot complete this description but represent *other oscillators* - other 'worlds'. The step from the free to the bound axiom is, then, a step that does not already occur during the formation of the free axiom. It is a unique and final step. By contrast, the usual system-constituitive quantification can be iterated as the example of euclidean geometry clearly shows.

The situation of rivalry between the instances quantified over in (2b) promotes the idea that the physical systems concerned are *independent possible worlds* - like Leibniz' monads. The independence means among other things, that – as already stated in III – we cannot infer from data belonging to one system the data of another system. In fact, in view of this independence the idea of coherence, dominating the formation of a single system, completely disappears from the scene as soon as it comes to the simultaneous consideration of several systems. Above all the independence means, then, that we may imagine physical systems as being capable of separate existence. There is a radical *fictionalism* at work in the thinking of modern physics, initiated by Galileo, that reaches far beyond the question of the universality of laws. To comprehend reality by means of our theories we outline possible worlds in the twofold sense that 1) we abstract from almost everything that constitutes the real world and 2) even that which remains is knowingly mistreated. We idealize, we neglect, we isolate, we simplify – in every case this means that we *replace* the real world in all its complexity by a more or less simple fiction. Electrodynamics in vacuo has as its object an electromagnetic field and a current density and nothing else. Quantum mechanics of the H-atom has as its object an H-atom or an ensemble of them and nothing else etc. In every case the rest is simply omitted. We act *as if* the respective object were the total universe, and if we do it rightly we are extremely successful with this method. This, by itself, is a highly non-trivial fact about our world that justifies the view of models of theories as being possible worlds.

The Possible-Worlds-View is in many cases somewhat concealed if the ratio of the number of constants of a theory to that of variables is in favour of the constants. The view, therefore, can be illustrated most properly by theories whose variables have (almost) displaced the constants. A good example for this situation is the theory of general relativity. If, in the context of Newtonian mechanics, we specify the shapes of bodies as spheres, cubes, etc. it goes without saying that we are talking of spheres, cubes, etc. *in absolute space*. As regards the latter there is no question of viewing *it* as being a space in something else. The situation changes when we do our physics on the basis of the Galileo group. Absolute space then disappears, and when we speak of a sphere we have to specify the reference system in which it is at rest. The space representing the reference system now is indeed part of something else, namely of *galilean spacetime*. And this time it is the latter of which we do not ask any more wherein we can find it. The situation is similar when we pass from absolute space to Minkowski spacetime.

But when we come to the theory of *general relativity* something more serious happens. Up to physics on the basis of the Galileo group or the Poincaré group there is a transworld identity between all possible worlds in form of spacetime, conceived as a unique entity. Everything that happens at all does happen in *this* spacetime. By contrast, in general relativity the possible worlds admitted include different spacetimes and by this very fact engage our phantasy all the more. The physicist will try to take all mystery from this by objecting that we here are dealing only with so many local applications. But a retreat in the sense in which Kant could still argue 'when we speak of several spaces we mean by this only parts of one and the same unique space' (and similarly for time, Kant 1781, B39 and B47) – a retreat like this does no longer come into consideration. For which spacetime could that be? Its metric and even its topology in the large depend on the energy and matter distribution in all of the spacetime continuum, i.e. it depends on entirely contingent circumstances that one would wish by no means to be determined by the theory. The only bond that ties together the local applications is the idea of the undetermined *existence* of a universe, well-determined by itself, but known to us only in fragments and housing all those applications.

V. Innerworldliness

Precisely when a theory allows for as much freedom to organize its possible worlds as does the theory of general relativity we feel confirmed in our intuition of expressing lawlike behaviour in this way. On the other side, these circumstances forcefully contrast with a further aspect of the Possible-Worlds-View that brings us back on earth, or rather: in our world. They contrast with the requirement, indispensable for any empirical science, that the systems decreed to be physically possible by a theory, irrespective of their being viewed theoretically as so many *different* possible worlds, yet should be realized *in one and the same,* namely our, *world* at least approximately and in sufficient number. I need not emphasize that we would not have the physics that we do have if this requirement were not fulfilled to an astonishing extent, and this was already pointed out as a non-trivial fact about our world. It is this fact that allows us to replace, in theory, the physics of our world by a physics of possible worlds so attractive because of its simplicity. And it is, therefore, the same fact that gives the natural law in its counterfactual version such a distinguished position. One is really inclined to believe that, if two moons of the planet Jupiter were removed from our world the rest would still behave according to Newton's gravitational law. Even modal realism could appeal to that fact (cf. Lewis 1986). The question to be treated in conclusion, however, is: Do the two factors of coherence and generality whose combination in a physical law was analysed still remain compatible if they are to be realised explicitly in the same world?

For an answer we have to bring in the idea of a physical system being *composed of* two or more subsystems. It is important to realize that this idea is *not* already contained in the theory form (2b), let alone (2a). They tell us nothing about the existence of a 'universe' in which two or more arbitrarily chosen instances of (2) can exist as subsystems. But physics knows of techniques to treat this question. Very roughly, two systems <a;x> and <a;z> have the composition <a;x,z>, and if both are instances of (2) the composite system is required to satisfy

$$\Sigma'(a;x,z) \equiv \Sigma(a;x) \wedge \Sigma(a;z) \tag{4}$$

This is essentially the way in which composite systems are treated in classical (Hamiltonian) mechanics and, with a noteworthy exception, also in quantum mechanics.

In both cases physical systems are described by a phase space (Hilbert space), a Hamilton function (– operator), and a solution of the Hamilton equations (the Schrödinger equation). The composition of two systems described in this way leads to a system having the cartesian product (Kronecker product) of the phase spaces (Hilbert spaces) of the subsystems as its phase space (Hilbert space), the direct sum of their Hamilton functions (– operators) as its Hamilton function (– operator) and the Cartesian (Kronecker) product of their solutions of the Hamilton equations (the Schrödinger equation) as solution of its Hamilton equations (Schrödinger equation). The total *independence* of the original systems in the composite system, as it is evident from the mere conjunctive decomposition (4), is expressed in more detail by the following consequences of the composition rules in the mechanical theories:

1) The lattice of contingent properties (the algebra of observables) is the free product of the lattices of properties (algebras of observables) of the original systems. The *free product* of two structures of a given kind is formed in such a way that the original structures are parts of the product structure while *no new relations* beyond this are introduced.

2) The direct sum of the Hamilton functions (– operators) means that the subsystems have *no interaction*.

Thus we see that the rules for building composite systems are such that the original systems appear as totally independent when seen as subsystems of the total system. This is in complete harmony with their being instances of one and the same law. But it is generally *not* in accordance with the idea of coherence applied to the total system. In both, quantum and classical, mechanics *dynamical reasons* are responsible for this. If in classical mechanics we have a dynamical law for a many particle system of a certain kind this means that we know a series H_n of Hamilton functions for which in all interesting cases

$$H_m + H_n \neq H_{m+n} \tag{5a}$$

for all $m,n \geq 1$ holds. Consequently, the composition of two systems in which they remain independent *never* leads to the interaction required by the dynamical law. An impressive case is gravitation. In its first treatment in Kepler's laws the planets are completely independent of each other. Accordingly, an exact innerworldly formulation of the laws is possible. In (5a) we have equations throughout. But Kepler's laws turned out to be empirically false and were replaced by Newton's theory. Thereby the situation changed so radically that not only all inequalities hold in (5a) but – as already mentioned – no subsystem of a Newtonian system is itself a Newtonian system. No two of the infinitely many solutions of the gravitational equations can be realized in *one* spacetime – *if* the theory is applied in all strictness.

With respect to interactions the situation in quantum mechanics is the same as in classical mechanics. But a new and analogous difficulty occurs that concerns already the *quantum mechanical states.* Just as it is hard to find systems in our world that, strictly speaking, would not interact with each other we rarely meet with quantum mechanical states of compound systems that are separable, i.e. are the product of independent states of the components. The normal situation is again characterized by an inequality

$$\phi_1 \otimes \phi_2 \neq \Phi \tag{5b}$$

where Φ is a state of the compound system and ϕ_1 and ϕ_2 are any two states of the subsystems. In such a situation Φ does not provide us with any definite information about the states of the subsystems. We only know the state of one of them if we know it of the other one. This is a new kind of coherence totally alien to classical physics. And again we see it as a rival of independence. The inseparability of states plays an important role in the direct empirical confirmation of quantum mechanics as deviating from classical mechanics (cf. van der Merwe 1993).

Against the foregoing argumentation it may be objected that the difficulties for an innerworldly interpretation of (2) in connection with interaction and inseparability do not have any *practical* importance. All fundamental interactions have a finite range allowing for practically independent and yet internally interacting systems. Similarly, we can prepare practically separable systems showing all the features of inseparability internally. And both possibilities are in accordance with the corresponding theories. Kepler's laws, for instance, follow from Newton's theory in good approximation if certain conditions are satisfied. All this has to be admitted, let alone the overwhelming number of cases where we find the independence in question simply by experience.

On the other hand, we have to remember that we are investigating a matter of principle. Theories about gravitation and the mechanism of compound systems are of a fundamental character. Such theories are not proposed only to say afterwards that they are to be taken *cum grano salis* anyway. And if such theories show us that the galilean method, successful as it is, in the last analysis not only misses the factual constitution of the universe but also violates its laws then this deserves to be recognized and understood. The question whether there are any *strictly* universal laws is a matter of principle. If we deny this – and there is every reason to do so – then natural laws are fictions whose ontological status could only be saved by taking refuge to modal realism. And this would be a desperate decision. We have not yet achieved a satisfactory understanding of natural laws (cf. Cartwright 1983).

References

Armstrong,D.M. [1983], *What is a law of nature?* Cambridge University Press: Cambridge 1983

Blanshard,B. [1939], *The nature of thought.* Allen&Unwin: London 1939

Blanshard,B. [1962], *Reason and analysis.* Allen&Unwin: London 1962

Cartwright,N. [1983], *How the laws of physics lie.* Clarendon: Oxford 1983

Einstein,A. [1929], Über den gegenwärtigen Stand der Feldtheorie. In: *Festschrift für A.Stodola.* Füssli: Zürich 1929, pp. 126-32

Feynman,R. [1965], *The character of physical law.* M.I.T. Press: Cambridge,Mass., 1965

Hempel,C.G. [1965], *Aspects of scientific explanation and other essays in the philosophy of science.* Free Press: New York 1965

Hund,F. [1972], *Geschichte der physikalischen Begriffe.* BI: Mannheim 1972

Kant,I. [1781], *Kritik der reinen Vernunft.* 2d Ed. 1787. Meiner:Hamburg 1990

Lewis,D. [1986], *On the plurality of worlds.* Basil Blackwell: Oxford 1986

Ludwig,G. [1978], *Die Grundstrukturen einer physikalischen Theorie.* 2d Ed. Springer: Berlin etc. 1990

Merwe,A. van der [1993], *Bell's theorem and the foundations of modern physics.* Ed. by A. van der Merwe et al. World Scientific: Singapore 1993

Russell,B. [1912], On the notion of cause. *Proc. Arist.Soc. 13*, pp.1-26

Scheibe,E. [1989], Coherence and contingency: Two neglected aspects of theory succession. *Nous, 23*, pp.1-16

Scheibe,E. [1991], General laws of nature and the uniqueness of the universe. In: *Philosophy and the origin and evolution of the universe.* Ed. by E.Agazzi and A.Cordero.Kluwer: Dordrecht 1991, pp.341-60

Sneed,J.D. [1971], *The logical structure of mathematical physics.* Reidel: Dordrecht 1971

Stegmüller,W. [1969], *Wissenschaftliche Erklärung und Begründung.* 2d Ed. *Erklärung, Begründung, Kausalität.* Springer: Berlin etc. 1983

Stegmüller,W. [1973], *Theorie und Erfahrung. 2.Halbband: Theorienstrukturen und Theoriendynamik.* Springer: Berlin etc. 1973. (English Edition: *The structure and dynamics of theories.* Springer: New York etc. 1976)

Suppe,F. [1974], *The structure of scientific theories.* Ed. with a critical introduction by F.Suppe. University of Illinois Press: Urbana,Ill.,etc. 1974

Weizsäcker,C.F.v. [1971], *Die Einheit der Natur.* Hanser: München 1971. (English Edition: *The unity of nature.* Farrar etc.: New York 1980)

Wittgenstein,L. [1921], *Logisch-philosophische Abhandlung.* Annalen der Naturphilosophie 1921. (English Edition: *Tractatus logico-philosophicus.* Routledge&Kegan Paul: London 1922)

Hermann Haken

Laws and Chaos

I. What is Chaos?

There is a drawing by the famous artist Escher. In its middle it shows a beautiful, highly symmetric transparent crystal, around which there is trash: broken bottles, opened and emptied cans, and other things. Escher called his drawing *Order and Chaos*. Quite clearly, he identifies the beautiful crystal with order, the trash with chaos. This is a static arrangement, however. In modern mathematics and physics, the notion *chaos* refers to dynamics. An example may be provided by a satellite photograph of the earth atmosphere (Fig. 1). In one part one may see well-ordered cloud streets, in some other part a giant vortex. Here the cloud streets stand for order, the vortex for chaos. But all the time the air is moving. These patterns are produced by a continuous partly turbulent motion of air.

In this article we shall be concerned with dynamic chaos. The word chaos denotes quite irregular processes. On the other hand, we connect with the word 'law' regularity, order. Thus the title of this article *Laws and Chaos* may, at first sight, represent an antinomy. However, as we shall outline below, chaotic processes obey certain laws. In this article I do not intend to give too many details, but I shall rather concentrate on the essential aspects. I shall elucidate them by means of typical examples. At the same time I will try to avoid too many formulas. Since the field of chaos research has become enormous over the past one or two decades, this article will be far from being exhaustive.

In physics, mathematics, and a number of other fields, we must clearly distinguish between microscopic and macroscopic chaos, which we shall treat separately.

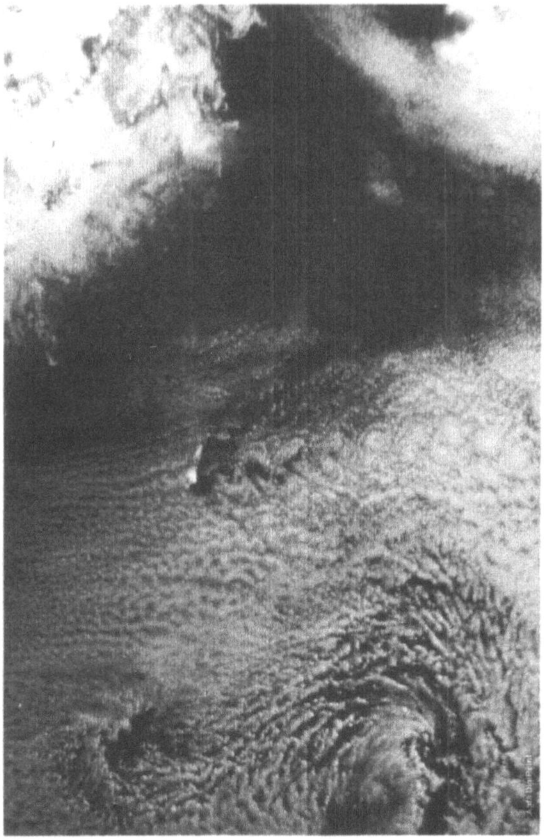

Fig. 1: A satellite photograph of the earth atmosphere: Dynamic order and chaos

II. Microscopic Chaos

Let us start with a field of physics that has seemingly nothing to do with chaos, namely thermodynamics. This field deals with concepts, such as heat, temperature, volume, or pressure and establishes relationships between them on a phenomenological basis. For instance, one finds for a so-called ideal gas a relationship between pressure p, volume V, and temperature T, in form of the equation

$$pV = RT, \tag{1}$$

where R is a constant. This law can be easily checked by experiments that occur in daily life. For instance, when we let the air out from a tyre, the air feels cooler than the surrounding. According to the law (1) the following happens: When we open the valve, the pressure rapidly drops, but the volume in the tyre remains the same. Thus according to that law, the temperature T of the gas must decrease. A central concept in thermodynamics is that of entropy. The entropy of any gas, liquid or solid body can be determined by a formula that relates the change of entropy S with absolute temperature T and the reversibly added or subtracted heat dQ. According to this, when a zero of entropy at the absolute temperature $T = 0$ is given, namely $S = 0$, one may, at least in principle, determine the entropy of any body, liquid, or volume of gas. A central law of thermodynamics states that in a closed system, i.e. where no exchange of matter or energy with the surrounding occurs, the entropy increases till its maximum value. These laws are based on macroscopically measurable quantities and are derived from macroscopic phenomena. They are thus called *phenomenological laws*.

In modern physics one wants to trace these phenomenological laws back to fundamental laws that are given, for instance, by mechanics, or, to be more precise, by quantum mechanics. To this end, Boltzmann introduced the hypothesis that a gas is composed of individual particles, the atoms or molecules. At first, this hypothesis was strongly fought against by famous scientists, such as Wilhelm Ostwald. According to the atomic view, the temperature T is directly related to the average-kinetic energy of these particles and the pressure is directly related to the momentum transfer of these particles on the walls. And what is the nature of the motion of these particles? The first hint may be derived from an observation by the biologist Brown. When a small particle is immersed in a liquid and its motion is observed under a microscope, one finds a zigzag motion. This irregular motion is produced by the random pushes that this particle receives from the molecules of the liquid. In this way, the following picture of the motion of the molecules of a gas or liquid emerges, where we shall concentrate on a gas. In a gas,

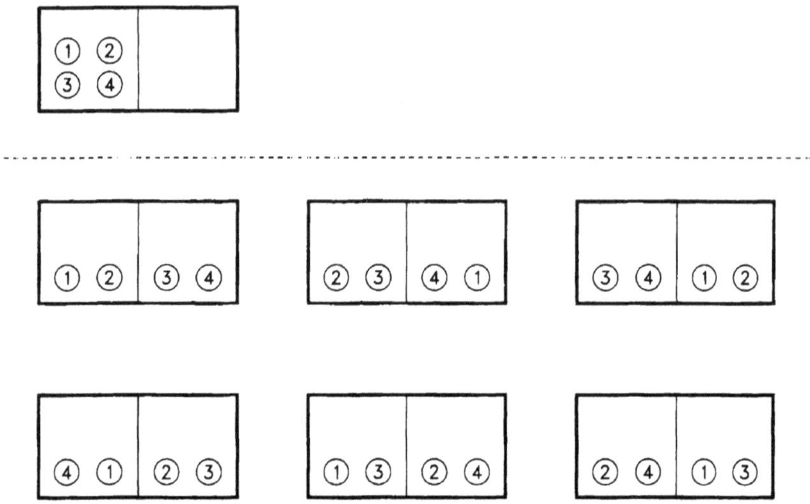

Fig. 2: Visualization of the determination of the number of realizations W

the molecules fly some distance, collide, change their direction, collide again, and so on. In this way, an irregular motion starts that is called *microscopic chaos*. Will it be possible to follow up the individual paths of the molecules and derive their motion by means of the basic laws of mechanics or quantum mechanics? Even the largest supercomputers will not be able to perform this task. Rather Boltzmann's ingenious idea was to introduce a new concept, namely to relate the macroscopic entropy with a microscopic quantity, the number of realizations W, by means of the formula $S = k \ln W$, where k is Boltzmann's constant and W the number of realizations. How W can be determined, may be visualized by an example where four balls are distributed over two boxes (Fig. 2). For instance, there is only one realization W, if the balls are in one box, but there are six possibilities W, when the balls are distributed evenly over the two boxes. According to Boltzmann a system in thermal equilibrium acquires a macroscopic state that is connected with the greatest number of realizations. In the present case, this would be the entropy S belonging to W = 6. Our picture is oversimplified, because it ignores the motion of the atoms, but in a simi-

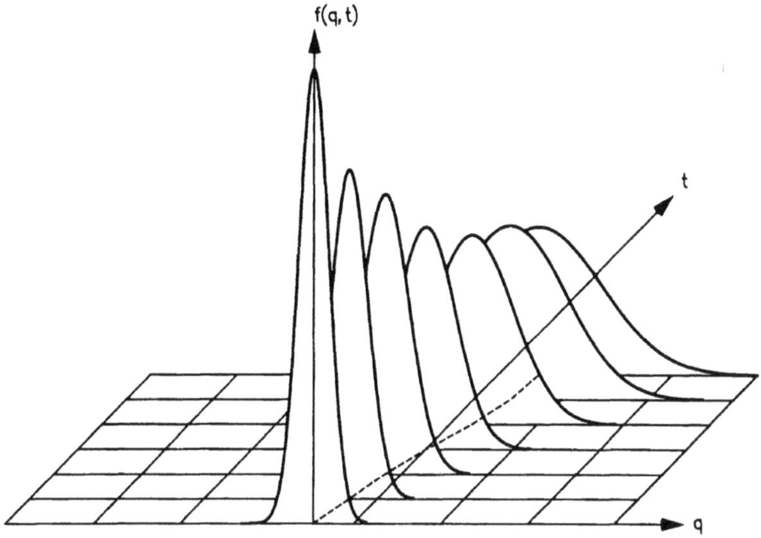

Fig. 3: The probability distribution $f(q,t)$ is shown as a function of a variable q that may, for instance, be identified with the velocity of a gas molecule. If this function is given initially at time t equal zero, it may change in the course of time, plotted backwards in this figure, and may then acquire a constant shape. It is this shape that is determined by the maximum entropy principle

lar way, one may consider the distribution of the balls over different velocities. Then the balls move and collide, and thus all the time will change their velocities and positions. Instead of tracing the motion of the individual molecules, one may look at their ensemble and ask how probable it will be that one will find a molecule with a given velocity in a specific velocity interval. Thus the decisive step by Boltzmann is the introduction of a probability distribution (Fig. 3). For large numbers of molecules one may express $ln\ W$ by means of that probability distribution. The postulate that systems in thermal equilibrium have the largest number of microscopic realizations can now be formulated as a requirement for the probability distribution p. But in all these cases, a central requirement has to be added. Namely we know that, for instance, a gas has its specific temperature, or, in other words, its molecules have a specific average-kinetic energy. When one maximizes the expression for the entropy under the constraint of a given mean kinetic energy, one finds the probability distribution for the velocities of the molecules. This is a function that had been derived previously by Maxwell and that is called the *Maxwell-Boltzmann distribution function*. According to Boltzmann, a distribution function can be derived for any other system equally well.This has led to the formulation of the maximum entropy principle. The probability distribution of any system can be found by maximizing the entropy under given constaints. Quite generally speaking, in physics these constraints are given by conservation laws for the number of particles, for energy, for total momentum, and so on. There are other ways to derive the distribution function from microscopic laws, but the one I described here is considered the *King's way*. It contains three ingredients:

1) the microscopic relationship between, for instance, the kinetic energy of a particle and the velocity, v, in the case of a gas
2) the maximum entropy principle
3) the law of large numbers.

These three ingredients are of quite a different nature. 1) stems from the microscopic laws of mechanics, or, in a refined fashion, from quantum mechanics. 3) is a purely mathematical relationship that holds for large numbers. (For experts we mention that it is Stirling's formula.) 2) is something that, if I am really honest, must be called "given by inspiration". There are a number of arguments, of course, that have the rank, in my opinion, of plausibility arguments, that can be traced back to Gibbs and have been reformulated in a clear-cut fashion by Jaynes.These plausibility arguments are based on the concept of making unbiased guesses: If one has no experimental control over a system other than by a few macroscopic quantities, such as total energy, volume, and so on, one has to assume that the system can realize and does realize all the states that are consistent with these constraints. But to be sure, this is a postulate. It is a remarkable fact that can hardly be overstressed that the laws of the whole

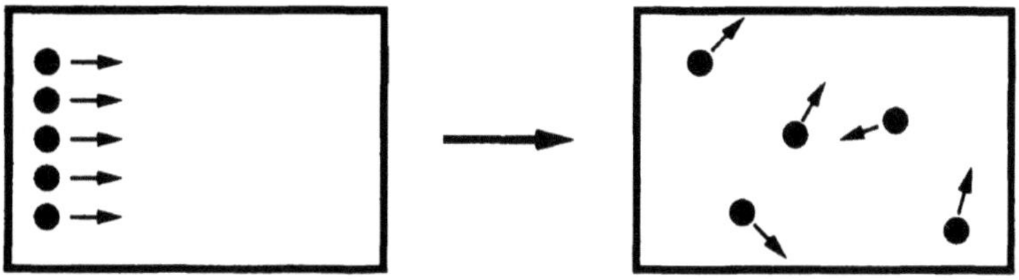

Fig. 4: A completely ordered state (l.h.s.) of gas molecules reaches its microscopically chaotic state after very short time being caused by the slighest perturbation

statistical mechanics, that deals with the relationship between microscopic laws and the phenomenological laws of thermodynamics, contain these three ingredients.

In view of these statements, it may not be surprising that, at least in my opinion, and of others, too, another fundamental problem of thermodynamics has not yet been solved satisfactorily from a microscopic point of view, namely that of irreversibility and the increase of entropy. To this end, let us consider the following thought model: A box may contain a number of gas molecules. Prepare an initial state so that all the molecules are in a plane and may run in just one direction (Fig. 4). Since there is only one state (W=1), the entropy is equal to zero. But the slightest perturbation will cause the molecules to collide and after a very short time complete microscopic chaos will be established. The entropy has reached its maximum value. But now the puzzle starts: Why do the gas molecules never come back to their initial state? According to the laws of microscopic mechanics or quantum mechanics, the motion is entirely reversible. Thus if we let a movie run in the reverse direction, it would be a physically acceptable process; but such reversions are never observed in nature. A number of tricks have been applied to resolve this puzzle. One of these tricks exists in the assumption that in reality the gas interacts with the surroundings so that it is coupled to a so-called heat bath. Eventually, the whole universe is such a heat bath and more recent ideas indicate that the expansion of the universe generates the irreversibility. But honestly speaking, at the moment there is no rigorous mathematical proof of irreversibility that is generally agreed upon. The majority of physicists would reject any attempt to modify the basic microscopic laws of physics in order to introduce irreversibility this way. This is an attempt that is actually made but that does not find support from others and, in my opinion, is not leading into the right direction.

III. Deterministic Chaos

As we have seen in the preceding section, microscopic chaos refers to the motion of many individual particles, say, atoms or molecules that occurs in a random fashion. Deterministic chaos, that we are going to consider now, is of quite a different nature. To elucidate the basic concepts occurring in this rather new field, let us start with a simple example from mechanics, namely the pendulum. A pendulum swings back and forth between two points of maximum elongation, where it comes to rest and reverses its direction of motion. In mathematics and physics, this kind of motion is represented in a so-called phase plane (Fig. 5), where one plots the velocity v along the ordinate and the position x along the abscissa at each instant of time. As one may derive from the laws of mechanics, when the pendulum moves one obtains an ellipse in the x-v plane, that is also called phase plane. But no real pendulum follows this law because of friction. In fact, the pendulum will slow down its motion and its elongation becomes smaller and smaller; so in a way the ellipse spirals down to, what is called a stable focus. By contrast, the undamped pendulum has the advantage that its motion does not finish and that the time of its swinging back and forth is a constant that is independent of the maximal elongation. In other words, such a pendulum is an ideal clock. But because of friction, a real pendulum comes to a stop. In order to construct a clock, one has to overcome the friction by driving it. In earlier times this was made by weights that by means of a transfer mechanism gave little pushes to the pendulum.

When one drives the pendulum more strongly by a periodically acting force, it may, eventually, show a quite irregular mo-

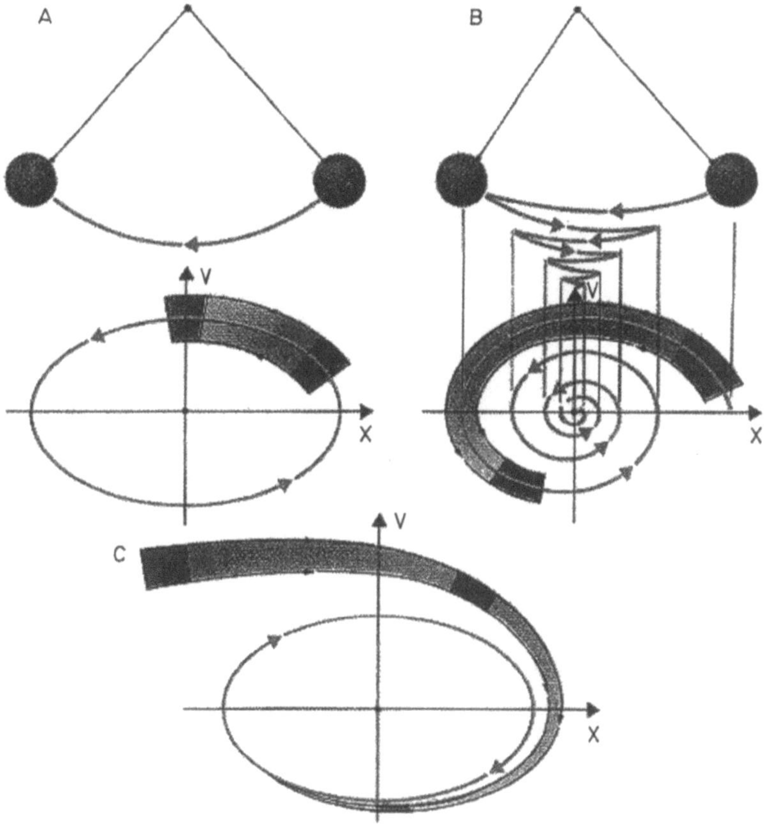

Fig. 5: Left upper side: The periodic movement of a pendulum is plotted as a trajectory in the *x*-*v*-phase plane. *x* is the position, *v* the velocity of the pendulum at each moment. Upper right: The same for a damped pendulum. Lower part: When the pendulum is both damped and weakly driven so that the damping can be adequately compensated, the trajectory reaches a stable ellipse, a so-called limit cycle

tion, a motion that is called *deterministic chaos*. Again one may plot the motion in a phase plane, but then one finds a crossing of trajectories. When one wants to follow up the motion of the pendulum in more detail, one may plot the position of the wheel of the driving motor along a third axis in the third dimension. Then one finds a trajectory in this three-dimensional space that never intersects and never leaves a certain region. We have the typical example of a so-called chaotic attractor (Fig. 6).

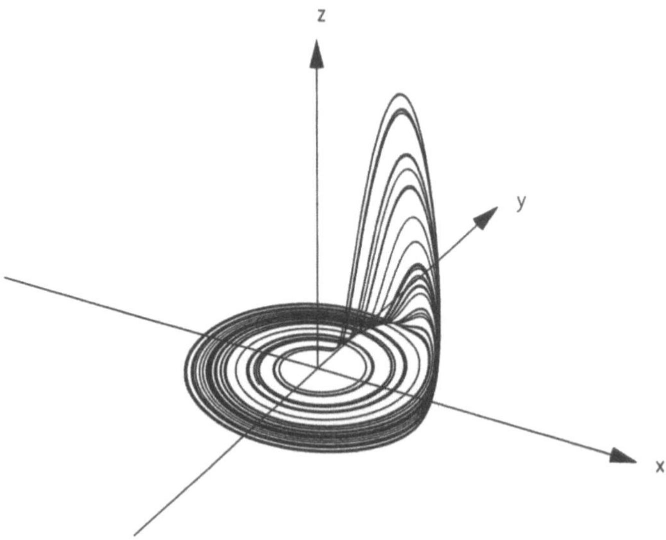

Fig. 6: a) The probably simplest examples of a chaotic attractor in three dimensions (Rössler attractor)

Historically, chaotic motion was conceived by Poincaré around the turn of the century, when he considered the so-called three-body problem, in which three celestial bodies are subject to the gravitational force and follow up an entirely irregular motion. Chaos was rediscovered by the meteorologist Lorenz, who analyzed equations of fluid dynamics by means of a highly simplified model.

A chaotic attractor has a number of quite typical features in contrast to the undamped pendulum, where the trajectory was a closed ellipse. In the case of chaos, the trajectory never ends. If it starts outside the region of the chaotic attractor, it will enter it and will remain in it forever. It is as if we are sticking a needle through a ball of wool. In this way, the trajectory cannot fill the whole space completely. The chaotic attractor possesses, as one says, a fractal dimension. When we look at two closely neighbouring trajectories, they will depart very quickly. This is called the sensitivity against initial conditions. The reason becomes clear when we look at two mechanical examples. In the case of

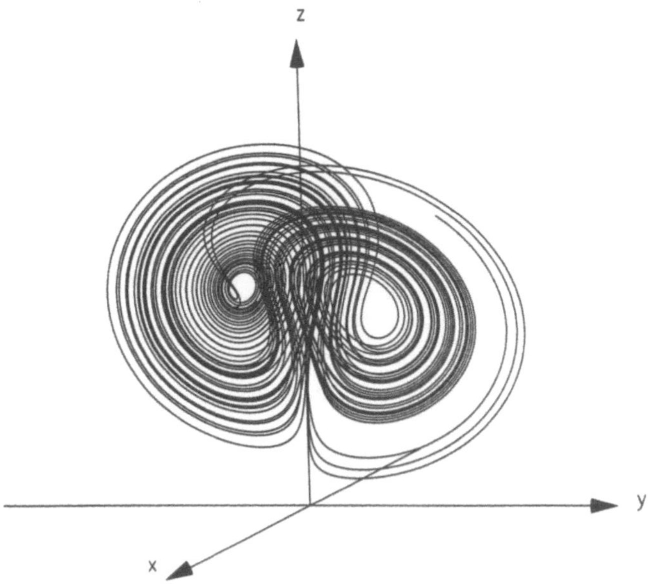

Fig. 6: b) The trajectories of the Lorenz attractor

the pendulum, which swings so strongly that it reaches its upper point, there are two ways of falling down, namely to the left- or right-hand side. Thus when we plot the resulting trajectory, we shall find one that will describe a motion to the left-hand side, whereas when the initial condition is just a little shifted to the right-hand side, the evolving trajectory will go to the right-hand side. Another example is provided by a steel ball falling on a vertical razor blade. Depending on the initial position of the center of gravity with respect to the razor blade, the ball will fall to the left or to the right. This sensitivity is typical for all chaotic attractors. From it the name deterministic chaos becomes evident. Both, the pendulum and the steel ball, follow the deterministic laws of Newtonian mechanics. On the other hand, the motion becomes entirely chaotic, as we may easily imagine when we think of a mechanism by which the steel ball is brought up again and again. When we plot its trajectory, it will follow a random path. The speed with which the trajectories separate from each other is measured by Lyapunov exponents. In this

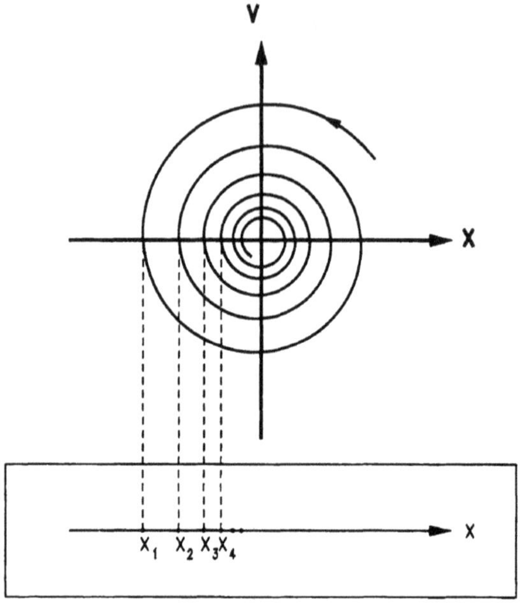

Fig. 7: An example for the construction of the Poincaré map in the plane. A spiraling trajectory causes the sequence of cross sections x_1, x_2, x_3, ...

way a chaotic attractor can be characterized by its fractal dimension and by its Lyapunov exponents, where on mathematical grounds relationships between Lyapunov exponents and dimensions of some specific kind can be established. In order to find more detailed laws, specific models have been developed. An important way to such a model was paved by Poincaré's idea of the return map. To this end consider a trajectory that is periodic or close to a periodic orbit in a three-dimensional space. Then according to Poincaré, we insert a plane that is more-or-less orthogonal to the trajectories so that the trajectories pass through that plane. We may plot the coordinates of the intersection points along a line (Fig. 7). If the trajectory is an ellipse, all the points coincide. We may think of a case, however, in which the points change. Then one may ask for a relationship between subsequent points, x_{n+1} and x_n. A nontrivial example is the so-called logistic map in which one has the relationship

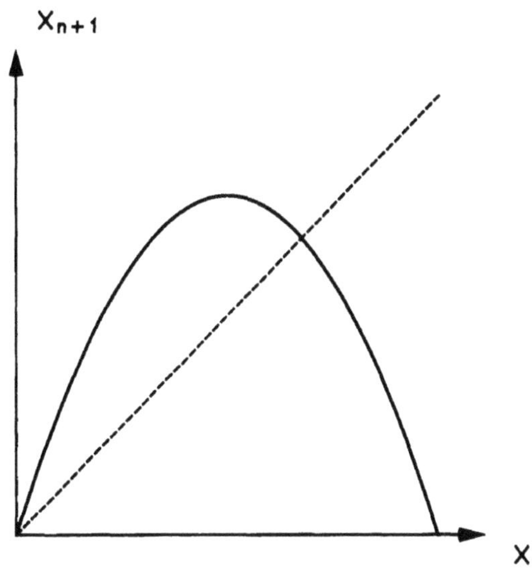

Fig. 8: The mapping of x_n on x_{n+1}. The map is represented by a parabola. The dotted line represents the diagonal

$$x_{n+1} = rx_n(1-x_n). \qquad (2)$$

The general curve that represents this relationship is a parabola that is drawn in Fig. 8. By some simple geometric considerations, one may show that the subsequent points may show quite different kinds of behavior depending on the control parameter value r. After some transients, the solution of equation (2) is a constant (Fig. 9). But when one increases r beyond a critical value, the points x_n (Fig. 10) alternate. When the value of r is further increased, suddenly the points vary within a period of 4 (Fig. 11), then after a period of 8, etc. Quite evidently, the period is doubling each time. Eventually, in the limiting case, an entirely irregular, i.e. chaotic, motion is found. Because each time the period doubles, one speaks of a *period doubling sequence*. When one plots the values of x_n versus r, one obtains a diagram of Fig. 12. A number of surprising features, or, if one wishes, laws have been found. For instance the onset of period doublings as a function of the control parameter r obeys a simple law, namely

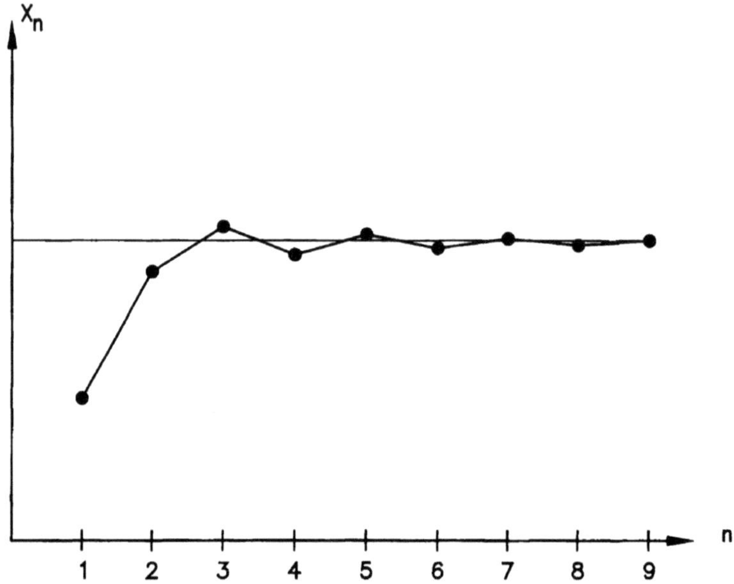

Fig. 9: When the index n increases, the sequence of points x_n tends to a fixed point for a small enough control parameter r

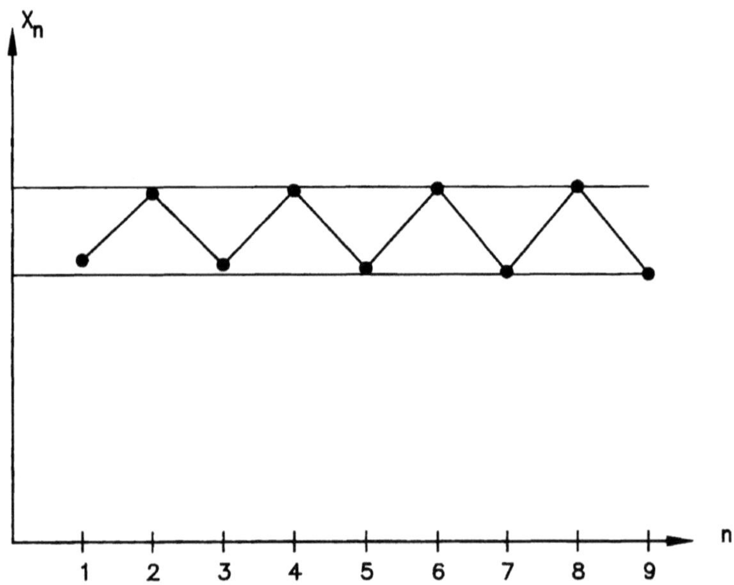

Fig. 10: Periodic change of x_n with period 2

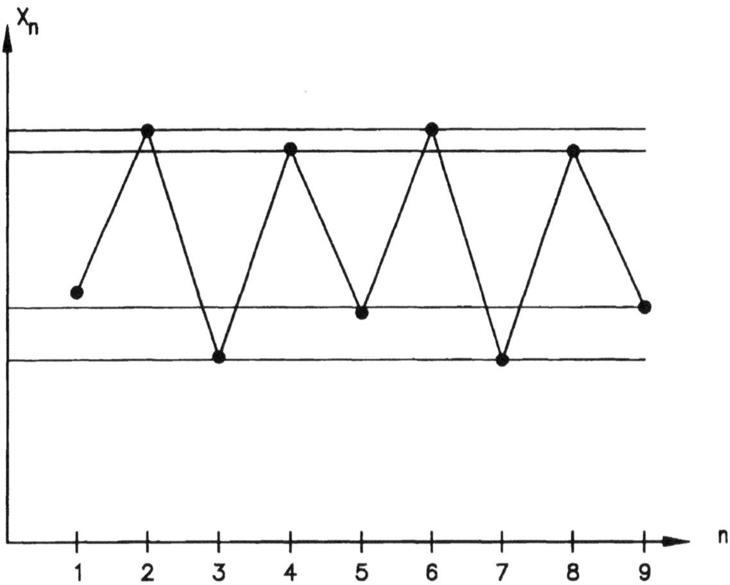

Fig. 11: The same as Fig. 10 with the period 4 of x_n for a somewhat bigger control parameter as that of Fig. 10

$$\lim_{n\to\infty} \frac{r_{n+1} - r_n}{r_{n+2} - r_{n+1}} = \delta = 4.6692$$

Another highly typical feature is that of self-similarity. An example is provided in the following way: When we cut a square around a bifurcation point, that is a branching point, where two new lines start, and enlarge it, it acquires the same form as the original picture. Or, in other words, when we magnify a picture, say part of Fig. 12, and consider it afterwards, we cannot say which magnification factor we used.

The period doubling route to chaos has been found in a number of physical systems, such as fluid dynamics. But what is the nature of these laws? In a way, they are a mixture between phenomenology and mathematics. For instance, when a variable of a physical system shows the behavior as plotted in Fig. 8, one may deduce on purely mathematical grounds all the features or, if you wish, laws of self-similarity, etc.. In simple cases, such as simple mechanical devices, the properties of the map of Fig. 8 are, of course, a consequence of the physical laws that, in turn,

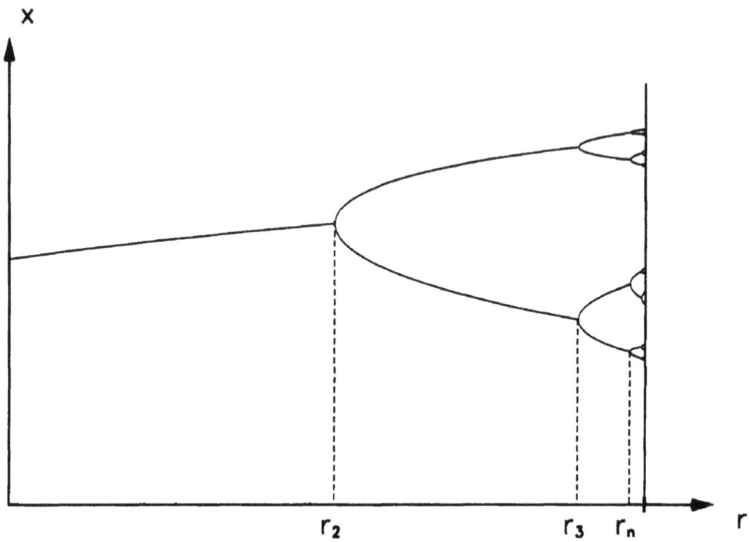

Fig. 12: Along the abscissa the control parameter r is plotted, whereas along the ordinate the points x_n are shown. For instance, for a small value of *r*, there is only one value x_n, namely the one to which the points tend in Fig. 9. When r is increased beyong a critical point, the points x_n start a jumping between two values, compare Fig. 10. After that the jumping occurs between four different points. If the neighbourhood that belongs to a branching point is enlarged, the same Fig. results for large enough n.

are described by differential equations of Newtonian mechanics. But the laws of chaos apply to far more complex systems also, as we shall see below. There are other routes to chaos, for instance by so-called intermittency (Figs. 13-15) and by bifurcation from a torus, but we shall not dwell here on the details, because the mechanism, how then laws are established, is precisely the same as the one just described with respect to period doubling. For a while it was believed that these three routes, namely by period-doubling, by intermittency, and by bifurcation from a torus are the only ones.

As we now know, there are infinitely many routes to chaos so that the universality of the three just mentioned routes has gone lost. The examples I provided occurred in systems that contain only few degrees of freedom. A pendulum, for instance, has as its variables position and velocity, so in the nomenclature

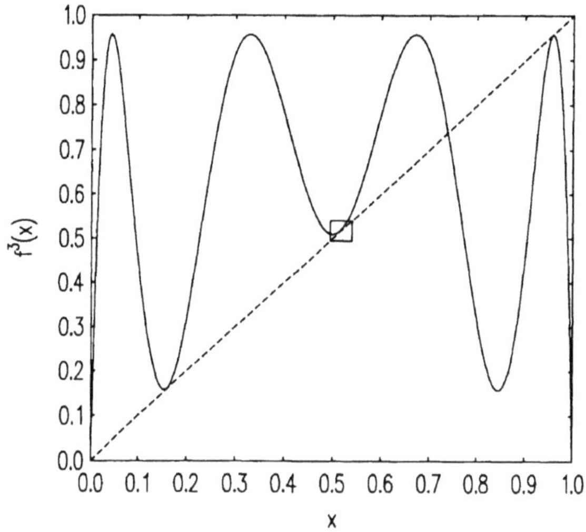

Fig. 13: When the map of Fig. 8 is applied three times, the map shown in this Figure results.

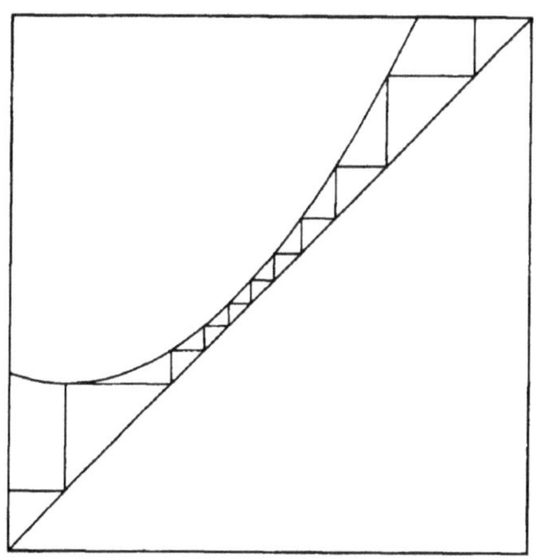

Fig. 14: An enlargement of the square of Fig. 13. The steps represent a trajectory where the points jump from the diagonal to the upper curve, back to the diagonal, up to the upper curve, a.s.o. Quite clearly, a very narrow channel results in which the jumps of the points x_n are very small. These are zones of quiet behavior of the system, whereas outside this region big jumps occur, indicating a turbulent state

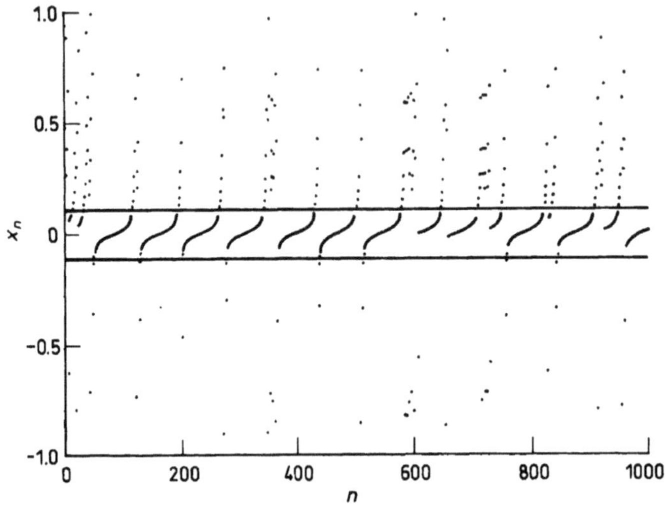

Fig. 15: Here the points x are plotted versus n; one may quite clearly see a quiet zone outside of which there are outbursts, again and again

of dynamic systems theory two degrees of freedom. Indeed, the results obtained so far refer to systems that can be described by few variables only and the most pertinent results have been obtained so far mainly for three degrees of freedom. Actually, three degrees of freedom are the minimum number necessary for the occurrence of chaos. On the other hand, typical features of deterministic chaos have been found in complex systems. Such systems are composed of very many components described by correspondingly very many variables.

What is then the relationship between complex systems and deterministic chaos?

IV. Chaos in Complex Systems

The answer why the behavior of complex systems may follow the laws of deterministic chaos can be found by the interdisciplinary field of synergetics. Synergetics originated from physics, where it dealt with open systems, i.e. systems whose states are maintained by an influx of energy and/or matter. An example is provided by the light source laser that produces light

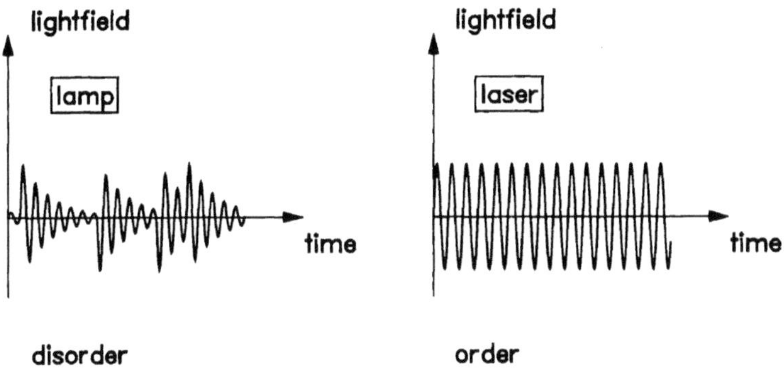

Fig. 16: The light field versus time. L.h.s.: The incoherent light of a lamp. R.h.s.: The coherent light of a laser

with properties that are quite different from those of light from conventional lamps (Fig. 16). While in usual lamps emitted light consists of wave trains that are independent of each other and may be visualized as Italian noodles representing microscopic chaos, the laser light, on the contrary, consists of a practically infinitely long, highly ordered wave. To elucidate this difference, let us consider a gas in a tube, where the individual atoms are excited by an electric current and start emitting light. While in a lamp the emission acts of the individual atoms are entirely independent of each other, in the laser these acts must be highly correlated so to produce the ordered laser light, or, to use a modern term, the atoms must self-organize in the laser. The phenomenon of self-organization is ubiquitous not only in open systems in physics, but in many other fields ranging from chemistry over biology to economy and sociology. In the present context the question arises: What are the basic principles underlying self-organization? To cut a long story short: When external conditions, for instance in the laser the electric current, are changed, the formerly disordered state may become unstable. Close to an instability point, specific collective modes tend to grow, compete with each other and one or several of these collective modes win the competition. All the other degrees of freedom of the system are enslaved by these modes that are called *order parameters.* As can be shown mathematically and has been verified by numerous individual experiments, in fluid dynam-

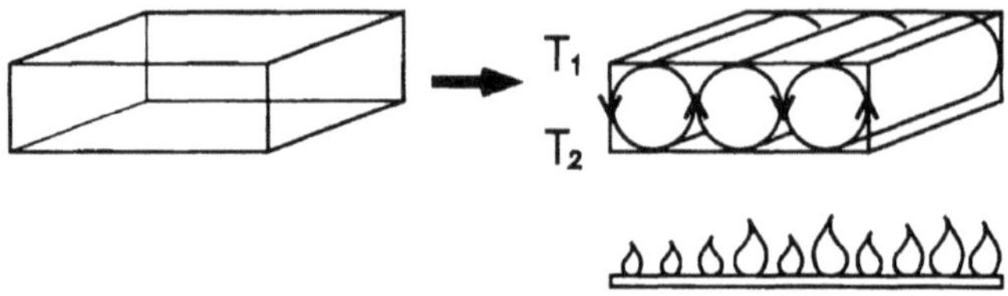

Fig. 17: Formation of rolls in a fluid heated from below as an example of pattern formation governed by a single order parameter

ics (Fig. 17), plasma physics, chemistry, etc. at instability points only few order parameters appear. They obey a closed set of equations among which are those that characterize deterministic chaos. In this way, the gap between systems with many degrees of freedom and those with few degrees of freedom giving rise to deterministic chaos is bridged. Close to instability points, the dynamics of the total system is governed by few quantities, the order parameters, which, in turn, may obey equations of deterministic chaos.

V. The Hierarchy of Laws

Since this volume is dedicated to the question of the nature of laws in nature, let us discuss the results of our foregoing analysis. It is rather generally agreed upon that in physics there is one set of laws that have been distilled from many experimental phenomena and that refer to elementary particles and to fields that provide the interaction between these particles, such as the electric and magnetic field, to mention an example. While research is going on to discover new kinds of elementary particles at higher and higher energies and to develop a unified theory of particles and fields, this article is concerned with another direction of research. Here we assume the laws of, say, atomic physics, as fundamentally given and we were concerned with the derivation of new laws at higher macroscopic levels, say, the motion of atoms in a gas, the behavior of fluids, and so on. The question is whether these laws at a higher macroscopic level follow immediately and directly from the laws at the microscopic level. While there are a few examples where the macroscopic laws can be derived from the microscopic laws step by step, in most cases additional assumptions or phenomenological facts must be incorporated. This is in particular the case with respect to chaos theory in physics. In general, these laws appear as a mixture of basic laws, experimental facts, and mathematical laws of the form: if such and such a set of experimental facts is present, then such and such predictions or explanations can be made. A lucid example for this is the application of the period doubling cascade. Probably not all researchers may agree to my point of view, but I hope that I have given, at least, a stimulus for further discussions of this fascinating subject.

References

Though Chaos Theory and Synergetics are rather young fields of scientific research, there is a huge literature. The quotation of two standard references may suffice here:

H. Haken, *Synergetics*, An Introduction, 3rd ed., Springer, Berlin (1983)
H.G. Schuster, *Deterministic Chaos*, An Introduction, 2nd revised ed., VCH, Weinheim (1988)

Paul Davies

Algorithmic Compressibility, Fundamental and Phenomenological Laws

1. Introduction

As a starting point let us agree at least on the following statements:

- There exists a real external world which contains certain regularities. These can be understood, at least in part, by a process of rational enquiry called the scientific method.
- Science is not merely a game or charade. Its results capture, however imperfectly, some aspect of reality. Thus these regularities are *real* properties of the physical universe and not just human inventions or delusions.

In making these assumptions one has to eschew extreme idealistic philosophies such as those in which the mind somehow imposes the regularites on the world in order to make sense of it. Unless one accepts that the regularities are in some sense objectively real, one might as well stop doing science.

As science progresses, so some regularities become systematized as *laws* or deductions from them. At this epoch the laws found in our textbooks image only imperfectly the actual regularites. Two points of view can be detected among practicing scientists regarding the ontological status of these laws. The first is that there exist 'real' laws, or 'the correct set' of laws, to which our current theories are only an approximation. As science progresses so we converge upon the 'true' laws of the universe, which are regarded as eternal, timeless, and transcendent of the physical states.

By contrast, some scientists deny that there are any 'true' laws 'out there,' existing independently of scientific enquiry. What we call laws, they maintain, are simply our attempts to cope with the world by ordering our experiences in a systematic way. The only laws are *our* laws, and they are to be judged solely on utilitarian grounds, i.e., they are neither true nor false, but merely more or less useful to us. My impression is that many scientists who practice what one might loosely call applied science incline to the latter philosophy, while those engaged in 'fundamental' research, for example, on quantum cosmology or the unification programme, adopt the former position. We can sidestep the issue of whether the laws of nature are discovered or invented for the moment if we view the world algorithmically. (Chaitin [1987])

2. Algorithmic Compressibility

The existence of regularities may be expressed by saying that the world is *algorithmically compressible* (Solomonoff [1964]). Given some data set, the job of the scientist is to find a suitable compression, which expresses the causal linkages involved. For example, the positions of the planets in the solar system over some interval constitute a compressible data set, because Newton's laws may be used to link these positions at all times to the positions (and velocities) at some initial time. In this case Newton's laws supply the necessary algorithm to achieve the compression.

It will be helpful to consider the idea of algorithmic compressibility in a little more detail. The subject of algorithmic information theory was created in the 1960s in the Soviet Union by Andrei Kolmogorov and in the United States by Gregory Chaitin of IBM. The essence of the idea hinges on a very simple question: what is the shortest message that can describe a system to a certain level of detail? Obviously a simple system can be described easily, but a complex system can't. (Try describing the structure of a coral reef in the same number of words needed to describe an ice cube.) Chaitin and Kolmogorov suggest that the complexity of something is defined as the length of the shortest possible description of that thing.

Let's see how this works for numbers. There are simple numbers, such as 2 or pi, and complicated ones, such as a string of 1s and 0s generated by coin tossing (heads = 0, tails = 1). What sort of descriptions can we give that will define such numbers uniquely? One strategy is simply to write them out in decimal or binary form (pi can only be given to a particular approximation, because it has an infinite decimal expansion). But this is not, obviously, the most economical description. The number pi, for example, would be better described by giving a formula that can be used to calculate it to any desired degree of approximation. If the numbers concerned are regarded as the output of a computer, then the shortest description of a number will be the shortest programme that will enable the computer to output that number. Simple numbers will be generated by short programmes, complex numbers by long programmes.

The next step is to compare the length of the number with the length of the programme that generates it. Is it shorter? Has a compression been achieved? To make this more precise, suppose that the output of the computer is expressed as a string of ones and zeros, such as

1011010111000101001101010001 ...

(where '...' denotes 'and so on, perhaps forever'). This string will have a certain information content, which is measured in 'bits.' We want to compare the amount of information in the output with the information content of the programme itself. To take a simple example, suppose the output is

101010101010101010101010101010.

This could be generated by the simple algorithm "Print 01 fifteen times." A very much longer output string could be generated by the programme 'Print 01 one million times.' The second programme is scarcely more complicated than the first, yet it produces vastly greater output information. The lesson is that, if the output contains any patterns, then these may be compactly encoded in a simple algorithm that can be very much shorter (in terms of bits of information) than the output itself. In this case the string is said to be algorithmically compressible. If, conversely, a string cannot be generated by an algorithm significantly shorter than itself, it is algorithmically incompressible. In this case the string will possess no regularities or patterns whatsoever. It will just be a haphazard collection of 1s or 0s. In this way the amount of algorithmic compression attainable can be seen to be a useful measure of the simplicity or structure present in the output, with low compressibility being a measure of complexity. Simple, regular strings are highly compressible, whereas complex, patternless ones are less so.

Algorithmic compression provides a rigorous definition of randomness: a random sequence is one that cannot be algorithmically compressed. It may not be easy to tell merely by looking whether a given string is compressible. It might possess patterns of great subtlety built into it in a cryptic way. Every codebreaker knows that what looks at a glance like a random jumble of letters might in fact be as a structured message; to tell, all you need is the code. The infinite decimal expansion (and its binary counterpart) of the number pi shows no obvious patterns at all over thousands of digits. The distribution of digits passes all the standard tests for randomness. From a knowledge of the first thousand digits alone there is no way of predicting what the thousand-and-first will be. Yet in spite of this pi is *not* algorithmically random, because a very compact algorithm can be written to generate its expansion.

Chaitin points out that these ideas of mathematical complexity can be convincingly extended to physical systems: the complexity of a physical system is the length of the minimal algorithm that can simulate or describe it. At first sight this approach seems to be rather arbitrary, because we have not yet specified which make of computer is to be used. It turns out, however, that it doesn't really matter, because all universal computers can simulate one another. Similarly, the computer language we choose to work in – LISP, BASIC, FORTRAN – is irrelevant. It is a straightforward matter to write instructions to translate one computer language into another. It turns out that the extra programme-length needed to convert the language and to run the programme on another machine is typically a very small correction to the total programme-length. So you don't have to worry about how the computer you use is actually made. This is an important point. The fact that the definition of complexity is machine-independent suggests that it captures a really existing quality of the system, and is not merely a function of the way we choose to describe it.

Although the universe is complex, it is clearly not random. We observe regularities. The sun rises each day on schedule, light always travels at the same speed, a collection of muons always decay with a half-life of two-millionths of a second, and so on. These regularities are systematized into what we call laws. The laws of physics are analogous to computer programmes. Given the initial state of a system (input), we can use the laws to compute a later state (output).

The information content of the laws plus initial conditions is generally far less than that of the potential output. Of course, a law of physics may look simple when written down on paper, but it is usually formulated in terms of abstract mathematics, which itself needs a bit of decoding. Still, the information required to understand the mathematical symbols is limited to a few textbooks, whereas the number of facts described by these theories is unlimited. A classic example is provided by the prediction of eclipses. Knowing the position and motion of the Earth, sun, and moon at one time enables us to predict the dates of future (and past) eclipses. So one set of input data yields many output sets. In computer jargon, we may say that the data set of eclipses has been algorithmically compressed into the laws plus initial conditions. Thus the observed regularities of the universe are an example of its algorithmic compressibility. Underlying the complexity of nature is the simplicity of physics.

Interestingly, one of the founders of algorithmic information theory, Ray Solomonoff, concerned himself with precisely these sorts of questions. Solomonoff wanted to find a way of measuring the relative plausibility of competing scientific hypotheses. If a given set of facts about the world can be explained by more than one theory, how do we choose between them? Can we assign some sort of quantitative 'value' to the contenders? The short answer is to use Occam's razor: you pick the theory with the least number of independent assumptions. Now, if one thinks of a theory as a computer programme, and the facts of nature as the output of that programme, then Occam's razor obliges us to pick the shortest programme that can generate that particular output. That is to say, we should prefer the theory, or programme, which offers the greatest algorithmic compression of the facts.

So can we conclude that cosmic complexity can all be compressed into a very short 'cosmic programme'? Although there are many conspicuous examples of algorithmic compression in nature, not every system can be thus compressed. There is a class of processes, the importance of which has only recently been recognized, known as chaotic. These are processes that exhibit no regularities. Their behaviour appears to be entirely random. Consequently they are not algorithmically compressible. It used to be thought that chaos was rather exceptional, but scientists are coming to accept that very many natural systems are chaotic, or can become so under certain circumstances. Some familiar examples include turbulent fluids, dripping taps, fibrillating hearts, and driven pendula.

Even though chaos is rather common, it is clear that on the whole the universe is far from being simple. It possesses a subtle kind of complexity that places it partway between simplicity on the one hand and randomness on the other. One way of expressing this quality is to say that the universe has 'organized complexity,' a topic that I have discussed at length in my book *The Cosmic Blueprint.* There have been many attempts to capture mathematically this elusive element called organization. One is due to Charles Bennett of IBM, and involves something he calls 'logical depth.' This focuses less on the quantity of complexity or the amount of information needed to specify a system, and more on its quality, or 'value.' Bennett explains:

> A typical sequence of coin tosses has high information content, but little message value; an emphemeris, giving the positions of the moon and planets every day for a hundred years, has no more information than the equations of motion and the initial conditions from which it was calculated, but saves its owner the effort of recalculating these positions. The value of a message thus appears to reside ... in what might be called its buried redundance – parts predictable only with difficulty, things the receiver could in prinicple have figured out without being told, but only at considerable cost in time, money and computation. In other words, the value of a message is the amount of mathematical or other work plausibly done by its originator, which its receiver is saved from having to repeat. (Bennett [1987: 297])

Bennett invites us to think about the state of the world as having coded information folded up in it, information about the way the state was achieved in the first place. The issue is then how much 'work' the system has to do – i.e., how much information processing went on – to reach that state. This is what he refers to as logical depth. The amount of work is made precise by defining it in terms of the time taken to compute the message from the shortest programme that will generate it. Whereas algorithmic complexity focuses on the length of the minimal programme to yield a given output, logical depth is concerned with the running time for the minimal programme to generate that output.

Simple patterns are logically shallow, because they may be generated rapidly by short and simple programmes. Random patterns are also shallow, because their minimal programme is, by definition, not much shorter than the pattern itself, so again the programme is very short and simple: it need only say something like 'Print pattern.' But highly organized patterns are logically deep, because they require that many complicated steps be performed in generating them.

One obvious application of logical depth is to biological systems, which provide the most conspicuous examples of organized complexity. A living organism has great logical depth, because it could not plausibly have originated except through a very long and complicated chain of evolutionary processes. Another example of a deep system can be found in the complex patterns generated by cellular automata such as the so-called game of Life invented by the mathematician John Conway. In all cases the rule used is very simple, so from the algorithmic point of view these patterns actually have low complexity. The essence of Life's complexity lies not with the rules, but with their repeated use. The computer has to work very hard applying the rule again and again before it can generate deeply complex patterns from simple initial states.

The world abounds in deep systems, which show evidence of enormous 'work' in fashioning them. Murray Gell-Mann once remarked to me that deep systems can be recognized because they are the ones we want to preserve. Shallow things can easily be reconstructed. We value paintings, scientific theories, works of music and literature, rare birds, and diamonds because they are all hard to manufacture. Motorcars, salt crystals, and tin cans we value less; they are relatively shallow.

So what can we conclude about the cosmic programme? For centuries scientists have loosely talked about the universe being 'ordered,' without having a clear distinction between the various types of order: simple and complex. The study of computation has enabled us to recognize that the world is ordered both in the sense of being algorithmically compressible, and in the sense of having depth. The order of the cosmos is more than mere regimented regularity, it is also organized complexity, and it is from the latter that the universe derives its openness and permits the existence of human beings with free will. For three hundred years science has been dominated by the former: the search for simple patterns in nature. In recent years, with the advent of fast electronic computers, the truly fundamental nature of complexity has been appreciated. So we see that the laws of physics have a twofold job. They must provide the simple patterns that underlie all physical phenomena, and they must also be of the form that enables depth – organized complexity – to emerge. That the laws of our universe possess this crucial dual property is a fact of literally cosmic significance.

3. Fundamental and Phenomenological Laws

Physicists usually distinguish between fundamental and phenomenological laws. This distinction, is, however, rather subtle and intuitive. Quantum electrodynamics, or QED, which describes the interaction of electrons and photons at the microscopic level, involves fundamental laws, but Ohm's law is regarded as phenomenological. The difference stems from the fact that QED is a completely general theory, applying to electrons and photons under essentially all circumstances, whereas Ohm's law refers to a very specific and complicated physical system (an electrical circuit).

It is often claimed that phenomenological laws can in principle be reduced to, or explained by, fundamental laws. Thus might Ohm's law be derivable, ultimately, from QED. In practice this can almost never be achieved. Nor is the claim strictly correct as it stands. True, Ohm's law is consistent with QED, and QED is almost certainly necessary for the validity of Ohm's law, but it is not sufficient. In addition to QED one needs to make all sorts of additional assumptions about constraints, boundary conditions and statistics that are not logically part of QED at all.

This distinction is crucial, I believe, when one confronts the puzzle of whether the laws have independent existence. A helpful analogy here is with the concepts of hardware and software in computing. The laws of physics correspond to software, while the physical universe is the hardware. (Granted, this stretches the use of the word hard quite a bit, as included in the definition of the physical universe are nebulous quantum fields and even spacetime itself.) The issue can be stated thus: Is there an independently existing 'cosmic software' – a computer programme for a universe – encapsulating all the necessary laws? Can this software exist without the hardware?

The discussion of these issues has become particularly pertinent since the proposal a few years ago by James Hartle of the University of California at Santa Barbara, and Stephen Hawking of Cambridge University, for a 'law of initial conditions.'

(Hartle & Hawking [1983]). To explain this proposal, I must first recap on what is meant by an ordinary law of physics. It is a simple though not very fancy law that every pitched baseball follows a parabolic path (neglecting air resistance and the curvature of the Earth). The law is general, but it does not suffice to fix precisely the shape of every baseball trajectory. Although each path is a parabola, there can be low, shallow parabolas and high, curved parabolas. The law determines the class of curves, but the actual curve in any given case also depends on the initial conditions. For the baseball, these conditions are the position of the pitcher and the angle and speed of projection. Given these, the path is determined uniquely.

Laws, by their very nature, refer to classes of process. By contrast, initial conditions must be determined by additional criteria; in the case of the baseball, by appeal to the physiology (and ultimately the psychology) of the pitcher. Physicists justify the use of certain initial conditions by investigation of the wider physical environment. There is, however, a fundamental problem when it comes to cosmology – the study of the Universe as a whole – for then there is no wider environment to consider. The initial conditions that accompanied the birth of the cosmos in a big bang must, it would seem, be accepted as given. They cannot be explained by prior processes because the big bang represents the origin of time; there were no prior processes.

Once the cosmic initial conditions are treated as given, they are on the same footing as the laws of physics. So rather than treat them as outside the scope of science (as is usually the case), you can consider certain simple mathematical formulations or statements that purport to describe these initial conditions. That is, instead of shrugging off the initial conditions as random or arbitrary or determined by a supernatural agency, you can make a case that (like the laws of physics) they should be the manifestation of a simple and elegant mathematical scheme. As with the laws of physics, you cannot prove in advance that a given scheme is the correct one; the fruitfulness of any given proposal must be judged against observation.

Hartle and Hawking propose a particular 'law of initial conditions.' They have formulated the law in the context of the quantum theory of cosmology, so it is expressed in terms of a wavefunction for the Universe. The details need not worry us. The essential point is that any such law of initial conditions, appended to those laws of physics, must share the property of transcendence. This is another legacy of theology. A transcendent God is one whose existence is logically independent of the Universe. Transcendent laws are laws that have independent existence; they are 'out there' in some abstract realm. They depend on the existence of the physical Universe for their manifestation but not for their existence.

Few scientists would be willing to suppose that the laws of physics are merely human inventions. To be sure, they are formulated by humans, but the physicist is motivated by the belief that the laws of physics reflect some aspect of reality. Without the connection with reality, science is reduced to a meaningless charade. On the other hand, most physicists are also reluctant to concede that the laws of physics have an independent existence.

The dilemma is nicely illustrated by the case of Ohm's law, a statement about the relationship between current, voltage and resistance in an electrical circuit. The idea that Ohm's law has a timeless, transcendent existence, and has been 'out there,' lying in wait, for aeons until somebody built an electrical circuit is surely ludicrous. And yet it is equally hard to believe that Ohm's law just happened to come into existence when the first circuit was constructed. The truth could be that Ohm's law is a secondary, as opposed to fundamental, law and that the primary laws to which some people claim it may be reduced (such as the laws of electrodynamics) are indeed timeless and transcendent.

There are, however, difficulties even with this position. Not all physicists believe that laws, such as Ohm's law, which refer to the collective properties of large assemblies of particles, can be completely reduced to more fundamental laws, at least without some auxiliary assumptions. Take an extreme example: Mendel's laws of genetics. There would seems to be no prospect of ever deriving them from the fundamental laws of physics alone, even in principle.

In my experience, physicists who study fundamental problems ususally assume the transcendent existence of the fundamental laws of physics. In particular, quantum cosmologists, who set themselves the goal of explaining how the Universe came to exist from nothing, are obliged to make this assumption. No law of physics can explain how the Universe came to exist unless that law already existed. If the laws of physics came into existence with the Universe, they could not explain how the Universe came into existence. And by definition a law of initial conditions cannot come into existence after the origin of the Universe.

However, the fact that the universe as a whole evolves, and that the hardware of the cosmos is not a fixture, has encouraged the speculation that the laws of physics might also evolve with time. Dirac proposed that the strength of the gravitational force gradually fades with time, although his formula for this has been ruled out by observation. However, this is not really what I mean by an *evolving law,* because Dirac merely replaced the usual law of gravity with another law that described how the first law systematically changed.

A more profound idea stems from the discovery that at higher and higher energies (or temperatures), what were once considered to be inviolate laws might in fact be violated. For example, it was long ago discovered that the old law of conservation of matter is violated when new particles and antiparticles are created in high energy processes. More recently, according to certain unified field theories the law of baryon conservation (that the total number of protons and related particles must remain constant) might fail slightly at ultra-high energies. It has been suggested that if one could drive the energy up without limit, one by one all our cherished laws would fail, leaving only chaos. These laws would be seen, not as eternal transcendent truths, but as incidental low-energy constraints on the behaviour of matter.

In principle we cannot drive up the energy (or temperature) of the universe without limit, but in the context of the big bang theory this happens naturally. The creation event itself can be regarded as a state of essentially infinite temperature. According to the chaos theory, then, the well-ordered universe we see is a frozen relic of primeval chaos. There was no lawfully supervised coming-into-being out of nothing. Everything came, to use the words of John Wheeler, 'out of higgledy-piggledy'. (Wheeler [1983: 398-404]).

Wheeler has elevated what he calls the fundamental mutability of the physical world to the status of a principle; in fact, the only principle. 'The only law', writes Wheeler, 'is that there is no law'. He has been led to this extreme position by the study of gravitational collapse, which predicts the existence of so-called spacetime singularities. These are boundaries or edges to spacetime where all physics seemingly breaks down. Wheeler favours a cosmology in which the entire universe eventually collapses in a 'big crunch', like the big bang in reverse, culminating in an all-encompassing singularity. Thus, space and time, as well as all matter and energy and all physical structures, would be destroyed. Reasoning that all laws of physics are formulated within the context of spacetime, Wheeler concludes that no law can survive the crisis of gravitational collapse.

If the physical world is not after all underpinned by eternal laws, where does its apparent lawfulness originate? Wheeler has an astonishing proposal, a sort of hyper-Kantian extreme. We the observers create that lawfulness by our very observations. Appealing to the oft-cited melding of observer and observed that emerges from quantum physics (and which I shall discuss shortly), he ventures: 'Could it be that the observership of quantum mechanics is the ultimate underpinning of the laws of physics?' In this 'participatory universe' *we* are the lawmakers, not by passively projecting our mental forms onto the physical world, but actively, by our very observations shaping that world into lawfulness.

The idea that the laws of nature have somehow evolved out of lawlessness is not new to philosophy. Friederich Nietzsche conjectured that 'the origin of the mechanical world would be a lawless game'. But just as biological order arises out of molecular chaos, so might the mechanical world 'ultimately acquire such consistency as the organic laws seem to have now'. Moreover, the existing laws would have emerged by a sort of natural selection from among a vast set of alternatives: 'All our mechanical laws would not be eternal, but evolved, and would have survived innumerable alternative mechanical laws'. The neuroscientist Richard Gregory recently expressed to me a similar belief, that the laws of physics have arisen from a type of Darwinian selection.

The shortcoming of this idea is that selection is meaningful only in the context of an ensemble of identical systems. Selection can occur in a biological species, for example, but what meaning has selection in the case of a unique universe? One possibility is to conjecture the existence of an ensemble of universes. Those with 'well-adapted' laws will 'prosper' in the sense of evolving in an orderly, interesting way to states of greater and greater organization and complexity until they spawn conscious beings who write articles about the meaning of it all. The universes less well served by their laws will degenerate into chaos and anarchy, and go unobserved. The selection mechanism here is known as the anthropic principle, and it has been much discussed in recent years. It implies that we live in a universe with the lawlike properties it has not because these laws are eternal and transcendent but because we have selected such a universe by our very presence. What we mistakenly call 'the' universe is in fact but one among a vast totality of generally uninhabitable universes. The lawfulness of our universe is no surprise, for lawfulness is a prerequisite of habitability and hence cognizability. (See Barrow/Tipler [1986])

Another imaginative proposal has been put forward by the controversial biologist Rupert Sheldrake. He also believes that there are no eternal laws. Instead he draws an intriguing analogy with human laws. The traditional timeless laws of physics he compares with statute law. But what if the laws of nature are more like common law? In other words, what we take to be absolutely inviolable, transcendent truths are, according to Sheldrake, really trends or habits, accumulating over the aeons out of an intially lawless state by repetitious application and mutual reinforcement. Let us look at the consequences of an assumption of evolving laws.

4. Evolving Laws

If one accepts the idea of evolving laws then the distinction between fundamental and phenomenological laws fades away. As new and ever more elaborate physical systems come into existence with time, so might the laws that apply to them. Thus one can regard life as an emergent quality, and the concommitant laws of biology as emergent alongside it. What were once regarded as primary laws, such as the laws of particle mechanics, would differ from 'secondary' laws, such as Ohm's Law or the laws of superconductivity, by temporal priority alone. The laws of mechanics were there first because isolated particles predated electric circuits. But the former would not be more fundamental than the latter, at least in the deep sense considered hitherto.

If I interpret him correctly, this seems to be the position of Ilya Prigogine, who, when asked, 'Can the laws of nature undergo transition?' replied: 'The laws of nature may indeed depend on the state of nature ... Can you speak of laws of biology if there is no life? Obviously not. In other words what we call laws or regularities, depend on their realization.' And elsewhere: 'There is no longer a single fundamental level. Our level, the level of macroscopic beings, is not less fundamental or more fundamental than the microscopic level ... there is no longer an absolute level of description. There are various levels, all interconnected in a much more complex fashion.'

Prigogine's suggestion that the laws of nature may depend on the state points up a key assumption that runs through all of orthodox science. Rooted in the very concept of 'law' is a conceptual separation between the abstract lawful statement itself and the state of the physical system to which the statement is applied. To take a simple example, consider again the motion of a ball under Newton's laws. The laws of mechanics make no reference to the actual state of the ball. They are the same laws whether the ball is here or there, or moving this way or that. (If I may be permitted a technical aside for clarity, the state of a mechanical system is represented by a point in phase space. But the dynamics is contained in the Lagrangian. Essential to all theories of mechanics is that the functional form of the Lagrangian is independent of the location of the representative point in phase space .)

As far as we know this separation of 'software' (laws) from 'hardware' (actual physical state) is justified for all simple systems. But I have a sneaking feeling it may fail for complex systems. I have argued elsewhere that the act of observation in quantum mechanics entails a 'hardware/software entanglement'. The biologist Robert Rosen has developed an elaborate mathematical description of complex systems in which there is no separation of dynamical laws and states. If one were to pursue this as far as cosmology, it would imply that the universe is a sort of gigantic feedback system where the laws that apply today depend on the details of cosmic history.

These evolutionist ideas are unlikely to find favour among particle physicists, cosmologists and unifiers, whose thinking is strongly influenced by the theory of relativity. Central to the evolutionary theme is the dimension of time. The laws of nature, like the cosmos itself, are given a history. An arrow of time is built into the operation of physical laws. Time is thus singled out from spacetime and ascribed a metaphysical significance that it lacks in relativity theory, where space and time are on an equal footing, and spacetime itself is part of the dynamical system. Indeed, Hawking and coworkers are busy deemphasising the significance of time, even to the extent of trying to redefine it as space in disguise.

5. Models and Reality

Now let me declare my own hand. I have always taken a somewhat positivist approach to the nature of reality. It seems to me that, ultimately, we have to work entirely with the (common) facts of experience. Physicists are in the business of making models to relate these facts. In the course of this they invent certain concepts such as 'energy' or 'atom'. At rock bottom, these concepts are merely code-words that encapsulate in a simplified way certain complicated mathematical properties of these models. However, they become so familiar to us that we tend to promote them to the status of real, independently-existing entitites. When we do that we then try to imagine them 'out there', possessing certain objectively real qualities. We then run into the trouble that these hypothetical 'things' simply cannot be imagined. Whatever mental image you may have of an atom, it is wrong! Certainly it is not a little solar system of orbiting particles, nor is it a packet of pulsating waves. It is nothing that we can visualise.

What, then, is an atom? I would say that it is an abstract concept – a Platonic Idea if you like – that helps us relate certain types of observations in a systematic way. 'Atomic theory' is actually simply a set of algorithms for effecting these relations (or more accurately, explicating correlations), and the 'atom' codifies the greater content of these algorithms. It is an abstract entity that belongs to the model; it does not really exist as an objective, independent entity.

The reader might be concluding that I am denying the reality of the external world. What I am denying is not the reality of the world, but the adjective 'external'. In modern physics, subject and object are indivisible. The 'real world' consists not so much of 'things out there' observed by us, but of connections – specifically, connections between observations. The reality is the connections. John Wheeler makes this point in characteristically extreme form: 'What we call "reality," that vision of the universe that is so vivid in our minds, we plaster in between a few iron posts of observation by an elaborate labor of imagination and theory'. He summerizes this in a pithy quote by Torny Segerstedt: 'Reality is theory'.

You might also believe that I am subscribing to the 'cultural' viewpoint, by arguing that atoms, and the laws of atomic physics, belong only to our models and not to 'the real world'. Well, I agree that there is an element of truth in this, because our models are undeniably products of our cultural milieu. But it is not the whole truth. Unless one is prepared to retreat into the lonely fastness of solipsism, one has to concede that the discoveries of science reveal that there is something orderly about the world, something not totally the result of our imaginings. The very fact that our observations can be fitted into mathematical models and algorithms surely tells us something about reality that we had no right to expect *a priori*.

Let me try to make this more precise. When we say that there are laws of nature, we mean that physical systems share certain common properties and patterns of behaviour wherever and whenever they are. Thus the motion of an electron in a hydrogen atom on Earth is the same as that on a distant quasar, and was the same billion of years ago as today (there is good observational evidence for this). Likewise, the paths of falling bodies on Earth and on a far-flung galaxy belong to the same geometrical class. When you think about it this coherence across space and time is a surprising and mysterious fact.

It is tempting to ask the question 'How does one electron (or falling body) know what the other is doing?' The orthodox answer is that it doesn't; both comply with the laws of physics. But this must mean that these laws have some sort of objective existence that transcends space and time. Or so it seems to me.

There is another way of expressing this, using the concept of computer algorithms. If we could take a 'God's-eye-view' and perceive the whole of space and time at once, together with the intricate tracery that is the collective history of all the particles of matter, we should notice that this texture is not random but patterned. The paths traced out in spacetime (what physicists call world lines) are correlated and ordered and have a preponderence for particular shapes. One could give an account of this panorama by dividing spacetime into minute cells and using binary arithmetic to register whether a cell was traversed by a particle path or not. The results would be a collosal array of 0s and 1s, representing the total information content of the history of the universe.

It is then meaningful to consider a computer algorithm that would generate the same sequence of 0s and 1s. If the algorithm has the same information content as the sequence it generates, one says the information is incompressible. This would be the case of the universe were completely random. In fact, the cosmic information is highly compressible. (When a computer is used to predict, say, eclipses it is utilizing – on a very small scale – precisely the compressiblity of cosmic information.) This compressibility we codify into 'laws of physics', which certainly have a cultural content. But the fact of compressibility is not explained by our culture.

(I should mention, as an aside, that the information content of the world, whilst enormously compressible, is not quite as compressible as was once thought, owing to the existence of so-called chaotic systems.)

So I find myself uneasily poised between two contending factions. On the one hand I am convinced that there are some trancendent eternal truths that constrain the nature of the physical world, and I believe that the coming-into-being of the universe was a lawful event and not the genesis of law. However, I am suspicious that what are conventionally regarded as 'fundamental laws' (e.g. the Lagrangian of some collection of fields or particles or strings) are altogether too narrow to encompass the full richness of physical phenomena, especially the organizational properties of complex systems. Thus I see the need for laws at all levels of description, and dislike the designation 'fundamental' applied to the bottom level alone. I agree wholeheartedly with Anthony Leggett when he wonders: 'Does the mere presence of complexity or organization or some related quality introduce new physical laws?' – and goes on to suggest that 'quantum mechanics breaks down in the face of increasing complexity. I must therefore concede that as the universe evolves, so new laws emerge.

Bibliography

Barrow, J.D./F.J. Tipler [1986], *The Anthropic Cosmological Principle.* Oxford: Oxford UP.

Bennett, Ch. [1987], 'Dissipation, Information, Computational Complexity and the Definition of Organization.' In: *Emerging Syntheses in Science* (ed. D. Pines), Boston: Addison-Wesley.

Chaitin, G.J. [1987], *Algorithmic Information Theory.* Cambridge: Cambridge University Press.

Hartle, J.B./S.W. Hawking [1983], 'The Wave Function of the Universe.' *Phys. Rev. D 28* (1983), p. 2960.

Solomonoff, R.J. [1964], 'A Formal Theory of Inductive Inference. Part I.' *Information & Control 7*,p.1.

Wheeler, J.A. [1983], 'On recognizing "law without law"', *Am.J.Phys.*, *51*(5),p.398.

Francisco J. Ayala

The Distinctness of Biology

The Issue

The similarities and differences between the biological and physical sciences, and between organisms and inanimate nature, are matters of considerable interest to philosophers of science. One question is whether the laws formulated in biology can be shown to be special cases of more general physical laws. This is the issue of (theoretical) reductionism that has received considerable attention over the last three decades (e.g., Simpson 1964; Koestler and Smithies 1969; Polanyi 1968; Schaffner 1969; Ayala and Dobzhansky 1974; Hull 1974; Grene and Mendelsohn 1976; Wimsatt 1979; Rosenberg 1985).

The discovery of the structure and mode of replication and action of the hereditary material, DNA, seems to open up the possibility of reducing the laws of genetics to the laws of chemistry (e.g., Hull 1974). Similarly, it seems to some authors that it's only a matter of time before the function of the brain processes, and even the mind, might be understandable as logically-derived consequences of certain laws of physics and chemistry concerning, for example, electrical and chemical transmission of signals.

Some authors have claimed, on the contrary, that even if some parts of biology can be reduced to physics and chemistry now or might be reduced in the future, a complete reduction is impossible in principle because biological disciplines employ patterns of explanation that do not occur in the physical sciences (Simpson 1964; Ayala 1968).

Historical explanations are sometimes mentioned as distinctive of the biological and social sciences. They are not, however, because they also occur in some physical sciences, such as astronomy and geology. Also, although historical explanations do play a role in evolutionary theory, the theory is primarily concerned with causal explanations of evolutionary processes.

I will argue, however, that biology is distinct from the physical sciences in that it uses patterns of explanation, and makes recourse to laws, that do not occur in, nor can be reduced to those formulated in the physical sciences. Specifically, I shall attempt to show that teleological explanations constitute patterns of explanation that apply to organisms while they do not apply to any other kind of objects in the natural world. I shall further claim that, although teleological explanations are compatible with causal accounts, they cannot be reformulated in nonteleological form without loss of explanatory content. Consequently, I shall conclude that teleological explanations cannot be dispensed with in biology.

Copernicus to Darwin

The publication in 1859 of Darwin's *On The Origin of Species* opened a new era in the intellectual history of mankind. The discoveries of Copernicus, Kepler, Galileo, and Newton had gradually led to a conception of the universe as a system of matter in motion governed by natural laws. The earth was found to be not the center of the universe but a small planet rotating around an average star; the universe appeared as immense in space and in time; the motions of the planets around the sun could be explained by simple laws – the same laws that accounted for the motion of objects in our planet. These and many other discoveries greatly expanded human knowledge. But the conceptual revolution that went on through the seventeenth and eighteenth centuries was the realization that the universe obeys immanent laws that can account for natural phenomena. The workings of the universe were brought into the realm of science – explanation through natural laws. Physical phenomena could be reliably predicted whenever the causes were adequately known.

Darwin completed the Copernican revolution by drawing out for biology the ultimate conclusions of the notion of nature as a lawful system of matter in motion. The adaptations and diversity of organisms, the origin of novel and highly organized forms, even the origin of man himself could now be explained by an orderly process of change governed by natural laws.

Before Darwin, the origin of organisms and their marvelous adaptations was most frequently attributed to the design of an omniscient Creator (see Hull 1989, ch. 2). God had created the birds and bees, the fish and corals, the trees in the forest, and best of all Man. God had given man eyes so that he might see, and had provided fish with gills to breathe in water. Philosophers and theologians often argued that the functional design of organisms evinced the existence of an all-wise Creator. In the thirteenth century St. Thomas Aquinas had used the argument-from-design as his "fifth way" to demonstrate the existence of God. In the nineteenth century the English theologian William Paley argued in his *Natural Theology* (1802) that the functional design of the human eye provided conclusive evidence of an all-wise Creator.

It would be absurd to suppose, wrote Paley, that the human eye by mere chance "should have consisted, first, of a series of transparent lenses ...; secondly of a black cloth or canvas spread out behind these lenses so as to receive the image formed by pencils of light transmitted through them, and placed at the precise geometrical distance at which, and at which alone, a distinct image could be formed ...; thirdly of a large nerve communicating between this membrane and the brain." The eight Bridgewater Treatises, written by eminent scientists and philosophers, were published between 1833 and 1840 to set forth "the Power, Wisdom, and Goodness of God as manifested in the Creation." The mechanisms and vital endowments of man's hand gave, according to Sir Charles Bell, author of one Bridgewater Treatise, incontrovertible evidence that the hand had been designed by the same omniscient Power that had created the world.

Adaptation

The apparent strength of the argument-from-design to demonstrate the existence of a Creator is obvious. Wherever there is function or design we look for its author. A knife is made to cut and a clock is made to tell time; their functional designs are contrived by a knifemaker and a watchmaker. The structures, organs, and behaviors of living beings are directly organized to serve certain functions. Thus the functional design, or teleology, of organisms and their features would seem to argue for the existence of a designer. Darwin showed that the directive organization of living beings could be explained as a result of a natural process – natural selection. There was no need to resort to a Creator or other external agent. The origin and adaptation of organisms were thus brought into the realm of science.

Darwin recognized that organisms are adapted to certain ways of life and their parts adapted to perform certain functions; i.e., that they are teleologically organized. Fish are adapted to live in water, kidneys are designed to regulate the composition of blood, the hand of man is made for grasping. Darwin accepted the facts of adaptation and then provided a natural explanation for those facts. His greatest achievement is that he explained scientifically the teleological features of living beings.

Darwin provided ample evidence for the occurrence of evolution, and for this he is appropriately given credit. More revolutionary, however, was that he extended the Copernican revolution to the world of living things. The origin and adaptive nature of organisms could now be explained, like the phenomena of the inanimate world, as the result of natural laws manifested in natural processes. Darwin's theory was opposed in certain religious circles, not so much because he proposed the evolutionary origin of living things (this had been proposed before, even by Christian theologians), but because he excluded God from the explanation of the adaptation of organisms. The Roman Catholic church's opposition to Galileo in the seventeenth century was similarly motivated.

The central argument of the theory of natural selection was eloquently summarized by Darwin in *Origin of Species*:

> As more individuals are produced than can possibly survive, there must in every case be a struggle for existence, either one individual with another of the same species, or with the individuals of distinct species, or with the physical conditions of life ... Can it, then, be thought improbable, seeing that variations useful to man have undoubtedly occurred, that other variations useful in some way to each being in the great and complex battle of life, should sometimes occur in the course of thousands of generations? If such do occur, can we doubt (remembering that more individuals are born than can possibly survive) that individuals having any advantage, however slight, over others, would have the best chance of surviving and of procreating their kind? On the other hand, we may feel sure that any variation in the least degree injurious would be rigidly destroyed. This preservation of favorable variation and the rejection of injurious variations, I call Natural Selection.

Darwin's argument addresses precisely the problem of explaining the adaptive nature of organisms. Darwin argues that adaptive variations ("variations useful in some way to each being") must occasionally appear, and that these are likely to increase the reproductive chances of their carriers. Over the generations, favorable variations will be preserved, injurious ones will be eliminated. In one of the few places where he uses the term "adaptation" or its cognates in *Origin of Species*, Darwin adds: "I can see no limit to this power [natural selection] in slowly and beautifully *adapting* each form to the most complex relations of life." Natural selection was proposed by Darwin primarily to account for the adaptive organization of living beings; it is a process that maintains and promotes adaptation. Evolutionary change through time (anagenesis) and evolutionary diversification (cladogenesis) are not directly promoted by natural selection, but they often ensue as by-products of natural selection fostering adaptation. "One problem of great importance ... is the tendency in organic beings to diverge in character as they become modified ... The solution [to this problem] is that the modified offspring of all dominant and increasing forms tend to become *adapted* to many and highly diversified places in the economy of nature" (Darwin 1958).

Teleology

According to the dictionary, teleology (from the Greek *telos* = end) is "the use of design, purpose, or utility as an explanation of any natural phenomenon" (*Webster's Third New International Dictionary* 1966). An object or a behavior is said to be teleological, or telic, when it gives evidence of design or appears to be directed toward certain ends. For example, the behavior of human beings is often teleological. A person who buys an airplane ticket, reads a book, or cultivates the earth is trying to achieve a certain end: getting to a given city, acquiring knowledge, or getting food. Objects and machines made by people also are usually teleological: a knife is made for cutting, a clock is made for telling time, a thermostat is made to regulate temperature. Similarly, features of organisms also are teleological: a bird's wings are for flying, eyes are for seeing, kidneys are constituted for regulating the composition of the blood. The features of organisms that may be said to be teleological are those that can be identified as adaptations, whether they are structures like a wing or a hand, or organs like a kidney, or behaviors like the courtship displays of a peacock. Adaptations are features of organisms that have come about by natural selection because they serve certain functions and thus increase the reproductive success of their carriers.

Inanimate objects and processes (other than those created by men) are not teleological because they are not directed toward specific ends, they do not exist to serve certain purposes (Binswanger 1990). The configuration of sodium chloride depends on the structure of sodium and chlorine, but it makes no sense to say that that structure is made up so as to serve a certain end. Similarly, the shape of a mountain is the result of certain geological processes, but it did not come about so as to serve a certain end. The motion of the earth around the sun results from the laws of gravity, but it does not exist in order to satisfy certain ends or goals.

We may use sodium chloride as food, a mountain for skiing, and take advantage of the seasons, but the use that we make of these objects or phenomena is not the reason why they came into existence or why they have certain configurations. On the other hand, a knife and a car exist and have particular configurations precisely in order to serve the ends of cutting and transportation. Similarly, the wings of birds came about precisely because they permitted flying, which was reproductively advantageous. The mating display of peacocks came about because it increased the chances of mating and thus of leaving progeny.

The previous comments point out the essential characteristics of telic phenomena, i.e., phenomena whose existence and configuration can be explained teleologically. I may now propose the following definition. "Teleological explanations account for the existence of a certain feature in a system by demonstrating the feature's contribution to a specific property or state of the system." Teleological explanations require that the feature or behavior contribute to the existence or maintenance of a certain state or property of the system. Moreover, and this is the essential component of the concept, the contribution *must be the reason why the feature or behavior exists at all.*

The configuration of a molecule of sodium chloride contributes to its property of tasting salty and therefore to its use as food, not vice versa; the potential use of sodium chloride as food is not the reason why it has a particular molecular configuration or tastes salty. The motion of the earth around the sun is the reason why seasons exist; the existence of the seasons is not the reason why the earth moves about the sun. On the other hand, the sharpness of a knife can be explained teleologically because the knife has been created precisely to serve the purpose of cutting. Motorcars and their particular configurations exist because they serve transportation, and thus can be explained teleologically. (Not all features of a car contribute to efficient transportation – some features are added for aesthetic or other reasons. But as long as a feature is added because it exhibits certain properties – like appeal to the aesthetic preferences of potential customers – it may be explained teleologically. Nevertheless, there may be features in a car, a knife, or any other man-made object that need not be explained teleologically. That knives have handles may be explained teleologically, but the fact that a particular handle is made of pine rather than oak might simply be due to the availability of material. Similarly, not all features of organisms have teleological explanations.)

Many features and behaviors of organisms meet the requirements of teleological explanation. The hand of man, the wings of birds, the structure and behavior of kidneys, the mating displays of peacocks are examples already given. In general, as pointed out above, those features and behaviors that are considered adaptations are explained teleologically. This is simply because adaptations are features that come about by natural selection. Among alternative genetic variants that may arise by mutation or recombination, the ones that become established in a population are those that contribute more to the reproductive success of their carriers. The effects on reproductive success are usually mediated by some function or property. Wings and hands acquired their present configuration through long-term accumulation of genetic variants adaptive to their carriers. How natural selection yields adaptive features may be explained with cases where the adaptation arises as a consequence of a single gene mutation. One example is the presence of normal hemoglobin rather than hemoglobin S in humans. One amino acid substitution in the beta chain in humans results in hemoglobin molecules less efficient for oxygen transport. The general occurrence in human populations of normal rather than S hemoglobins is explained teleologically by the contribution of hemoglobin to effective oxygen transport and thus to reproductive success. A second example, the difference between peppered-gray moths and melanic moths is due to one or only a few genes. The replacement of gray moths by melanics in polluted regions is explained teleologically by the fact that in such regions melanism decreases the probability that a moth be eaten by a bird. The predominance of peppered forms in nonpolluted regions is similarly explained.

Not all features of organisms need to be explained teleologically, since not all come about as a direct result of natural selection (Binswanger 1990). Some features may become established by random genetic drift, by chance association with adaptive traits, or in general by processes other than natural selection. Proponents of the neutrality theory of protein evolution argue that many alternative protein variants are adaptively equivalent. Most evolutionists would admit that at least in certain cases the selective differences between alternative ammino acids at a certain site in a protein must be virtually nil, particularly when population size is very small. The presence in a population of one amino acid sequence rather than another adaptively equivalent to the first, would not then be explained teleologically. Needless to say, in such cases there would be amino acid sequences that would not be adaptive. The presence of an adaptive protein rather than a nonadaptive one would be explained teleologically; but the presence of one protein rather than another among those adaptively equivalent would not require a teleological explanation.

Teleology: Internal and External; Determinate and Indeterminate

In the previous section some man-made objects and adaptive traits of organisms served as examples of teleological phenomena. We may now explore several kinds of teleological phenomena (Ayala 1968, 1970). Actions are *purposeful* when the end-state or goal is consciously intended by an agent. Thus, a man mowing his lawn is acting teleologically in the purposeful sense; a lion hunting deer and a bird building a nest have at least the appearance of purposeful behavior. Objects resulting from purposeful behavior exhibit *external* (or *artificial*) teleology. A knife, a table, a car, and a thermostat are examples of systems exhibiting artificial teleology: their teleological features were consciously intended by some agent.

Systems with teleological features that are not due to the purposeful action of an agent but result from natural process exhibit *internal* (or *natural*) teleology. The wings of birds have a natural teleology; they serve an end, flying, but their configuration is not due to the conscious design of any agent. We may distinguish two kinds of natural teleology: *determinate*, bounded or necessary, and *indeterminate*, unbounded or contingent.

Determinate natural teleology exists when a specific end-state is reached in spite of environmental fluctuations. The development of an egg into a chicken, or of a human zygote into a human being, are examples of determinate natural teleological processes. The regulation of body temperature in a mammal is another example. In general, the homeostatic processes of organisms are instances of determinate natural teleology. Two types of homeostasis are usually distinguished – physiological and developmental – although intermediate conditions exist. Physiological homeostatic reactions enable organisms to maintain certain physiological steady states in spite of environmental shocks. The regulation of the concentration of salt in blood by the kidneys, or the hypertrophy of muscle owing to strenuous use, are examples of physiological homeostasis. Developmental homeostasis refers to the regulation of the different paths that an organism may follow in the progression from fertilized egg to adult. The process can be influenced by the environment in various ways, but the characteristics of the adult individual, at least within a certain range, are largely predetermined in the zygote.

Indeterminate or unbounded teleology occurs when the end-state served is not specifically predetermined, but rather is the result of selection of one from among several available alternatives. For teleology to exist, the selection of one alternative over another must be deterministic and not purely stochastic. But what alternative happens to be present may depend on environmental and/or historical circumstances and thus the specific end-state is not generally predictable. Indeterminate teleology results from a mixture of stochastic (at least from the point of view of the teleological system) and deterministic events.

Many features of organisms are teleological in the indeterminate sense. The wings of birds require teleological explanations: the genetic constitutions responsible for their configuration came about because wings serve to fly and flying contributes to the reproductive success of birds. But there was nothing in the constitution of the remote ancestors of birds that would necessitate the appearance of wings in their descendants. Wings came about as the consequence of a long sequence of events, where at each stage the most advantageous alternative was selected among those that happened to be available; which alternatives were available at any one time depended at least in part on chance events.

In spite of the role played by stochastic events in the phylogenetic history of birds, it would be mistaken to say that wings are not teleological features. As pointed out earlier, there are differences between the teleology of an organism's adaptations and the nonteleological potential uses of natural inanimate objects. A mountain may have features appropriate for skiing, but those features did not come about so as to provide skiing slopes. On the other hand, the wings of birds came about precisely because they serve for flying. The explanatory reason for the existence of wings and their configuration is the end they serve – flying – which in turn contributes to the reproductive success of birds. If wings did not serve an adaptive function they would have never come about, and would gradually disappear over the generations (Binswanger 1990).

The indeterminate character of the outcome of natural selection over time is due to a variety of nondeterministic factors. The outcome of natural selection depends, first, on what alternative genetic variants happen to be available at any one time. This in turn depends on the stochastic processes of mutation and recombination, and also on the past history of any given population. (What new genes may arise by mutation and what new geneic constitutions may arise by recombination depend on what genes happen to be present – which depends on previous history.) The outcome of natural selection depends also on the conditions of the physical and biotic environment. Which alternatives among available genetic variants may be favored by selection depends on the particular set of environmental conditions to which a population is exposed.

Objections and Responses

Some evolutionists have rejected the teleological mode of explanation in evolutionary biology because they have failed to recognize the various meanings that the term "teleology" may have (Pittendrigh 1958; Mayr 1965, 1974; Williams 1966; Ghiselin 1974). These biologists are correct in excluding certain forms of teleology from evolutionary explanations, but they err when they claim that teleological explanations should be excluded altogether from evolutionary theory. In fact, they themselves often use teleological explanations in their works, but fail to recognize them as such, or prefer to call them by some other name, such as "teleonomic." Teleological explanations, as explained above, are appropriate in evolutionary theory, and are recognized by most evolutionary biologists and philosophers of science who have thoughtfully considered the question (Beckner 1959; Nagel 1961; Simpson 1964; Dobzhansky 1970; Ayala 1968, 1970; Wimsatt 1972; Hull 1974). Which kinds of teleological explanations are appropriate and which ones are inappropriate with respect to various biological questions may be briefly specified.

According to Mayr (1965) teleological explanations have been applied to two different sets of biological phenomena. "On the one hand is the production and perfection throughout the history of the animal and plant kingdoms of ever-new and ever-improved DNA programs of information. On the other hand is the testing of these programs and their decoding throughout the lifetime of each individual. There is a fundamental difference between end-directed behavioral activities or developmental processes of an individual or system, which are controlled by a program, and the steady improvement of the genetically coded programs. This genetic improvement is evolutionary adaptation controlled by natural selection." The "decoding" and "testing" of genetic programs of information are the issues considered, respectively, by developmental biology and functional biology. The historical and causal processes by which genetic programs of information come about are the concern of evolutionary biology. Grene (1974) uses the term "instrumental" for the teleology of organs that act in a functional way, such as the hand for grasping; "developmental" for the teleology of such processes as the maturation of a limb; and "historical" for the process (natural selection) producing teleologically organized systems.

In the terminology that I have proposed, organs and features such as the eye and the hand have determinate (and internal) natural teleology. These organs serve determinate ends (seeing or grasping) but have come about by natural processes that did not involve the conscious design of any agent. Physiological homeostatic reactions and embryological development are processes that also have determinate natural teleology. These processes lead to end-states (from egg to chicken) or maintain properties (body temperature in a mammal) that are on the whole determinate. Thus, Mayr's "decoding" of DNA programs of information and Grene's "instrumental" and "developmental" teleology, when applied to organisms, are cases of determinate natural teleology (Mayr prefers to use the term "teleonomy" for this type of teleology). Human tools (such as a knife), machines (such as a car), and servomechanisms (such as a thermostat) also have determinate teleology, but of the artificial kind, since they have been consciously designed.

I propose that it is appropriate to consider natural selection as a teleological process, but only in the sense of indeterminate natural teleology. It is not consciously intended by any agent, nor is it directed towards specific or predetermined end-states. Yet the process is far from random or completely indeterminate. Among the genetic alternatives available at any one time, natural selection favors those that increase the reproductive success of their carriers in the particular environmental circumstances in which the organisms live. Reproductive success is mediated by some adaptive function, say flying, that is in turn determined by the genetic variants that are favored by natural selection.

Some authors exclude teleological explanations from evolutionary biology because they believe that teleology exists only when a specific goal is purposefully sought. This is not so. Terms other than "teleology" could be used for natural (or internal) teleology, but this might in the end add more confusion than clarity. Philosophers as well as scientists use the term "teleological" in the broader sense, to include explanations that account for the existence of an object in terms of the end-state or goal that they serve (Binswanger 1990).

It is important, for historical reasons, to reiterate that the process of evolution by natural selection is not teleological in the purposeful sense. Thomas Aquinas and the natural theologians of the nineteenth century erroneously claimed that the directive organization of living beings evinces the existence of a Designer. The adaptations of organisms can be explained as the result of natural processes without recourse to consciously intended end-products. There is purposeful activity in the world, at least in man; but the existence and particular structures of organisms, including humans, need not be explained as the result of purposeful behavior.

Some scientists and philosophers who held that evolution is a natural process erred, nevertheless, in seeing evolution as a determinate, or bounded, process. Lamarck (1809) thought that evolutionary change necessarily proceeded along determined paths from simpler to more complex organisms. Similarly, the evolutionary philosophies of Bergson (1907), Teilhard de Chardin (1959), and the theories of *nomogenesis* (Berg 1926), *aristogenesis* (Osborn 1934), *orthogenesis*, and the like are erroneous because they all claim that evolutionary change necessarily proceeds along determined paths. These theories mistakenly take embryological development as the model of evolutionary change, regarding the teleology of evolution as determinate. Although there are teleologically determinate processes in the living world, like embryological development and physiological homeostasis, the evolutionary origin of living beings is teleological only in the indeterminate sense. Natural selection does not in any way direct evolution toward any particular kind of organism or toward any particular properties.

Teleology and Causality

Teleological explanations are fully compatible with causal explanations (Nagel 1961; Ayala 1970). It is possible, at least in principle, to give a causal account of the various physical and chemical processes in the development of an egg into a chicken, or of the physicochemical, neural, and muscular interactions involved in the functioning of the eye. (I use the "in principle" clause to imply that any component of the process can be elucidated as a causal process if it is investigated in sufficient detail and in depth; but not all steps in almost any developmental process have been so investigated, with the possible exception of the flatworm *Caenorhabditis elegans*. The development of *Drosophila* fruitflies has also become known in much detail, even if not yet completely.) It is also possible in principle to describe the causal processes by which one genetic variant becomes eventually established in a population. But these causal explanations do not make it unnecessary to provide teleological explanations where appropriate. Both teleological and causal explanations are called for in such cases.

One question biologists ask about features of organisms is "What for?" That is, "What is the function or role of a particular structure or process?" The answer to this question must be formulated teleologically. A causal account of the operation of the eye is satisfactory as far as it goes, but it does not tell all that is relevant about the eye, namely that it serves to see. Moreover, evolutionary biologists are interested in the question why one particular genetic alternative rather than others came to be established in a population. This question also calls for teleological explanations of the type: "Eyes came into existence because they serve to see, and seeing increases reproductive success of certain organisms in particular circumstances." In fact, eyes came about in several independent evolutionary lineages: cephalopods, arthropods, vertebrates.

There are two questions that must be addressed by a teleological account of evolutionary events. First, there is the question of how a genetic variant contributes to reproductive success; a teleological account states that an existing genetic constitution (say, the gene coding for a normal hemoglobin beta chain) enhances reproductive fitness better than alternative constitutions. Then there is the question of how the specific genetic constitution of an organism enhances its reproductive success; a teleological explanation states that a certain genetic constitution serves a particular function (for example, the molecular composition of hemoglobin has a role in oxygen transport).

Both questions call for teleological hypotheses that can be empirically tested. It sometimes happens, however, that information is available on one or the other question but not for both. In population genetics the fitness effects of alternative genetic constitutions can often be measured, while the mediating adaptive function responsible for the fitness differences may be difficult to identify. We know, for example, that in *Drosophila pseudoobscura* different inversion polymorphisms are favored by natural selection at different times of the year but we are largely ignorant of the physiological processes involved. In a historical account of evolutionary sequences the problem is occasionally reversed: the function served by an organ or structure may be easily identified, but it may be difficult to ascertain why the development of that feature enhanced reproductive success and thus was favored by natural selection. One example is the large brain of man, which makes possible culture and other important human attributes. We may advance hypotheses about the reproductive advantages of increased brain size in the evolution of man, but these hypotheses are difficult to test empirically.

Teleological explanations in evolutionary biology have great heuristic value (Binswanger 1990). They are also occasionally very facile, precisely because they may be difficult to test empirically. Every effort should be made to formulate teleological explanations in a fashion that makes them readily subject to empirical testing. When appropriate empirical tests cannot be formulated, evolutionary biologists should use teleological explanations only with the greatest restraint (see Williams 1966).

References

Ayala, F.J. [1968]: 'Biology as an autonomous science.' *Amer. Sci. 56*, pp 207-221.

Ayala, F.J. [1970]: 'Teleological explanations in evolutionary biology.' *Phil. of Sci. 37*, pp 1-15.

Ayala, F.J./T. Dobzhansky (Eds.)[1974]: *Studies in the Philosophy of Biology.* London: Macmillan, and Berkeley: University of California Press.

Beckner, M. [1959]: *The Biological Way of Thought.* New York: Columbia University Press.

Berg, E.S. [1926/1969]: *Nomogenesis or Evolution Determined by Law.* London. Boston: MIT Press (reissued).

Bergson, H. [1907]: *L'Évolution Créatrice.* (Engl. Transl. *Creative Evolution.* New York: 1911.)

Binswanger, H. [1990]: *The Biological Basis of Theological Concepts.* Los Angeles, California: The Ayn Rand Institute Press.

Darwin, C. [1859]: *On the Origin of Species by Means of Natural Selection.* London: Murray.

Darwin, C. [1958]: *The Autobiography of Charles Darwin, 1809-1882*, Nora Barlow (Ed.). London: Collins.

Dobzhansky, T. [1970]: *Genetics of the Evolutionary Process.* New York: Columbia University Press.

Ghiselin, M.T. [1974]: *The Economy of Nature and the Evolution of Sex.* Berkeley: University of California Press.

Grene, M. [1974]: *The Understanding of Nature.* Essays in the Philosophy of Biology. Boston: Reidel.

Grene, M./E. Mendelsohn (Eds.)[1976]: *Topics In The Philosophy Of Biology.* Dordrecht, Holland: D. Reidel.

Hull, D. [1974]: *Philosophy of Biological Science.* Englewood Cliffs, New Jersey: Prentice-Hall.

Hull, D. [1989]: *The Metaphysics of Evolution.* Albany, New York: State University of New York Press.

Koestler, A./J.R. Smythies [1969]: *Beyond Reductionism.* London: Hutchinson.

Lamarck, J.B. [1809/1963]: *Zoological Philosophy*, translated by H. Elliot. New York: Hafner (reprinted).

Mayr, E. [1965]: 'Cause and effect in biology.' In: D. Lerner (Ed.), *Cause and Effect.* New York: Free Press 1965, pp. 33-50.

Mayr, E. [1974]: 'Teleological and teleonomic, a new analysis.' In: R.S. Cohen/M.W. Wartofsky (Eds.), *Boston Studies in the Philosophy of Science, XIV.* Boston: Reidel 1974, pp 91-117.

Nagel, E. [1961]: *The Structure of Science.* New York: Harcourt, Brace, and World.

Osborn, H.F. [1934]: 'Aristogenesis, the creative principle in the origin of species.' *Amer. Nat. 68*, pp. 193-235.

Paley, W. [1802]: *Natural Theology*. London.
Pittendrigh, C.S. [1958]: 'Adaptation, natural selection, and behavior.' In: A. Roe/G.G. Simpson (Eds.), *Behavior and Evolution*. New Haven, Connecticut: Yale University Press 1958, pp. 390-416.
Polanyi, M. [1968]: 'Life's Irreducible Structure.' *Science 160*.
Rosenberg, A. [1985]: *The Structure of Biological Science*. New York: Cambridge University Press.
Schaffner, K.F. [1969]: 'The Watson-Crick Model and Reductionism.' *British J. Phil. Sci. 20*.
Simpson, G.G. [1964]: *This View of Life*. New York: Harcourt, Brace, and World.
Teilhard de Chardin, P. [1959]: *The Phenomenon of Man*. New York: Harper.
Webster's Third New International Dictionary. 1966.
Williams, G.C. [1966]: *Adaptation and Natural Selection*. Princeton, New Jersey: Princeton University Press.
Wimsatt, W.C. [1972]: 'Teleology and the logical structure of function statements.' *Stud. Hist. Phil. Sci. 3* , pp. 1-80.
Wimsatt, W.C. [1979]: 'Reduction and reductionism.' In: P.D. Asquith/H. Kyburg, Jr. (Eds.), *Current Research in Philosophy of Science*. East Lansing, Michigan: Philosophy of Science Association, pp. 352-377.

IV. Historical Views

Jane E. Ruby

The Origins of Scientific "Law"

1. Introduction

Apparently without reservation, for over three hundred years scientists have called the intelligible, measurable, predictable regularities they find in nature 'laws.' In contrast, in the thirteenth century Thomas Aquinas, followed by Francisco Suárez in the seventeenth, declared the use of 'law' for the inanimate metaphoric; in the fifteenth Lorenzo Valla found it ridiculous to speak of law with reference even to animals; and in the sixteenth Peitro Pomponazzi argued that because natural events are not responses to words or persuasion, it is meaningless to speak of 'law' in nature.[1] As late as 1682 the chemist Robert Boyle, again assuming that law is necessarily prescriptive, said that he could not 'conceive how a body devoid of understanding and sense, truly so called, can moderate and determinate its own motions, especially so as to make them conformable to laws, that it has no knowledge or apprehension of.' He explained that if he sometimes did not 'scruple to speak of the laws of motion and rest that God has established among things corporeal, and to call them ... as men are wont to do, the laws of nature,' it was 'for brevity's sake, or out of custom.'[2]

How, then, did the purely descriptive 'law' of modern science arise? In keeping with the assumption Boyle voices, the

1 Aquinas, *Summa theologiae*, 2-1. 91.2 ad 3; 90, 1; Suárez, *Tractatus de legibus ac Deo legislatore*, 1.1.2 (Madrid, 1971), I, 12; Valla, *Elegantiae*, 4. 48, *Opera omnia*, (2 vols.; Turin, 1962), I, 139; Pomponazzi, *De fato, de libero arbitrio et de praedestinatione* (Lucani, 1957), 132.

2 *A Free Inquiry into the Vulgarly Received Notion of Nature. Works* (6 vols.; London, 1772), V, 170.

nineteenth- and twenthieth-century scholars who have offered explanations of modern scientific 'law', have all found its origin in the metaphor of divine legislation, with the prescriptive connotations subsequently disappearing.[3] Edgar Zilsel and Joseph Needham, the two who haven written most extensively on the question, are in almost complete agreement as to how and when the use arose.[4] Zilsel says that by 'changing the divine laws of the Bible into geometric prescriptions', Kepler became 'the first naturalist to use the law-metaphor'; but he credits Descartes, who combined the biblical idea of God's legislating for nature with the 'basic idea of physical regularities and quantitative rules of operation ... [he] took over from the superior artisans of his period', with having 'created the modern concept of natural law.'[5] Earlier approximations – 'prescientific natural law' – Zilsel found only in Ovid, Seneca, the Christian orator Arnobius (c. 300 A.D.), Richard Hooker and Shakespeare.[6] Saying 'there is no doubt that the turning-point occurs between Copernicus (1473-1543) and Kepler (1571-1630)', Needham accepts Zilsel's suggestion that even had a genuinely scientific

3 Besides Zilsel and Needham (see note 4), Thomas Erskine Holland, *Elements of Jurisprudence* (Oxford, 1880), 16, cited in the article 'Law' in the eleventh edition of the *Encyclopaedia Britannica;* Franz Borkenau, *Der Übergang vom feudalen zum bürgerlichen Weltbild* (Paris, 1934), 21; A.R. Hall, *The Scientific Revolution, 1500-1800* (Boston, 1954), 171-2; Ray Bashkar, *Dictionary of the History of Science* (Princeton, 1981), 229. Holland gives the alternative of legislation 'by an abstraction called Nature.'

4 Zilsel, 'The Genesis of the Concept of Scientific Law,' *The Philosophical Review,* 51 (1942), 245-67. Needham, 'Human Laws and the Laws of Nature in China and the West,' *Journal of the History of Ideas,* 12 (1951), 3-32, 194-231, incorporated, with minor changes, into his *Science and Civilisation in China* (Cambridge, Eng., 1954-), II (1956), 518-83. Needham's Hobhouse Memoria Trust lecture of the same title, published in London in 1951, is included, with some minor changes, in his *The Grand Titration: Science and Society in East and West* (Toronto, 1969). Needham says that he learned of Zilsel's account only after having prepared the first draft of his (*JHI,* 12 [1951], 18, n. 46; *Science,* 2, 533, note a).

5 'The Genesis,' 265-69.

6 'The Genesis,' 260.

oulook developed earlier, its expression in terms of law could not have done so, for only where political absolutism prevails is God's law understood as sufficiently rigorous and comprehensive to provide a suitable designation for the regularity of physical phenomena.[7]

Although prima facie, the explanation of scientific 'law' as arising from the idea of divine legislation is highly plausible, it is for the most part mistaken. The idea of legislation by God or Nature does account for much of ancient use of 'law' for natural phenomena. However, the modern use emerged through different processes at different times in three distinct fields, in only one of which the idea of divine legislation had any part. In all three it appeared before 1543.

As A. C. Crombie pointed out a few years after Needham wrote, in the thirteenth century Roger Bacon already used 'law' *(lex)* in optics in a modern way.[8] On learning of Bacon's 'law,' Needham accommodated it to his and Zilsel's thesis, pronouncing it, together with the concept of nature it represented, an abortive anticipation of the sixteenth-century development.[9] However, not only is the premise of a thirteenth century absolutism remarkable but the assumption that Bacon's 'law' was derived from the idea of divine legislation is quite unfounded. So, too, is the supposition that it was stillborn.

In the fifteenth century, 'law' in a recognizably modern sense appeared in mathematics. Again, the idea of divine legislation played no part. Only in astronomy, where 'law' was used in a modern way at least by 1514 and where the most important inspiration lay in ancient uses for physical phenomena, can the theological idea be said to have contributed to the development.

---

7 Zilsel, 'The Genesis,' 277-79; Needham, *JHI* 12 (1951), 27, 30; *Science*, 2, 540, 542.

8 Crombie, 'The Significance of Medieval Discussions of Scientific Method for the Scientific Revolution,' *Critical Problems in the History of Science* (Madison, 1959), 89. Cf. his *Medieval and Modern Science* (2 vols; Garden City, New York, 1959), 24.

9 *The Grand Titration*, 310

Even so, in the first modern uses of 'law' in astronomy there is no vestige of the idea of divine legislation.

2. 'Law' in Optics

Roger Bacon (c. 1210 – c. 1292) used 'law' in what he called 'perspective' and we call 'geometric optics'. He wrote repeatedly of the 'law(s) [*lex*, *leges*] of reflection', the 'law(s) of refraction(s)', and the 'law(s) of multiplication of species', and of all of these collectively as 'laws of nature.'[10] To Bacon, multiplication of species, that is, the emanation of corporeal simulacra or images from an object, each 'multiplying' itself from point to point in a straight line through a medium, was, although known only through perspective, in the theory the principle of all natural actions.[11] Despite some resemblance to the idea of a field of force, the concept is scarcely modern. However, apart from

10 There are four fragments of Bacon's *Opus tertium* in print, the three of interest here edited by J. S. Brewer, *Opera quaedam hactenus inedita* (London, 1859), 3-310; F. A. Gasquet, *English Historical Review,* 12 (1897), 497-517; and Pierre Duhem, *Un fragment inédit de l'Opus tertium* (Quaracchi, 1909), repr. Frankfurt, 1964. These fragments are designated below as 'Brewer,' 'Gasquet,' and 'Duhem.' 'Law' is used of reflection, refraction, and multiplication of species in the *Opus majus* (2 vols. and suppl. vol.; Oxford, 1897), pars 4, dist. 2, cap. 2 (Vol. I, p. 113); dist. 3, cap. 1 (I, 122); dist. 4, cap. 2 (twice), 4, 5, 7, 8 (I, 130, 132, 136, 137-8, 142, 148); pars 5-1, dist. 6, cap. 2 (twice) (II, 36, 37-38); dist. 7, cap. 1 (II, 49); also in Brewer, 37, 114; Duhem, 78, 90 (twice), 147; Gasquet, 512, 513; in *De Erroribus medicorum,* in *Opera hactenus inedita,* (16 vols.; London, 1905-40), IX, 167; and in *De Speculis,* published with the *Perspectiva* (Frankfurt, 1614), 196 (twice). References to 'laws of nature' occur in the *Opus majus,* pars 4, dist. 2, cap. 2 (I, 117); Duhem, 78 (twice); Brewer, 113, 114, 118. Cf. in the *Opus majus,* pars 4, dist. 2, cap. 2, *leges rerum materialium et corporalium.*

11 Description, *Opus majus,* pars 4, dist. 2, cap. 1 (I, 111-112); pars. 5-1, dist. 6, cap. 2; dist. 7, cap. 1; dist. 8, cap. 2 (II, 37, 49, 57-8). Known only through perspective: Brewer, 37. Principle of all actions: *Opus majus,* pars. 4, dist. 4, caps. 1, 2, 3, 5, 7 (I, 127, 130, 134, 137-38, 142); Brewer, 37; Gasquet, 512.

Bacon's exception of the 'tortuous' passage of light through the optic nerve as 'not following the laws of nature' or 'following laws above nature', his use of 'law' in 'perspective' for regularities in nature is indistinguishable from ours.[12]

Crombie suggested that Bacon's 'law' expresses 'a program for mathematicising physics,' a shift from forms to laws begun by Robert Grosseteste (1175-1253).[13] The Platonizing Grosseteste had declared that 'through the power of geometry one can arrive at the causes of all natural effects' and that those causes 'must be expressed by means of lines, angles and figures.'[14] Although more eclectic, Bacon spoke repeatedly to the same effect.[15] The one time in his optical writings he recognized the validity of physical and metaphysical explanations, he went on to say that he was omitting them from his treatment and proceeding by means of geometric lines alone.[16]

Greek and Arabic writers on optics anticipated Grosseteste and Bacon in their geometric approach, and some of the Arabs, al-Kindi in particular, in the theoretical extension of the principles of perspective to all of nature.[17] What was novel was Bacon's use of 'law.' In the medieval Latin translations of Greek

---

12 See 'laws of nature' in note 10 above. In the absence of any other suggestion of animism Bacon's single reference to the 'laws which nature observes [*servat*] in the bodies of the world' (Duhem, 78) is almost certainly figurative.

13 See note 8 above.

14 *De Natura locorum. Die philosophischen Werke. Beiträge zur Geschichte der Philosophie und Theologie des Mittelalters* (henceforth *Beiträge),* ed. C. Baeumker, IX (Münster i. W., 1915), 65; *De lineis angulis et figuris, seu de fractionibus et reflexionibus radiorum, ibid.,* 59-60.

15 The first four chapters of Part IV of the *Opus majus* argue the importance of mathematics. Cf. also pars 4, dist. 4, cap. 8 (I, 143-4); Brewer, 36, 110; Gasquet, 512.

16 Gasquet, 513. Cf. *Opus majus,* pars 4, dist. 4, cap. 8 (I, 143).

17 David C. Lindberg, *Theories of Vision from al-Kindi to Kepler* (Chicago and London, 1976), 11ff and 119. Graziella Federici Vescovini, *Studi sulla prospettiva medievale* (Turin, 1965) (henceforth Vescovini), 40, 44-47. Lynn Thorndike, *History of Magic and Experimental Science* (8 vols; New York, 1923-58), I, 646.

and Arabic works dealing with perspective and in passages in Augustine, the only Latin thinker who wrote on vision prior to Grosseteste, the term does not occur.[18] Nor is there any mention of 'law' in the surviving Greek originals or in the Arabic.[19]

Bacon's 'law' is the more remarkable in that use of the term for physical phenomena was out of fashion in his day. In the twelfth century writers on nature such as William of Conches and Bernard Silvester had used 'law' freely.[20] In the metaphysical work by Hermann of Carinthia, *habitudo,* defined as our means of knowing '*qua lege* [literally, by what law] things occur,' is one of the five 'essences' that enable us to know what will happen when and where – the others being matter, form, space, and time; and in the poems of Alan of Lille a personified Nature lays down laws.[21] In the thirteenth century, however, reference to 'law' all but vanished from writings on nature.

Probably one reason was the displacement of the Roman and early Christian writings that had inspired twelfth-century uses by translations of Greek and Arabic works in which 'law' rarely occurs. Although twelfth-century translations used in the thir-

18 *L'optique de Claude Ptolémée, dans la version latine d'après l'arabe de l'émir Eugène de Sicile* (Louvain, 1956). A.A. Björnbo and S. Vogl, 'Alkindi, Tideus und Pseudo-Euklid, Drei optische Werke,' *Abhandlungen zur Geschichte der mathematischen Wissenschaften,* 26.3 (Leipzig and Berlin, 1912), 1-176. *Opticae thesaurus Alhazeni Arabis* (Basel, 1572), 1972. Augustine, *De Trinitate,* 11 *(Patrologia latina* [henceforth *PL*], 42: 983ff). *De Musica,* 6, (*PL* 32: 1174). I have not seen the translation of Euclid's *Optics* used in the thirteenth century, but given the absence of *lex* from other early translations, it is unlikely that the word occurs there.

19 Of Greek works only Euclid's *Optics,* the *Catoptrics* mistakenly ascribed to him, and Theon of Alexandria's recensions of the two works survive. Arabists tell me there is no word corresponding to 'law' in Arabic.

20 William, *Glossae super Platonem* (Paris, 1965), 214-16. Bernard, *De Mundi universitate* (Frankfurt a.M. [1876] 1964), 8, 9, 11, 16, 19, 35, 40, 44, 55; *Mathematicus* (Paris, 1895), 15, 16, 17, 23, 26, 19, 30, 31, 37. Cf. Therry of Chartres, *De Sex dierum operibus,* 45, ed. N. Häring, *Commentaries on Boethius by Thierry of Chartres and His School* (Toronto, 1971).

21 Hermann, *De Essentiis* (Comillas [Santander], 1946), 33. Alanus de Insulis, *De Planctu naturea* (*PL* 210: 448C, 452C-D); *Anticlaudianus* (491B).

teenth introduced an occasional *lex*, in the Greek originals a word for 'law,' *nomos*, or any other, seldom occurred, and in the Arabic never.[22] Another possible reason was the rejection of the idea that non-rational creation can share in 'law.' The *Summa* attributed to Alexander of Hales (d. 1245), compiled some time before 1250, recognized three kinds of natural law: for rational creation, for non-rational but sentient, and for non-rational and non-sentient.[23] However, in 1250-51 Bonaventura limited natural law to the sentient, defining it, in the words ascribed to the Roman jurist Ulpian, as what nature has taught all animals, namely cohabitation, procreation, and the rearing of the young; and some time after 1265 Aquinas, following Cicero's reading of law as a function of reason, limited natural law to the human realm and declared the participation of the rest of creation in eternal law to be only a metaphor *(per similitudinem)*. [24]

Whatever the explanation, none of the scientifically oriented thinkers of the thirteenth century whom Bacon cites or praises – Peter of Maricourt, Jordanus of Nemours, Campanus of Novara, or Grosseteste – spoke of 'law.' At the beginning of the centruy Alfred of Sarashel did so, and later John Pecham (1240-92) was to draw on Bacon in saying that in certain circumstances vision proceeds according to the 'law of spirits' *(spirituum)* rather than the 'law of diaphanousness.'[25] Otherwise Bacon was, to my knowledge, the only thirteenth-century writer on natural phenomena to use *lex*.

How did he come to do so? Besides Needham's explanation, one other has been offered. Matthias Schramm, explaining not the word 'law' but the phrase 'laws of nature,' found its

22 On the Greek, see p. 354 below; on the Arabic, note 19 above.

23 *Summa theologiae*, 3-2.1.3.4 (Quaracchi, 1924-), IV. 338-39.

24 Bonaventura, *In primum Sententiarum Petri Lombardi*, 4.33.1., *Opera omnia* (10 vols.; Quaracchi, 1882-1902) IV, 747-48. Ulpian (*Digesta*, 1.1.1). For Aquinas, see note 1 above. Cicero *De Republica* 3.22.

25 Alfred, *De Motu cordis*, 15, *Beiträge*, XXIII, 77. David Lindberg, *John Pecham and the Science of Optics* ...(Madison, 1970), 124. Cf. *Tractatus de perspectiva* (Bonaventura, N.Y. 1972), 52. On Pecham's indebtedness to Bacon, Lindberg, *John Pecham*, 10-1, 14-15, 26-27.

source in the 'law of universal nature' of Avicenna's *Sufficientia* echoed by Bacon in his discussion of vacuum.[26] However, the phrase comes not from Avicenna (980-1037), who wrote in the *Sufficientia* simply of 'universal nature' but from Gundisalvis's twelfth-century translation, and it cannot be the source of the 'laws of nature' of Bacon's opics.[27] As Schramm says, 'universal nature is not yet nature as a whole whose unity lies in the laws of nature, but ... is ... superimposed on the manifold of special natures', limiting their tendencies and preventing discontinuity.[28] Although Bacon used the concept of a providential universal nature elsewhere, in his writings on perspective there is no trace of it. There is no mention of individual natures, or even of 'the law of nature.' There are only 'laws' and 'laws of nature'. In the world of discourse of Bacon's optics, nature is indeed a whole whose unity lies in its laws.[29]

Besides, Bacon was familiar with many examples of the phrase 'laws of nature' free of the providential note. He cites the first chapter of the *De Caelo,* where Aristotle says, in a use of 'law' exceptional in his work, that we have taken the number three from nature as if it were one of its laws.[30] Bacon refers often to the moral epistles and essays of Seneca, in which there is repeated reference to the laws of nature; and he knew well

---

26 'Aristotelianism: Basis and Obstacle to Scientific Progress in the Middle Ages: Some Remarks on A.C. Crombie's *Augustine to Galileo.' History of Science,* 2 (1963), 108. Crombie adopted Schramm's explanation immediately ('The Relevance of the Middle Ages to the Scientific Movement...,' *Perspectives in Medieval History* [Chicago, 1963], 54-5) and reiterated it in the article on Bacon he co-authored with J.D. North in the *Dictionary of Scintific Biography,* I (New York, 1970), 380.

27 *Sufficentia,* 1.7. Gundisalvi's translation: Avicenna, *Opera* (Venice, 1508), 17^{V}.

28 Cf. note 26 above.

29 As Bacon's thought matured, he would seem to have cast out the Avicennian concept. In the *Opus majus* he rejects as foolish the argument that the law of universal nature precludes the existence of any vacuum (pars 4, dist. 4, cap. 9 [I, 151]).

30 Aristotle, *De Caelo,* I 1, 268a, 13-14. Roger Bacon, *Opus majus,* pars 6, cap. 12 (II, 197).

works of St. Basil and St. Ambrose, in which the picturesque example set in adjurations to steadfastness is that of fish, which, following the laws of nature, either stay in one place or move as a school.[31]

What calls for explanation is Bacon's use not of the phrase 'laws of nature' but of such phrases as 'the law of refraction'. There were, to be sure, ancient precedents for the use of 'law' for specific phenomena, and Bacon was familiar with many of them. Yet it seems unlikely that they account for the 'law' of the optics. For one thing, unless one includes the astrological 'law' indicating the governing influence of the planets at the times of their ascendencies, Bacon echoes not a single earlier use. For another, the syntax of his 'law', with its genitive of process or action, is all but unprecedented.[32]

There is ample evidence that the explanation lies not in the earlier use of *lex* for physical phenomena, but in that of *regula* (rule), together with the long association of the two terms in extra-scientific disciplines. Although thirteenth-century scientific writers almost never used *lex*, some used *regula*, Grosseteste among them. In most of his works, Grosseteste indicates the universality of the principles of multiplication of species and of perspective with an 'every' *(omnis)* or 'it is necessary' *(necesse est)*, or by simple statement.[33] However, in the *De Natura locorum*, he uses *regula* several times.[34] Bacon himself uses the word frequently, on occasion interchangeably with *lex*. For example, in the same paragraph he writes of 'the law of multiplica-

31 Seneca, *Ad Lucilium epistulas morales*, 101.5; *De Providentia*, 1.2; *Ad Helviam de consolatione*, 6.8. Cf. *De Ira*, 2.27.2. Basil, *Hexaemoron*, 7.3-4 and 8.4 *(Patrologia graeca*, 29.155; 158). Ambrose, *Hexaemeron*, 5.10 *(PL* 14:217).

32 The closest approximation I have found is Seneca's speaking of the 'law of the speed' *(lex celeritatis)* of the planets *(Naturales quaestiones*, 7.12.5).

33 E.g., *De luce seu inchoatione formarum, Beiträge*, 53-55; *De lineis angulis et figuris*, 60-62.

34 *Beiträge*, 65; 66, thrice; 67, twice.

tions' and 'the rules of perspective' and elsewhere of 'the rules of refraction'.[35]

Originally meaning straightedge or ruler, *regula* had acquired early in its Roman use a second meaning of 'rule' in the sense of guideline or standard.[36] Similarly, *lex* was used not only of government or jurisdiction but also of principles laid down by authorities or developed by custom for the practice of various disciplines.[37] The Greeks had used *nomia,* a cognate of *nomos,* in much the same way, and *astronomia,* indicating not, as the medieval West thought, 'law of the stars', but a set of rules for their study and interpretation, just as *oikonomia* meant a set of rules for household management.[38] Among the Romans one finds *lex* in this disciplinary sense as early as Cicero, for example, in the question, 'Who does not know that the first law of history is not to venture to say anything false?'[39] Although in the case of natural phenomena it is not always possible to distinguish the use of *lex* for formulated principles from that for regularities in nature itself, in a few instances the use for principles seems reasonably clear. Seneca the Elder (53 B.C. - 39 A.D.) speaks of those who 'reduced the course of the seasons to a fixed law' *(ad ceram legem redegerunt);* Seneca the Younger (ca. 4 B.C. - 65 A.D.) of comets as 'not yet grasped in fixed laws *(nondum teneri legibus certis)*';' and Manilius (1st century A.D.), in his astrological poem, the *Astronomica,* of 'following a fixed law *(certa ... lege)* in seeking information about individual signs.'[40] In the second century *lex* and *regula* are coupled in much the same sense in the discipline of oratory. In the *Institutes* Quintilian says that what is published as a model 'must be ... brought

35 Respectively, Gasquet, 512, and Duhem, 95.

36 *Oxford Latin Dictionary* (Oxford, 1983), 1602.

37 *Thesaurus linguae latinae* (Leipzig, 1900-), VII, 2.8. 1248.

38 Ch. Daremberg and Edm. Saglio (eds.), *Dictionnaire des antiquités grecques et romaines* (6 vols. in 10; Paris, 1877), I, 476.

39 *De Oratore,* 2.62.

40 Seneca the Elder, *Suasoriae,* 1.4; the Younger, *Naturales quaestiones,* 7.25.3; Manilius, 3.390.

into conformity with rules and standards *(ad legem ac regulam compositum esse oportere).*"[41]

Both terms were to evolve further in meaning. At least by the fifth century there had appeared the adverb *regulariter,* meaning, as 'regularly' means today, occurring steadily or uniformly, or at fixed intervals.[42] Evidently at some point, *regula* itself, stripped of prescriptive connotations, took on the meaning of regularity (cf. today's 'as a rule') it has in Grosseteste and Bacon.

In much the same way *lex* shifted in meaning, as principles originally understood as set down by authorities or evolving through custom came to be perceived as more or less certainly, more or less completely, inherent in the nature of things. It became, accordingly, either ambiguously precriptive-descriptive, or purely descriptive, free from all connotations of authoritative precept. One finds clues to the development in St. Augustine's treatment of meter in the fourth century and in John of Salisbury's of grammar in the twelfth. After saying that 'all the meters already celebrated by the poets have had each one its author and inventor to keep us from transgressing certain fixed laws *[quasdam certas leges]* they laid down,' Augustine observes that there are some meters 'the ancient authors did not institute as if not already whole and fixed in the nature of things, but found by reasoning and made known through designation.'[43] Similarly John, using *lex* together with *regula* repeatedly for grammatical principles, says that 'while grammar developed to some extent, and indeed mainly, as an invention of man, still it imitates nature, from which it partly derives its origin.'[44] Bacon himself stands at the end of a similar evolution in the treatment of ac-

41 12.10.50, trans. H.E. Butler (4 vols; London, 1920-2), IV, 79.

42 Macrobius, *Somnium Scipionis,* 1.20.12; Martianus Capella, *De Nuptiis philologiae et Mercurii* (Leipzig, 1866), 9.898; *Digest,* 15.3.3.2.

43 *De Musica,* 5.1 *(PL* 32:1147).

44 The statement: *Metalogicus,* 1.14 *(PL* 199:840D); cf. 1.17 (847C). Translation by Daniel D.McGarry (Berkeley, 1955), 39. The use of *lex*: 1.15, 18 (843B, 845A, 849B, 851A,D).

cents. Those that everyone uses, he says, unlike those that are rare, are a function not of poetic license but of natural law.[45]

In the thirteenth century *lex* appeared in logic in designation of what Aristotle had called the most indubitable of logical principles, the principle of contradiction: that of two contradictory propositions, one must be false.[46] Prior to the early 1240s, while *regula* had been used for other logical principles, the principle of contradiction had been simply stated without rubric; but beginning with Peter of Spain, Lambert of Auxerre, and William of Shyreswood, the phrase *lex contradictionis* (sometimes *lex contrariorum)* appeared in almost every passage on the principle, including Bacon's, written about 1250, years before he first wrote on multiplication of species and perspective.[47] Thus, in a discipline Bacon pronounced basic to all knowledge, *lex* had taken on the descriptive meaning of universality and predictability.[48] Besides, it had appeared in a phrase paralleling in its genitive the *lex fractionis, reflexionis, multiplicationis* of his optics.

It seems reasonable to postulate that when for some reason an impulse arose to label the hitherto unlabeled principle of contradiction, its distinctive certainty suggested, very likely subconsciously, a term long synonymous with *regula* but more

---

45 *Opus majus*, pars 3, cap. 10 (Suppl. vol., 113). Cicero wrote of natural accents (*Orator*, 18.58), and grammarians distinguished between accents by nature and by position (cf. e.g., Bede, *De Arte metrica*, in *Grammatici latini* [7 vols.; Hildesheim, 1858-80], VII, 229). Bede wrote also of accents by poetic license *(ibid.*, 230) and Priscian of 'the general law of accents' *(generalem accentuum legem) (ibid.*, III, 132).

46 *Metaphysica*, 4.4 (1005b).

47 Peter of Spain, *Summulae logicales*, 1. BL MS. Royal 7F vii, 16, 17. Lambert of Auxerre, BN MS. 13966, 11^{V}; William of Shyreswood, *Sitzungsberichte der Bayerischen Akademie der Wissenschaften, Philosophisch-historische Abteilung* (1937), X, 35. That the term was just finding its way into common use at the time of the three would seem borne out by the fact that in the diagrams accompanying their texts the word *lex* does not appear, while in almost all later texts it does. Bacon, *Summule dialectices*, *Opera hactenus inedita*, XV, 244.

48 Brewer, 104.

forceful, just as 'law' is for us more forceful than 'rule.' It is significant that while in the centuries following the thirteenth both *lex* and *regula* were used for other logical principles, *lex* was almost invariably used either alone or in the phrase *lex sive* or *seu natura* (law or nature), present already in Bacon, for the pre-eminently certain principle.[49] Very likely Bacon's optical *lex* was similarly an expression, albeit again a scarcely conscious one, of his conviction of the certainty as well as of the importance of the principles for which it was used.[50] In his writings on perspective, as on logic, there is a subtle distinction between *lex* and *regula*. In the paragraph in which he uses both terms, *lex* is for the more fundamental principle of multiplication; and while he uses *lex* far more often than *regula* to designate the general principles of both multiplication and perspective, for variants under specific conditions he never uses *lex*, but instead either *regula* or its naturalized Greek equivalent, *canon*. [51]

In sum Bacon's optical 'law' was not in the least metaphorical. From the start it meant regularity pure and simple. It was used in the context of an approach in optics that, apart from the exclusion of a single physical phenomenon from the realm of natural law, resembled that of modern science – at least until the emergence of the Uncertainty Principle – in regarding nature as a set of intelligible, measurable, predictable regularities.

Neither the term nor the approach disappeared. There is substantial evidence of continuity in optics from Bacon to Kepler

49 Among those using *lex* alone were John Buridan, *Commentum Magistri Johannis Dorp super textu summularum Magistri Johannis Buridani* (n.p., 1499) and Walter Burleigh, *Super artem veterem* (Venice, 1497); among those using *lex sive natura*, Iodocus Trutuetter, *Summule totius logice* (Erfurt, 1501), Nicolas de Orbellis, *Logica brevis* (Venice, 1489), Johannis Versoris, *Super septem tractatus Magistri Petri Hyspani* (Nuremberg, 1495), and the authors of the *Modernorum summulae logicales* (Mainz, 1489).

50 He argues the certainty of mathematics, of which perspective is a part (*Opus majus*, pars 4, dist. 1, cap. 3 [I, 105-7]).

51 *Canon: Opus majus*, pars 4, dist. 3, cap. 1 (I, 120); pars 4, dist. 4, cap. 1, 2 (I,127, 130); pars 5-3, dist. 2, cap. 2; dist. 3, cap. 4 (II, 150, 165); Brewer, 115, Duhem, 95; Gasquet, 512.

(and therefore to Descartes and Newton). Although Pecham's *Perspectiva* was printed around 1482 and Bacon's only in 1604 and although there were far fewer manuscripts of Bacon's than of Pecham's, Bacon's had many readers.[52] In the fourteenth century Domenico Chivasso (or Clivaxo), Biagio Pelacani of Parma, and the anonymous author of *De Visione stellarum* all cited it; Pierre Dubois recommended study of the part of the *Opus majus* on mathematics and Thémon Juif referred to the *De Multiplicatione specierum,* both writings in which optical 'law' *(lex)* appears.[53] In the fifteenth century a treatise on astronomical physics cited Bacon's *Perspectiva;* Regiomontanus pointed to Bacon as an authority on optics; Ghiberti paraphrased the *Perspectiva* in his *Third Commentary* without acknowledgement; Pico della Mirandola owned part of the work; and Leonardo, although relying heavily on Pecham, mentioned Bacon.[54] In the sixteenth century, John Dee and Thomas Digges attributed their knowledge of lenses and mirrors primarily to manuscripts of Bacon's works; Maurolico, Benedictine abbot of Messina, drew directly on Bacon's *Perspectiva* in his *Photismi,* completed in 1567 but published only in 1611; and so did Giambattista Porta in his *De Refractione* of 1593.[55] Regiomontanus, Ghiberti,

52 Pecham, *Perspectiva communis* (Milan); Bacon (Frankfurt).

53 Domenico, *Questiones super perspectivam,* Vescovini, 206, note 38. Biagio, *Questiones perspectivae, Rinascimento,* 2d. ser., 1 (1961), 214; *Le Questiones de anima,* 2.13, Accademia Toscana di Scienze e Lettere, *Studi,* 30 (Florence, 1974), 117. *De Visione,* Vescovini, 199. Pierre Dubois, *De Recuperatione Terrae Sanctae* (Paris, 1891),1.79 (xlix)). *L'Oeuvre astronomique de Thémon Juif, maître parisien du xive siècle* (Geneva and Paris, 1973), 304.

54 Treatise: Thorndike, *A History of Magic,* IV, 145-46. Regiomontanus, *Opera collectanea* (Osnabrück, 1972), 49. Gezenius ten Doesschate, *De Derde commentar van L. Ghiberti in Verband met de middeleeuwische Optick* (diss. Utrecht, 1940). Pearl Kibre, *The Library of Pico della Mirandola* (New York, 1936),100, 207. Eugenio Garin, 'Sull'ottica di Leonardo,' *Gionarle critica della filosofia italiana,* ser. 3,10 (1956), 280-81.

55 Dee and Digges: Francis R. Johnson, *Astronomical Thought in Renaissance England* (Baltimore 1937), 178. For Roger Bacon's works in Dee's library, *The Private Diary of John Dee* (London, 1842), Cata-

Maurolico, and Porta all used *lex* of optical phenomena.[56] So did Henry of Langenstein in the fourteenth century and Bernardino Telesio in the sixteenth.[57]

Although Zilsel says that in the *Ad Vitellionem Paralipomena (Supplement to Witelo)* (1604) Kepler 'never used the term "law"', in fact he wrote there of the 'geometric laws' *(leges Geometricas)* under which light falls, the 'laws of repercussions' *(repercussuum legibus)*, the 'laws of density' *(leges densitatis)*, 'optical laws' *(legibus opticis)*, and the 'law of nature' *(lex ... naturae)* that there has always been refraction.[58] How Kepler arrived at this use of 'law' is not clear. There is no 'law' in the treatise of Witelo (born c. 1220), which he was presumably 'supplementing.' He did not know Bacon's *Perspectiva.* He knew Porta's *Magia naturalis,* which contains a brief passage on perspective with no *lex,* but tried in vain to produce a copy of the *De Refractione;* and although his views in the *Paralipomena* resemble those of Maurolico's *Photismi,* not only was Kepler scrupulous in acknowledgements but there are no parallels either in reference or in syntax in the use of *lex* in the two.[59] The only earlier work using *lex* one can be certain Kepler knew is Pecham's *Perspectiva,* with its single sentence contrasting the

---

logue of Books. Maurolico, *Photismi de lumine et umbra ad perspectiva* ...(Naples), trans. Henry Crew (New York, 1940). He refers to Bacon on 73, 74, and 76. Porta, *De Refractione optices parte* (Naples, 1593), the references to Bacon on 24, 79.

56 Regiomontanus, 50; Ghiberti, 12 (94), 14 (98); Maurolico, 74, thrice; 76; 78. Porta, [3] and 151.

57 Henry, *Questiones perspective,* q. 9, f. 72v (Vescovini, 184, n. 62). Telesio, *De Rerum natura iuxta propria principia* (Naples, 1586), 145.

58 Zilsel, 266, n. 77. Kepler, *Ad Vitellionem paralipomena quibus astronomiae pars optica traditur, Gesammelte Werke* (henceforth *GW)* (Munich, 1937-), II, 20, 71 (cf. 73), 88, 103, 152, 187, 196, 197; 132; also *Dioptrice* (1611), *GW,* IV, 331, 334, 339. Zilsel (275) makes the same error in the case of Newton's *Lectiones opticae* (London, 1729), cf. 1.1 23 (29-30).

59 On the attempt to get Porta's work, *Paralipomena, GW,* II, 181. On Kepler's scrupulousness, J. Pelseneer, 'Gilbert, Galilée, Kepler, Harvey, et Descartes: leurs relations,' *Isis,* 17 (1932), 205, and Vasco Ronchi, *Histoire de la lumière* (Paris, 1956), 104-5.

'laws of spirits' with the 'law of diaphanousness'.[60] Yet, not only does the passage seem anything but an adequate explanation for Kepler's varied uses of *lex* but the phrase 'law of density' appears to echo, at a remove, Bacon's *secundum legem incessus a subtiliora in densiora.* [61]

Certainly optical 'law' was in the air in Kepler's day. Thomas Harriot wrote of 'the law of refraction' in 1601, and the term appears in a work Marcantionio de Dominis (1560-1626) published in 1611 but probably wrote years earlier.[62] In any case, given Kepler's acquaintance with Pecham's *Perspectiva,* his optical 'law' is indebted to the thirteenth century; and in it as in that of Descartes and Newton, there is not the slightest suggestion of divine legislation. Descartes might contend that even mathematical verities are the work of God; but the *lex* of optics reached modern times before him, independent of the idea of divine legislation.[63]

3. 'Law' in Astronomy and Mathematics

Regiomontanus (1436-76) used *lex* not only in optics but also in astronomy and mathematics. In mathematics he was, so far as I have discovered, the first ever to use the term, in astronomy the first, with one exception, since the twelfth century. A century earlier Henry of Langenstein had used the phrase *secundum legem agencium naturaliter* ('in accordance with the law of things

60 Cf. note 25 above.

61 *De Multiplicatione specierum,* 2.4 (II, 476); cf. 213 (468-9).

62 Murièl Rukeyser, *The Traces of Thomas Harriot* (New York, 1971), 64-5. De Dominis, *De Radiis visus, et lucis vitris, perspectivis et iride tractatus* (Venice, 1611), 13-4, 33. On the date of composition, Carl Boyer, 'Kepler's Explanation of the Rainbow,' *American Journal of Physics,* 8 (1940), 366.

63 Letter to Mersenne, 15 April, 1630. *Oeuvres* (12 vols.; Paris, 1897-1910), I, 145. Cf. letters of 6 May, 27 May, and 27 November, 1630 (I, 149, 151, 182).

acting naturally') in his work on comets; but his *lex*, unlike that of Regiomontanus, did not find imitators.[64]

Regiomontanus's *numeraria lege* and *lex superioris divisionis* are phrases for which we have seen classical parallels and which we might translate with 'rule' but would find no difficulty rendering as 'by arithmetic law' and 'law of higher division'.[65] The *lex* of his more frequent ablative phrases, *hac lege* and *qua lege* (literally, 'by this law' and 'by what law'), the latter a phrase we have met already in Hermann of Carinthia, is another matter. Neither in directions for procedure, as for example, those in his *Tabulae directionum et profectionum* for calculating the right ascension of planets, nor in descriptions, such as those in his *De Triangulis* of the way the perpendicular falls given this or that premise, does it translate well with 'law'.[66] Nor does it in the *hac lege* of one of the 'conclusions' prefaced to his *Epitome* of Ptolemy's *Almagest:* 'Finally, the earth is not moved around some other axis, for then the altitude of the pole would change for us [even] when we do not move on the earth. Since this appears the case to no one, it is established *hac lege* that the earth cannot be in motion.'[67]

The reason is that the principles of procedure and the regularities for which Regiomontanus uses the phrases do not have the importance of those for which we use 'law.' *Qua lege* and *hac lege* translate more satisfactorily as 'in what way' and 'in this way' or as 'how' or other words to the same effect or, where they are used prescriptively, sometimes with *lex* construed as

---

64 *Questio de cometa*, 1368 (Hubert Pruckner, *Studien zu den astrologischen Schriften des Heinrich von Langenstein* [Leipzig/Berlin, 1933], 112).

65 *De Proportione sermo noster* (Cataniae, 1554), 49, 59, and *Oratio habita Patavii in praelectione Alfragani, Opera collectanea*, 49. In the latter work Regiomontanus speaks also of a *lex compositionis* that accounts for the word *geometria*.

66 *Tabulae* (Tübingen, 1559), iv, 3, 6 *(alia lege)*, 6^{V}. *Regiomontanus on Triangles*, ed. and trans. Barnabas Hughes (Madison, 1967), 1.43 (88), 1.49 (98), 1.51 (100, twice), 4.30 (254).

67 *Epytoma ... in almagestum ptolomei* (Venice, 1496), I, *conclusio* 5.

'rule.'[68] That Regiomontanus himself was not necessarily thinking in terms of 'law' appears from his repeated use of *quo pacto* interchangeably with *qua lege.* [69] While *qua lege* had no idiomatic meanings in ancient times, *quo pacto* (literally 'by what covenant') was, along with *quodomodo,* the classical Roman way of saying 'in what way' or 'how' in either the prescriptive or the descriptive sense.[70] This is not to say that it is impossible to use 'how' for a principle of sufficient generality and importance to merit the designation 'law'. Maurolico, in fact, parallels *qua lege* with both *quo pacto* and Bacon's *lex fractionum.* ' It is true', he wrote, 'that how *[quo pacto]* vision occurs or under what law of refraction or of spirits *[aut sub qua lege fractionum vel spirituum]* has never been easy to determine'; and 'How', he asks, 'by what law would they [visual rays] be refracted *[quo pacto, qua lege frangentur]?'* [71]

In all their ambiguity Regiomontanus's phrases became part of the scientific vocabulary. They recur, after other echoes, in the writings of Copernicus, who acquired the *Tabulae directionum* and the *Epitome* not long after their publication in 1490 and 1496 respectively, and the *De Triangulis* in 1539.[72] Copernicus uses *hac lege* and the synonymous *ea lege* in the *De Revolutionibus* of 1543 in his 'demonstrations' or models of the movements of the moon, Mercury, and Venus.[73] The model of those

68 Hughes translates one of the phrases in the *De triangulis* as 'by which rule,' (89), one as 'by what law' (255), three as 'in which way'or 'which way' (99, 101 twice).

69 *Quo pacto, Tabulae,* 8^V, 14^V; *hoc pacto,* a. 10^V, 13^V; *Regiomontanus on Triangles,* 1.41 (86), 4.30 (254).The 'conclusion' of the *Epistome* contains the clauses, *si aerea cum terra hoc pacto moveri putaveris,* and *terram hac lege moveri non posse constat.*

70 On *quo pacto,* Ethan A. Andrew, *A Copious and Critical Latin-English Lexicon* (New York, 1851), 1066.

71 *Photismi,* 74, 76.

72 Paul Czartoryski, 'The Library of Copernicus,' *Science and History. Studia Copernicana,* 16 (Wrocław, 1978), 366-7. Aleksander Birkenmajer, *Etudes d'histoire des Sciences en Pologne* (Wrocław, 1972), 602.

73 *Nicolaus Copernicus Gesamtausgabe,* II (Hildesheim, 1984), 4.1 (262); 5.25 (407); 5.22 (401).

of Venus reads in part: '...let us take N as center and KN as radius and describe a small circle, on which [the center of] the orbital circle of Venus is borne and moves *ea lege, ut* [literally, 'by the law that'] whenever the Earth falls upon diameter ACB ... the center of the orbital circle ... will always be at least distance ... and when ...'[74] In the *Commentariolus*, the brief account of his views he completed at least by 1514 and circulated in manuscript, one finds the kindred phrases *lege diversa* and *contraria illius lege* (literally, 'by a different law' and 'by a law the contrary of that of ...'), both of which translate better otherwise than with 'law'.[75]

Copernicus, however, also uses 'law' in one of the principal ways we do. In the *Commentariolus* one reads of the 'fixed laws of motion of the planets in the firmament' *(statae sub firmamento motuum leges)*, of a 'law of opposition' [of celestial movements] not yet sufficiently understood' *(lege nondum satis deprehensa)*, and in the *De Revolutionibus* of the movements *(motus)* of the planets, which, either circular or composed of many circular movements, maintain their regularities 'by a fixed law' *(certa lege)* and of a 'fixed law kept' *(certa lege observata)* in the movements of Venus and Mercury, while the three higher planets move 'by certain other laws' *(aliis quibusdam legisbus)*. [76] Here Copernicus echoes ancient Roman uses derived from the Stoic concept of the law of nature.

Unlike the Romans, the Greeks very seldom used 'law' in a way resembling that of modern science. For some two centuries before Stoicism emerged at the end of the fourth century, the Sophisitic understanding of law *(nomos)* as conventional or even arbitrary and therefore antithetical to nature *(physis)* pre-

---

74 5.22 (401). Translation by C.G. Wallis, *Great Books of the Western World*, 16 (Chicago, 1939), 780.

75 *Nikolaus Kopernikus Erster Entwurf seines Weltsystems...*, ed. Fritz Rossman (Munich, 1948), 19, 26. Edward Rosen translates *lege diversa* as 'with a different arrangement' and *contraria illius lege* as 'by contrast' *(Three Copernican Treatises*, [3d ed. rev., New York, 1971], 75, 85).

76 *Commentariolus*, ed. L. Rossman, 16, 25, 19. *De Revolutionibus*, 1.4 (10), 6.1 (459).

vailed; and when the Stoics broke with the Sophistic tradition, not only was their cosmological natural law at once physical and ethical, descriptive and prescriptive, but they almost never associated it with any physical phenomenon smaller than the cosmos itself.[77] All in all, unless one count Aristotle's statement about the supposed natural law of three – with which, incidentally, Copernicus, like Bacon was familiar – there survive from before the Christian era but two instances of use at all comparable to that of modern science: Plato's assertion that the sick body goes against the laws of nature, and the second-century Peripatetic Kritolaos's declaration that because of 'those immutable statutes, the laws of nature', the Myrmidons could never have sprung full-grown from the ground.[78]

Of the Romans who used 'law' of physical phenomena, Copernicus mentions only Martianus Capella, speaking of him as one of those who held that Venus and Mercury circle not the earth but the sun.[79] However, Martianus does not use 'law' in the section on astronomy in his *Marriage of Philology and Mercury,* where he sets forth this view but only in allegorical passages elsewhere in the work. One of these contains the Hermetic image Copernicus follows in speaking of the sun as 'sitting on a royal throne and governing the wheeling family of stars', but there are other possible sources of the image.[80] Though Copernicus mentions only Martianus, he was familiar with uses of 'law' in works by other Romans. It is certain that

77 On the development of the antithesis see W.K.C. Guthrie, *A History of Greek Philosophy* (6 vols., Cambridge, Eng., 1962-81), II, ch. 4 and G.B. Kerferd, *The Sophistic Movement* (Cambridge, Eng., 1981), ch. 10. For the Stoics see Ioannes ab Arnim, *Stoicorum veterum fragmenta,* (3 vols.; Leipzig, 1903-1905) and vol. 4 (index) by M. Adler (Leipzig, 1924); Zeno, I, 42-43; Cleanthes, I, 122-23; Chrysippus, III, 76-80. Of the three Cleanthes comes closest to modern scientific usage.

78 For Aristotle , see note 30 above. Plato *Timaeus* 83E. Kritolaos, *De aeternitate mundi,* 59.

79 *De Revolutionibus,* 1 10 (19).

80 2. 186. One of the most likely is Ficino's *De Sole,* in *Opera omnia* (Turin, 1959), I-2, 968.

he knew Pliny's *Natural History,* which commends Hipparchus and others for having rescued men from superstitions about eclipses and the dying of the moon 'through discovering their law' *(lege deprehensa),* a phrase that seems echoed in Copernicus's *lege nondum satis deprehensa.* [81] He must have known also Manilius's *Astronomica,* where *lex* is used astronomically as well as astrologically.[82] He shares with Manilius the phrase *diversa lege;* and the 'dance of the planets' *(siderum chorea)* of the last sentence of the *Commentariolus* appears to echo Manilius's line, 'the planets execute their various dances under the law of nature' *(sidera exercent varias naturae lege choreas).* [83]

Copernicus did not, however, follow Manilius in asserting that God brought the whole universe under law.[84] He speaks of God not as legislator but as the wisest and most orderly of craftsmen.[85] His only uses of the governmental and legislative ideas are purely figurative. It is the sun that governs the family of stars, and 'the motion of the earth' that, in addition to exercising command, 'prescribes law' *(terrae mobilitas exercet imperia legesque praescripsit).* [86] While God can be conceived literally as legislating, the motion of the earth cannot.

Sixteenth-century astronomers all drew on the works of Regiomontanus, and after 1543 almost all if not all, whether or not they accepted the heliocentric hypothesis, on the *De Revolutionibus.* For the most part they used *lex,* as Regiomontanus and Copernicus did, without reference to God. There is nothing theological in Caspar Peucer's definition of astronomy in 1551 as 'the science of the motions of the heavenly bodies and of the fixed and perpetual successions and laws of the motions' *(motuumque certis et perpetuis vicibus ac legibus),* or in his speaking of 'the fixed and unchanging law' *(stataque et immota lege)*

81 Pliny, 2.9 (54), *Commentariolus,* 14. Copernicus owned the *Natural History* (Czartoryski, 372).

82 E.g., 1.201, 479; 3.411; 4.327, 330.

83 Manilius, 3.468 and 1.670-71. *Commentariolus,* 19 and 28.

84 Manilius, 2.475

85 *De Revolutionibus, Praefatio* (4); 1.10 (21).

86 1.10 (21); 6.1 (453).

by which the celestial orbs sustain their course.[87] Nor is there any note of divinity in Giovanni Antonio Magini's (1551-1617) *oppositionis lege, hac lege,* and *qua lege,* or Michael Maestlin's (1550-1631) *hac motus lege* and *his legibus,* or of Konrad Dasypodius's (1532-1600) assertion that comets 'drift without laws' *(sine legibus feruntur).* [88]

Among major sixteenth-century astronomers only Tycho Brahe linked 'law' with God. However, when, in his *De Disciplinis mathematicis* of 1547 he wrote that 'the wondrous and perpetual laws of the celestial motions, so diverse and yet so harmonious, prove the existence of God', he would appear already to have known the *De Revolutionibus.* [89] In the *De nova Stella* of 1573 he not only spoke of the Copernican hypothesis but said that the new star was not accommodated by either the Alphonsine or Copernican tables.[90]

The only writers on astronomical phenomena who used 'law' of the skies in the Copernican way, yet presumably did not know Copernicus's writings, were Maurolico and Philip Melanchthon. Maurolico wrote in his *Cosmographia,* published in the same year as the *De Revolutionibus,* of the planets being 'in motion by law' *(in motu lege);* Melanchthon in 1531, in his strongly anti-Epicurean dedication of Sacrobosco's *Sphere,* of the 'laws *[leges]* of the great orbs and stars,' and in 1537 of man's knowledge of the 'laws of motion' *(motuum leges),* as proof

87 *Elementa doctrinae de circulis coelestibus et primo motu* (Wittenberg, 1551), A, E2^{V}. Cf. E9^{V}, H4^{V}, and Ta.

88 Magini, *lex oppositionis: Novae coelestium orbium theoricae* (Venice, 1589), 4; *qua lege* and *hac lege:* 4^{V}, 7, 8, 55^{V}, 87^{V}, 88; also *Ephemerides coelestium motuum* (Venice, 1582), 4, 50^{V}. Maestlin, *Epitome astronomiae,* written in 1588 (Tübingen, 162), 43: *De Dimensionibus orbium et sphaerarum coelestium,* in Kepler, *GW,* I, 141.32; 134.15 (trans. Anthony Grafton, *Proceedings of the American Philosophical Society,* 117 [1973], 532-58). Dasypodius, *Brevis doctrina de cometis, et cometarum effectibus* (Strasbourg, 1578), Ciii (3).

89 *Opera omnia* (15 vols.; Copenhagen, 1913-39), I, 152.

90 *Opera omnia,* I, 37.

of his immorality.[91] While Maurolius's *lex* may well be explicable through his acquaintance with both Bacon's and Roman uses of the term, it would seem remarkable for Melanchthon to have arrived at precisely Copernicus's phrase *motuum leges* without knowing, directly or indirectly, the *Commentariolus.* The phrase does not occur in Roman writings, nor in the Bible, where God legislates for nature only by placing restraints on the rain, the abyss, and the sea.[92] One must wonder whether someone in Melanchthon's circle in Wittenberg knew the *Commentariolus.* Some think it was while Rheticus, Copernicus's young champion, was teaching there that he learned of Copernicus's views.[93]

For formulated perceptions of regularities in nature Copernicus did not use 'law' but only 'hypothesis'. It was Rheticus who revived the use we have seen in the two Senecas and Manilius. He revived also the use for collective procedural principles found in the Greek *astronomia.* In his *Prima narratio,* the elegant account of Copernicus's views he published in 1540, he wrote that for the 'observations of all ages to harmonize in a consistent law' *(consentienti sibi invicem lege responderent),* his 'teacher' (that is, Copernicus) had to distinguish three different motions of the earth, and that such questions as whether the

91 Maurolicus, *Cosmographia* (Venice), 22. (He cites Manilius, 8^{V}.) Melanchthon, *Liber Iohannis de Sacro Busto, de Sphera* (Wittenberg), A iii; *Epistula,* published with *Rudimenta astronomica Alfragani* (Nuremberg), a 3.

92 Job 28:26; Prov. 8:27, 29.

93 Rosen, who says that it was first through Rheticus's *Prima narratio* of 1540 (see below, n. 94) that Melanchthon encountered Copernicus's theory, thinks Rheticus probably learned of it first in Nuremberg *(Three Copernican Treatises),* 394, 393. Fred Hoyle says, without documentation, that Rheticus heard a synopsis of the theory in Wittenberg *(Nicolaus Copernicus, an Essay on His Life and Work* [London, 1973], 31). Karl H. Burmeister, Rheticus's biographer, supports Rudolf Wolf's suggestion *(Geschichte der Astronomie* [Munich, 1877], 237) that Copernicus's theory was discussed first in the circle of Kaspar Cruciger and Erasmsus Reinhold in Wittenberg *(Georg Joachim Rheticus,* [2 vols.; Wiesbaden, 1967-81], I, 56).

sun moves are to be decided 'not by plausible opinions but by mathematical laws' *(non ex plausibilibus opinionibus, sed legibus mathematicis).* [94] He also distinguished 'law' from 'hypothesis' as we do, saying that Copernicus, 'having before his eyes the observations of all the ages together with his own, recalling ancient hypotheses ... and posing new ones, finally found it possible to write down the laws of astronomy' *(leges astronomiae demum conscribere).* [95]

4. Later Developments

By 1540, then, except for the association of scientific laws with the names of their discoverers, which, to my knowledge, occurs first with Ramus in 1567, all the familiar modern scientific uses of 'law' were in place.[96] They did not yet, to be sure, have their eventual currency. In the development of that, it would probably be agreed, Newton's *lex* was especially influential. Except for the laws of nature of Descartes's *Principia philosophiae*, the laws of motion of Newton's *Philosophiae naturalis principia mathematica*, as principles of mechanics explaining celestial and terrestrial phenomena alike, had a generality and a force unmatched in the uses so far dealt with.

It is primarily in prompting Newton's use that Descartes's is important to the story of scientific 'law'.[97] One might wonder whether Newton's phrase was not suggested by the association

94 *Georgii Joachimi Rhetici Narratio prima,* éd. critique, *Studia Copernicana* 20 (Wrocław, 1982), 4.28 (47); 8.74 (56). Roger Bacon once wrote of the 'laws of this science' *(leges istius scientie)* (Brewer, 37), but given the single occurrence of the phrase, one must wonder whether he was not thinking in terms of regularities in nature covered by the science rather than of formulated principles.

95 *Studia Copernicana* 20 (1982), 13.33-34 (69).

96 *'Leges Ptolomaei ... leges Euclidis.' Prooemium mathematicum* (Paris), 339, 401.

97 On the relationship of the two *Principia*'s, see I.B. Cohen, '"*Quantum in se est*"; Newton's Concept of Inertia in Relation to Descartes and Lucretius,' *Notes and Records of the Royal Society of London*, 19 (1964), 135,

of *lex* with motion by astronomers – Copernicus, Peucer, Brahe, Maestlin, Kepler.[98] Evidently, though, it was not. Newton's knowledge of astronomy seems to have come entirely from works by Wing, Streete, Gassendi, and Nicholas Mercator, in none of which the association occurs.[99] Nor would Newton have found it in the *De Motionibus naturalibus a gravitate pendentibus* of Borelli, whom he recognized as a precursor.[100]

Given Descartes's method and the place of God in his thought, it is probable that he re-invented the use of the term 'law' *(lex)* through the idea of divine legislation. He may also have re-created the concept of law by combining with the biblical idea 'that of physical regularities and quantitative rules of operation' inspired by contemporary artisans. However, just as the extensive use of 'law' prior to Descartes depended on something other than the Bible, so the earlier presence of the idea of

---

and Brian Ellis, 'The Origin and Nature of Newton's Laws of Motion,' *Beyond the Edge of Certainty,* ed. Robert G. Colodny (Englewood Cliffs, N.J., 1965), 29-68. Arguing that Newton's laws were primarily conceptual, Ellis writes (40), '... they were the delayed products of Descartes' wondering what sorts of laws would obtain in a world presided over by an immutable God.'

98 Kepler, *Epitome astronomiae Copernicanae, GW,* IV, 399.

99 Vincent Wing, *Astronomia instaurata...*(London, 1656); Thomas Streete, *Astronomia Caroline: a New Theory of the Celestial Motions* (London, 1661); Nicholas Mercator, *Institutionum astronomicarum libri duo, de motu astrorum communi et proprio secundum hypotheses veterum et recentiorum praecipuas* (London, 1676). The reference to ancients and moderns in Mercator's title recurs in the first sentence of Newton's *Principia.* On Newton's reading, I.B. Cohen, 'Newton,' *Dictionary of Scientific Beography,* X, 43; D.T. Whiteside, 'Isaac Newton: the Birth of a Mathematician,' *Notes and Records of the Royal Society of London,* 19 (1964), 58; Direction to Richard Bentley (c. 1691), in *Correspondence of Sir Isaac Newton and Prof. Cotes* (London, 1850), 274.

100 Alphonse Borelli (Regio Iulio, 1670). Walter Charleton, whom Newton also read and who speaks of 'Motion, its Causes, Kind, and Universal Laws' *(Leges),* draws on Descartes *(Physiologia Epicuro-Gassendo-Charletoniana* [London, 1654], 435). On Newton's use of Charleton, see Richard S. Westfall, 'The Foundations of Newton's Philosophy of Nature,' *The British Journal for the History of Science,* 1 (1962), 172, n. 5.

regularities and quantitative rules depended, at least primarily, not on contemporary example but on the example of the Greek mathematical school as counterposed to the Aristotelian.

At the beginning of the *Principia* Newton says, 'Since ... the moderns, having rejected [more literally, 'dismissed'] substantial forms and occult qualities, have undertaken to bring back the phenomena of nature to mathematical laws *[et recentiores, missis formis substantialibus et qualitatibus occultis, Phaenomena Naturae ad leges Mathematicas revocare aggressi sunt].* I have in this treatise cultivated mathematics so far as it relates to philosophy.'[101] In all probability Rheticus, who can be considered to speak for Copernicus, had in mind the same contrast in counterposing mathematical laws to 'plausible opinions.' The *De Revolutionibus* was, for all its novel hypotheses, in a very true sense a revision of the *Almagest,* at the beginning of which Ptolemy distinguishes mathematical science from physical. While the physical, he says, is expounded in terms of conjecture, the mathematical, properly pursued, gives 'certain and trustworthy knowledge with demonstrations both arithmetic and geometric resulting from indisputable procedures.'[102]

With Roger Bacon the case is less clear. Although in theory he held that multiplication of species accounts for all natural phenomena, in practice his mathematical approach was limited to optics, and even there, as we have seen, did not cover every phenomenon; and although he pronounced Aristotle's science incomplete, he did not reject it.[103] Yet, while Newton's moderns 'dismiss' substantial forms and occult qualities and undertake to bring back the phenomena of nature to mathematical laws, Bacon 'omits' physical and metaphysical explanations and uses geometric lines alone. Flawed though it was and limited in application, the concept of law, inspired by the mathematical tra-

101 (London, 1686), Brussels, 1965; trans. A. Motte, rev. by F. Cajori (Berkeley, 1960).

102 Trans. R. Catesby Taliaferro, *Great Books of the Western World,* 12 (Chicago, 1938), 5.

103 Cf. note 16 above.

dition and expressed in the term 'law', was present in the thirteenth century. The roots of the seventeenth-century scientific revolution run deep.[104]

104 For a philosophical discussion of the concept see Charles S. Peirce, 'The Laws of Nature and Hume's Argument Against Miracle,' *Values in a Universe of Chance. Selected Writings of Charles S. Peirce*, ed. Philip P. Wiener (Stanford, 1958), 289-321.

I want to express my appreciation to Paul Oskar Kristeller for his valuable suggestions and to F. Edward Cranz and I. Bernard Cohen for theirs.

Friedrich Steinle[1]

The amalgamation of a concept – Laws of nature in the new sciences

I. Views on Laws

According to a widespread view, one of the main characteristics of the scientific revolution of the 17th century is the switch from a search for causes to a search for laws of phenomena. An impressive number of what we now see as classical examples for laws of nature were established in this very period: Kepler's astronomical laws, Galileo's laws of free fall and of motion of pendulums, the law of inertial motion, the laws of impact, the sine-law of optical refraction, Boyle's law of the compression of air, Hooke's law of the spring of elastic bodies, and Newton's law of gravitation. There is, moreover, Galileo's famous statement that he is engaged in investigating *properties* ("passioni") of the accelerated motion of heavy bodies, not its *cause* (1638, 202)[2]. Taking a modern understanding of laws of nature as being mathematically formulated rules, which describe the regular behaviour of things, and thus enable predictions, we may count Galileo's statement as a supporting instance of the above mentioned view.

From an historical perspective, however, things are more complicated. A 17th century reader would most probably not agree with the former view, if she or he understood it at all.

1 Michael Hagner, Dominik Perler, Claudia Sanides, and Matthias Schramm gave valuable comments on earlier versions of this paper. David Hyder helped me in preparing the English text. I wish to thank them all.

2 In quoting 17th century authors, I refer to later editions of their works as noted in the respective entries of the list of references.

There is, to be sure, a process of fundamental change in what one is willing to accept as causes of phenomena. This process forms one of the main driving forces of the formation of the new sciences. But, in the course of this process, the concept of causes does not lose its importance, as the mentioned view might imply. The search for causes is one of the main goals of even the new sciences, although causes come to be, of course, understood in various new ways. I shall discuss a few prominent examples that illustrate this point. A more problematic issue, however, concerns the concept of laws. Many historians and philosophers of science use a rather more modern concept of laws of nature even when analyzing a period in which it first gets formed.[3] The reader – at least one who is not familiar with the sources – gets the impression that, even in the early stages of the new sciences, such a concept of laws of nature is well established and in common use.[4] But this is a misleading impression. Many of the above-mentioned "classical" examples of laws of nature are not called "laws" at the time they are established; this holds for free fall and optical refraction as well as for the compression of air. Some promoters of the new science do not have a concept of laws at all. Even if they have one, it is sometimes – as in the case of Galileo – not used in the course of scientific inquiry; often it is considerably different from what a modern reader has in mind. I will concentrate on these latter points in this paper.

There are various, and very different, approaches to forming and explicating a concept of laws in the 17th century. The questions of when and where "laws" become important, of the context in and the background against which the concept is first used, and, lastly, of what are the various factors involved in its formation and development, have been studied from sever-

3 This holds for classical studies as well as for more recent ones. To have only a few examples, one may look at Sabra [1967], Drake [1978 and 1980], Cohen [1980], Maull [1980], Fischer [1983], Damerow et al. [1992], Garber [1992] or Schuster [1993].

4 Drake goes even so far as to make the interlocutors of one of his fictive early 17th century dialogues use the term "law"; see his [1983:143].

al different points of view. Zilsel (1942) was the first to collect a lot of relevant material concerning the use of "laws of nature" and related concepts from antiquity to the 17th century. He attempts, moreover, to give a sociological explanation of why the concept became so prominent in the course of the 17th century. Schimank (1949) discusses some more philosophical aspects and presents source material extending from antiquity to the 19th century. Needham (1956, ch. 18) focusses on the differences between European and Chinese thought. Oakley (1961a, 1961b) gives a thorough study of the concept of "natural laws" from the middle ages to the 17th century. Posing serious objections against Zilsel's sociological explanation, he argues for the voluntaristic theological tradition as being one of the main sources of the rise of the concept. His argument is widened and elaborated by various studies: McGuire (1972, concentrating on Boyle), Milton (1981), Klaaren (1985) and Oakley himself (1984). The latter two studies concentrate again on Boyle. The variety of different traditions present in the middle ages is treated by Schramm (1962, 1981) and by Ruby (1986) [Reprinted in this volume, pp. 289-315] Loeck (1986) gives a thorough philosophical analysis of Descartes' concept.

Most of these accounts deal, however, more or less with *one* concept of laws of nature which seems to come in common use in the 17th century.[5] Galileo, Descartes, Boyle and Newton are often mentioned in the same breath. As I shall argue, however, there is no such common concept. There are, to be sure, important common aspects involved in all 17th century approaches. But the differences among these seem to me to be larger than previously recognized. They are most closely connected with the different approaches to a new science in general. Among the few first propagators of a concept of law there are, moreover, important differences regarding their intellectual and religious traditions, the theological authority they have to deal with, the patronage system, or the various needs of establishing their new form of natural philosophy. In order to give a more comprehen-

5 The most significant exception is Loeck (1986).

sive historical account, one would have to include, moreover, the wider usage of concepts of laws among lesser known figures, and even in everyday life.

Clearly I cannot undertake such a project in the course of this paper. What I shall do instead, is to pick out some isolated points of such a history, thereby making obvious the need for further research. I shall analyze only a few studies of a few authors in only a few respects. In dealing with Galileo, Descartes, Bacon, Boyle and Newton[6], I shall first give an account of their respective use of the term "law". My further analysis is guided by questions concerning the very content of what is adressed by their terminology. I shall deal with the role ascribed to laws within the whole project of science, with the relation between laws and causes, with the ways in which knowledge of laws is thought to be achieved, and with the question of whether and how considerations on instrumentalistic versus realistic views – to use our modern terms – come into play.

Concentrating thus on rather more philosophical aspects, I hope that even such a narrowly focussed analysis may reveal some essential similarities, as well as differences, between the various concepts of the 17th century, and between those concepts and modern ones. Moreover, it may possibly point to characteristic philosophical approaches, in the context of which certain problems inherent to modern concepts of law occur.

Another restriction concerns the material used. Due to the limited space I shall concentrate on material published by the authors themselves. Unpublished manuscripts and letters are used only if they provide background information essential for the understanding of the published writings. As is now widely accepted in history of science, published papers and private notes are sources of different kind and have to be evaluated differently. Published statements, received by the author's contemporaries, are adressed with a certain aim to a particular readership. The reason why an author wants them to be recognized

6 In the last section, I shall discuss the reasons for selecting these authors, see p. 354f

as his own ones may not necessarily be that he really identifies himself with them. This becomes important particularly in respect to the question of *how* knowledge of laws is thought to be obtained. My study is to be understood as giving an account of how the respective authors wish to present themselves rather than of how they actually proceed. The latter question would require another study.

II. The Inexorability of Nature

In Galileo's published writings the term "law" in the sense of law of nature occurs hardly at all. It is particularly remarkable that his most famous achievements – often brought forward as evidence for the switch from causes to laws – viz. the "laws" of free fall, of inertia, and of pendulous motion, are never termed "laws" by Galileo himself. He instead speaks of "some properties of accelerated motion" (1638, 202; cf. letter of 7 Jan. 1639; EN XVIII, 13), of the "definition of uniformly accelerated motion" (1638, 205), of the "proportion" or of the "rule" of the acceleration of falling bodies (ibid, 212 resp. 274), or the like. "Laws of nature" are only mentioned in a rather general context, namely in the course of a discussion of the mutual relation between the knowledge we may obtain by studying the Holy Scripture, and that obtained by studying "the book of nature". Galileo's aim is to separate clearly these two ways of achieving knowledge, for they have different goals. What links them is that both the Holy Scripture and the book of nature proceed from the Divine Word, the former as an inspiration of the Holy Spirit, the latter as a consequence of God's giving certain laws to nature. While the wording of the Holy Scripture, however, is often accommodated to the limited human abilities of understanding, nature itself, being "inexorable and immutable", does not care at all about human understanding (1613, 282-3 or 1615,

316.)[7]. Nature "never transgresses the bounds of the laws imposed to it" (ibid.), being a "most careful executor of the orders of God" (1613, 282 or 1615, 316). Nature's independence from human reasoning is again highlighted by Galileo some years later, when – in a rather polemical context – he emphasizes that nature "laughs about the constitutions and decrees of princes, emperors and monarchs, in virtue of which it would not alter even an iota of its laws and statutes" (1624, 538). Again the essential point seems to be that these laws are given by the one omnipotent regent of the world, and that nature obeys them in a manner so strict and immutable, that it defies comparison with human obedience of juridical law.

This point, as well as Galileo's very formulation,[8] shows a deep embeddedness of his considerations of laws of nature in a theological context. It makes clear a possible reason for his apparent indifference to extending these laws into the realm of particular scientific inquiry. The main point in separating the two mentioned ways of gaining knowledge is, of course, his interest to deal with particular scientific questions free from any theological considerations. Importing the terminology of "laws", once embedded in a theological context, into particular scientific reasoning, would probably counter this interest, holding the theological commitments still present where he thinks they should not be.

Thus the direct textual base in Galileo seems to be too narrow to permit a definite explication of his ideas on laws of nature. Given our frequent use of the term "Galileo's law of free fall", and the common view of this "law" as illustrating paradigmatically the "laws" of the new sciences, I shall, for a moment, go beyond the immediate textual evidence. I shall ask how we could

---

7 Both letters of which these passages are taken of were made public by Galileo. The latter one has even been published in print [in 1636], together with a latin translation; see Drake [1980:80].

8 Galileo refers here to passages of the Old Testament: Job 28, 26 and 38, 8-11, or Psalm 104, 9, for example. For further relevant passages, see Zilsel [1942:248].

further explicate Galileo's ideas, given his numerous considerations on what makes up the new sciences. To be sure, I shall not give this as Galileo's own view, since it is not clear whether he would endorse any such explication of his thoughts. My few observations can only give a strong, and in some way speculative, interpretation.

First, since nature follows its laws immutably, one should under equal circumstances expect equal phenomena. Thus laws entail regularities. Second, if one includes Galileo's view of the book of nature being written in mathematical language (1623, 232, for example), i.e. of the inner constitution of nature being organized mathematically, this should hold for the laws of nature as well: laws should have a mathematical form. One may therefore, and this is my third point, surmise that human beings should be able, at least in principle, to achieve knowledge of laws of nature, just as they can have knowlege of some of God's infinitely many mathematical truths (1632, 128-9, for example). The way in which we might achieve such knowledge is not given explicitly by Galileo.

Pursuing this interpretation one step further, one may ask whether his famous "law of free fall" agrees with what he says about laws. There seems to be no easy answer. Galileo gives his propositions in a mathematical form, and he is, of course, convinced that he has knowledge of the correct ones. But there is a problematic point. Galileo is well aware that falling bodies do not actually behave as his propositions on free fall say. These propositions are said to hold only under idealized conditions, not realizable at all in his time. Should one take as a law *of nature* a rule, which is said to hold only under conditions, which possibly never and nowhere do occur in nature? If one is to give a positive answer, the relation between laws of nature and actually observed phenomena appears to be quite complicated and needs to be explicated in detail, all the more as nature is said to obey its laws with the above-mentioned inexorability and immutability. I do not find such an explication in Galileo. What may become clear in these considerations, is that, even in Galileo's rudimentary idea of laws, there is a certain tension between what he calls the inexorability of nature in obeying its laws, the mathematical certainty aimed at in his general approach to science, and the actual variety of phenomena which, although they may occur regularly, do not allow of a direct mathematical treatment.

III. Metaphysical Certainty

The term "law" is found in Descartes much more frequently than in Galileo, and it is used essentially in discussions of physical questions. It refers exclusively to the rather small set of Descartes' three famous "laws of nature". Neither Galileo's propositions on free fall, nor Kepler's astronomical propositions, nor the sine-ratio occurring in the refraction of light are called "laws" by Descartes.[9] Sometimes Descartes uses the term "rule" synonymously with "law"[10]; but by "rule" he generally envisages a significantly wider scope than by "law".

Descartes introduces his three "laws of nature" in the second book of his *Principia Philosophiae* (1644), entitled "The Principles of the Material Things". It starts with a proof of the existence of matter (II, §1-3)[11], and continues with discussions of space and matter (§4-23), and of an appropriate definition of motion (§24-35). Then Descartes turns to considering the "causes of motion". The "universal and primary cause" is God Himself who has created matter in motion and rest and who preserves in the world a constant total quantity of motion, taken as the numerical sum of the quantities of motion of all bod-

9 Schimank [1949:142] claims quite the contrary without, however, giving any evidence. Loeck [1986] fails to consider this restriction. This is a significant flaw in an otherwise brilliant conceptual analysis. Regarding refraction, one may be in doubt about one passage of the *Dioptrique* which I shall discuss below (see note 19).

10 There are many such passages in the *Principia* and even more in his posthumously published *Le Monde*; see, e.g., [1664: 37-43].

11 In referring to the *Principia*, I give the number of the book and of the paragraph. If not noted otherwise, the English translations are taken from CSM.

ies (§21). As the "secondary and particular causes"[12], regarded as being the causes for the change of the state of motion of particular bodies, Descartes introduces the three "laws of nature". The first law states that "each and every thing, insofar as it is simple and undivided, always remains, insofar as it can, in the same state, nor is it ever changed except through external causes" (§37). This quite general conservation law is specified in the case of local motion by the second law, which states that "each and every part of matter, regarded by itself, never tends to continue moving in any curved lines, but only along straight lines." (§39)[13]. The third law, finally, gives a general rule of which motions will be the result of an impact of two bodies under certain conditions (§40).

I shall not analyse the content and various changes of these laws in detail here[14], but instead concentrate on the role they play qua "laws" within the whole of Descartes' project of science. Loeck (1986) has worked out in detail the centrality of laws of motion within Descartes' general approach. I agree with this analysis[15] and shall now discuss some aspects which are not in the centre of her study but which are of high importance for the further development of the concept of laws. Laws do not define what kind of entities act in the world – instead, those entities (matter) given, laws govern their action and mutual interaction. Thus there is a clear dichotomy between things ("res") and laws, the latter being the rules imposed on things by God. Another dichotomy becomes clear when Descartes considers the

---

12 Although Descartes uses the term "secondary causes", I do not find the term "secondary laws", contrary to what Funkenstein [1980: 186] suggests. Descartes does not regard the "first cause" as a "law", he rather names some more general "principles"; see below, p.327

13 Translations of Garber [1992: 211 and 218].

14 This has thoroughly been done by Garber in [1992: chs. 7., 8 and 9].

15 From the historical aspect, Loeck defends the idea that Descartes was the first to introduce a modern concept of laws into scientific explanation [e.g. p.81 and p.91]. Many historians since Zilsel, however, offer evidence refusing such a claim. Loeck does not discuss them in her argument [ch. III - 2.2].

possibility that the initial arrangement of matter may have been completely chaotic. He is convinced that even in this case the world must have changed, by the wisdom enclosed in the laws, to the well-ordered state we perceive now (1644, III, §47)[16]. Here he distinguishes clearly between an initial state and its development. While laws can not tell anything about which initial state has been realized, they do govern, and indeed completely determine, the manner in which all temporal change of this state takes place.

This distinction plays an important role in Descartes' concept of causes. When he considers physical causes of phenomena, he usually has in mind initial states, not laws.[17] As to such causes one may – so he says – form various different hypotheses, from which – by application of the laws of nature – conclusions can be drawn which are to be compared with what we find empirically as the actual state (III, §45-47 and IV, §204-206). This comparison may – as his own considerations on a chaotic or a well-ordered initial state of the world show – not be sufficient for deciding which of the hypotheses should be preferred, but it provides at least a necessary criterion for the possible acceptance of a hypothesis. In the course of this procedure, laws – being the governors of the action and interactions of matter – work as a sort of inference rule[18]. They are of a different kind of certainty. As will become clear, laws are obtained in a way fundamentally different from that by which causes in the sense of initial conditions are considered.

16 For an analogous consideration, see *Le Monde* [1664: 34], or his *Discours* [1637:43 and 45].

17 In the mentioned passage of the *Principia* in which the laws of nature are introduced as the "second, particular causes" of motion, Descartes is concerned rather with general, metaphysical causes than with particular physical ones. In the main body of what follows in the *Principia*, he concentrates, however, exclusively on the latter ones.

18 Descartes states, for example, in the French edition of the *Principia* that, starting from certain suppositions as causes, he was going to "deduce all effects apparent in nature by the only means of the laws explicated above" [III: 47; see AT IX-2, 124]. Loeck [1986: 115-7] elaborates this point in detail.

As to the way in which knowledge of laws can be achieved, Descartes is quite clear and decided. The derivation he offers of his three laws of nature is directly and exclusively founded on previous insights in the immutability of God which he regards as a certain truth (1644, II, §37-42; or 1637, 43).[19] The certainty of the laws thus derives directly from the certainty of some metaphysical and theological truths. It is, in this respect, comparable to the certainty of mathematical truth. As a warranty of the reliability of our knowledge of laws, Descartes argues that God, as he gave laws to nature, simultaneously gave us the correct notions by which we can grasp them (1637, 41; or letter to Mersenne of 15 April 1630, AT I, 145). Descartes emphasizes that, due to the clear derivation of these laws, one can be sure that they will be valid in every possible world (1664, 47; or 1637, 43).

Thus our knowledge of laws is so clear and so reliable, that Descartes does not see any need to confirm the laws by experience. He acknowledges, in contrast, that experience may appear even to contradict these laws as well as rules which are directly derived from them. This results, so he argues, from the great multitude of simultaneously acting factors which cannot be encompassed by the laws. Laws only tell what happens under idealized conditions which may, however, never be realizable in nature (1644, II, §52 and 53)[20]. This marks a significant difference to any concept of empirical laws. Descartes is not even ready to regard as laws those propositions whose derivation involves some hypothetical principles; the most prominent example may be the sine-proportion in optical refraction.[21] The place

19 I shall not analyze his derivations in detail here. My main point is Descartes' explicit insisting on the validity of his a priori derivation. See Larmore [1980: 8 ff].

20 The point becomes particularly clear in the French edition of 1647; see also Garber [1992: 358, n.12].

21 There is one passage of the *Dioptrique* which might appear problematic. After having demonstrated a sine-proportion for a moving ball, Descartes says that the action of the light follows "the same laws" as the motion of the ball [1637, 100]. An analogous claim is found in the letter to Mersenne

at which the laws of nature are introduced within his *Principia* becomes thus understandable. Their place is in the second book, which contains the (metaphysical) principles of the world, and which thus provides rather a general framework for a more detailed treatment of physical questions than such a treatment itself.[22] This is only done in the third and fourth book in which, however, no further laws of nature are introduced.

One should be careful not to equate Descartes' concept of laws with what he calls "principles", which concept in some way overlaps with the described concepts of laws and causes. Descartes distinguishes two different types of principles, according to the way we come to know them. There are, first, those of which a "clear" knowledge can be achieved. These are rather general ones; an example is his aforementioned proposition concerning the global conservation of the quantity of motion. Principles of the second type are those of which knowledge can be achieved only by speculation, by inferring certain

of 27 May 1638 [AT II, 143]. At first sight these passages may be interpreted as if he would count the sine-proportion as a law (see Zilsel [1942: 268]. A closer look, however, reveals that the term "law" again refers to the three laws of nature (as already Milton [1981: 179] observes, without giving further explication). In the course of his considerations of the motion of the ball, Descartes uses at least the first and second laws; and it is, of course, essential for his argument to state that these laws hold likewise for the propagation of light (cf. Garber [1992: 214, 225]). Similar problems may, moreover, occur with respect to the particular collision rules, given in *Principia* II, §46-52. Descartes insists on having certain demonstrations of these rules (§52 and 53), but does not present them explicitly. It is not really clear whether he thinks of a demonstration based exclusively on the three laws, or allowing other principles too. As to his terminology, he speaks almost everywhere of "rules". I know only one passage in which he uses the term "laws" (letter to Mersenne, August 1649, [AT V: 405]). Compared with the large number of negative instances, however, this one passage seems to me to be not significant to allow the conclusion that Descartes counts these rules as laws in the strict sense. For discussion and literature, see Garber [1992: 358, n.18].

22 In a letter to Mersenne of 26 April 1643, Descartes proves his first law by what he explicitly calls "metaphysics" [AT III: 648]; cf. his letter to Mersenne of 15 April 1630 [AT I: 144].

conclusions, and by comparing these ones with experience. Of these principles we can have, therefore, only what Descartes calls "moral certainty" (1644 IV, §204 and 205). They typically concern more particular propositions. Principles of the first type are something called "first causes" by Descartes; they depend directly on God and his attributes.[23] In this very respect laws can be regarded as a subclass of this type of principles. The main additional characteristic of laws consists in that they are to specify what happens in singular cases, whereas principles in general may give only global boundary conditions. Laws should enable to determine the outcome of particular interactions between bodies.[24]

Within such a concept of law the question of an instrumentalistic view simply does not arise. Laws, derived directly from indubitable attributes of God, are *per se* laws of nature itself. Their validity as inference rules for our achieving an understanding of nature is ancillary, in that it is merely a consequence of their being the rules governing nature.

To sum up, in Descartes we find a concept of law much more developed and explicated than the rudimentary hints found in Galileo. If Galileo is interpreted as strongly as I attempted one may, however, find some important similarities. For both Galileo and Descartes the concept of law is most intimately and inextricably connected with theological considerations concerning God's activity as legislator. In both cases, a mathematical form of laws is at least envisaged. Neither Galileo nor Descartes do explicitly discuss a way by which knowledge of laws can be achieved by starting from empirical studies; in contrast, both approaches involve the possibility that experience may apparently contradict what laws state; but neither discusses in detail how this difficulty should be overcome.

---

23 [1637: 64], see Larmore [1980: 10] and Rodis-Lewis [1993: 251].

24 This is also Garber's view of the role of laws; [1992: 201-2]. The distinction of laws and more global principles gets lost in the approach of Freudenthal and McLaughlin, since they use a more general concept of laws (Damerow et al., [1992: 70 and 90]). This concept differs significantly from Descartes' one.

Not all of these points are characteristic or even self-evident in the early period of the new sciences. This becomes clear by looking at other early concepts of laws, in particular those of another, different scientific movement which Kuhn has labelled as "Baconian" science. I shall first present some observations an Bacon himself, and then move to a closer analysis of what we find in the work of Boyle, one of the most prominent and most influental promoters of an empiristic program.

IV. Inquiring Into Forms of Nature

Francis Bacon makes use of the term "law" at several places in the *Novum Organum*[25], published in 1620. He mentions, for example, "laws of matter" (II, 4), "laws of action" (II, 17), and "laws of processes" (I, 51). In all these passages it is not a theological context which is prominent; Bacon's terminology of "laws and its paragraphs" (II, 2) makes one rather think of juridical or political laws.[26] This is not to say that Bacon's ideas do not involve theological commitments. But, in his discussions of laws of nature, theological considerations are less significant than one finds in Galileo or Descartes. I shall return to this point later.

In order to understand his point of view one has to include what he calls "forms", the investigation of which he regards as being the "work and aim of human knowledge [humanae scientiae]" (II, 1). Forms are in a close way linked to the "for-

25 I shall concentrate on this work since it "contains the core of his new philosophy of science" (Urbach [1987: 8]; cf. Krohn [1987: 61]). In my references, I give the number of the part of the work, together with the number of the aphorism. If necessary, I give additionally the page number of vol.I of the *Works*. The English text is taken, with some slight alterations, from Bacon [1960].

26 For a closer analysis of this point, see Krohn [1987: 131-3]. One must not forget that Bacon was a practicing barrister. Wheeler [1983] emphasizes that Bacon takes an empiristic approach to both juridical laws and natural ones.

mal causes" of the philosophical tradition. To know the "form" of a thing is, according to Bacon, to know in what it consists, or, in other words, to know "its essence and quiddity" (II, 20, p.262). The "thing differs from its form not otherwise than as the apparent differs from the real [existens]." (II, 13). As he himself emphasizes, however, his "forms" cannot be simply equated with the "forms" of the school philosophy (II, 17). When he, for example, characterizes forms as a "source of emanation" (II, 1), this is rather taken from alchemical thinking. The interplay of various different traditions makes it difficult to gain a sharp view of his own account. I shall not deal with the complicated task to explicate his concept of forms in detail[27], but rather concentrate on the relation between forms and laws. Bacon states that

in nature nothing really exists besides individual bodies, performing pure individual acts according to a law. In philosophy, however, this very law, and the investigation, discovery, and explication of it, is the foundation as well of knowledge as of operation. And it is this law with its clauses that I mean when I speak of forms, a name which I rather adopt because it has grown into use and become familiar (II, 2).

This passage, besides showing Bacon's nominalistic view, reveals an intimate connexion between forms and laws. It is somewhat explicated when Bacon claims that

Matter should be the object of our attention, its configurations and metaconfigurations [schematismi et metaschematismi] and pure action, and the law of action or motion; for forms are figments of the human mind, unless you will call those laws of action forms" (I, 51).

His insistance on laws being laws of action "or of motion", and his distinction between action and laws of action makes one think of laws of motion, analogous to what one has later in Descartes. That things are more complicated, however, becomes obvious of the way in which Bacon actually uses his concept of forms. Illustrating his proposed method of inquiry by the ex-

27 For a closer study, as well as for further literature, see, e.g. Krohn [1987], or Urbach [1987].

ample of the form of heat, he gives a first summary of his considerations, concluding "that the form or true definition of heat ... is, in few words, as follows: Heat is a motion, expansive, restrained, and acting in its strife upon the smaller particles of the bodies" (II, 20, p.266). Considering that shortly before he states that "the form of heat and the form of light is the same thing as the law of heat or the law of light" (II, 17), one may regard the quoted passage as giving the law of heat. But what is presented here is not what one would expect as a law of action, expressing a rule in which certain entities act and interact. If one tries to remain within the frame of Bacon's terminology of action and its laws, the "form" of heat given above seems more to concern the action itself – saying in what it consists – than to give a law of it.

Thus Bacon's "laws" should be taken quite generally as something which imposes a certain order on the material world,[28] be this a statical arrangement, or a set of rules governing movements. This corresponds well to the dichotomy, introduced by Bacon in the course of defining forms, between configuration [schematismus] and action [actus, processus] (II, 1 and II, 9, for example). It may roughly be interpreted as a distinction between statical arrangements and temporal changes of those.[29] Both configurations and actions are subsumed under the general concept of forms, and one should take the concept of laws likewise to be so broad as to embrace both; although Bacon's actual use of the term law is mostly combined with the aspect of action or motion.

That which Bacon terms "law" is, therefore, not very sharply defined. It may be that the concept has to remain as incomplete and aphoristic as Bacon's program itself. The analysis of further material may provide further information, but also further com-

28 This corresponds well to Bacon's rather wide use of the term "law" even in respect to human action. He criticizes, for example, the "empiricists" for their proceeding "without law", i.e. without a "regular system of operations" [1620, preface, p.128].

29 For a discussion see Krohn [1987: 119-123].

plication as does, for example, his idea of a "summary law of nature" which is not presented in the *Novum Organum*[30]. A 17th century reader may, due to the aphoristic character of Bacon's work, not have searched for a well-formed and explicated concept of "laws of nature", but rather have grasped the points which are prominent in Bacon's actual usage of the term.

Among these is the aspect of laws as governing motions or, more generally, actions, of great importance. Another point concerns the realm of inquiry in which laws are of most interest: Bacon emphasizes repeatedly that the actions (and configurations) to be examined are latent ones, hidden to the senses, and essentially located on a microscopic level (II, 1, 5, 6, 9, 17, among others). The laws which should primarily be discovered are laws of microscopic processes, not of everyday experience. In this context one may find, moreover, a sort of determinism. Bacon repeatedly emphasizes that any "nature" is present (or absent) just in case its form is present (or, respectively, absent) (II, 4 or 12); such a correspondence holds even for the degree of intensity (II, 13). Thus the apparent thing is completely determined by the underlying form or law. In respect to the later development of the concept of laws, it is, finally, remarkable that Bacon does not even hint to any connection between laws and their possible mathematical formulation or, more generally, between laws and quantitative investigation of phenomena. This has, on the one hand, to do with the close connection to the traditional concept of forms and, on the other, with Bacon's general indifference to mathematical reasoning in scientific context (I, 96, for example). This openness to the possibility of not quantitatively formulated laws left its traces in the early empiristic tradition of the Royal Society. Boyle, whom I shall discuss below, provides a good example. This point – among others – distinguishes Bacon's thinking from that of Galileo and Descartes.

---

30 See his *Advancement of Learning* [1605: 356], resp. its latin version *De Augmentis Scientiarum* [1623: 567]; or the fable of Cupido in *De Sapientia Veterum* [1609: 655]. For an analysis, see Krohn [1987: 122] or Urbach [1987: 67-8].

When considering how knowledge of forms should be achieved, Bacon proposes to start with the investigation of phenomena, provided by nature or even by human interaction with it. He is convinced of the possibility of a way leading from a systematic inquiry into phenomena to the knowledge of the laws of nature, i.e. of the latent processes of matter (I, 26 and 97-106, for example). As his own examples concerning particular phenomena show, the first step of such investigation includes the attempt to generalize the observed processes or manipulation rules. Such rules, once established, are regarded as "particular and special habits of nature [consuetudines naturae]", in contrast to the "fundamental and universal laws which constitute forms" (II, 5). This contrast reminds one immediately of what we now call the distinction between laws of nature and mere empirical rules. Bacon seems unwilling to regard a mere empirical regularity as a law. I find in Bacon three possible grounds for this distinction. First, laws concern "latent processes", while habits of nature deal with directly accessible phenomena. Since the latent processes are regarded as the fundamental ones, this difference entails a second one, concerning the scope of generality attributed to laws as opposed to habits of nature. Third, Bacon regards forms – and likewise laws – as being eternal and immutable (II, 9).[31] This may not necessarily be true of habits of nature. Thus, for instance, Bacon rejects many of his previously assembled observations at a certain stage of his discussion of the form of heat; for they do not, as he says, regard heat in itself, but only as it appears to us (II, 20, p.262). What he here adresses is the difference between those structures and regularities which only appear to the inquiring subject and those ones which should be regarded as really pointing to latent structures and processes, i.e. to laws. Habits of nature may possibly be interpreted in an instrumentalistic way.

But such a difference renders the problem of getting from habits to laws of nature particularly acute. Bacon proposes the procedure of tables which, however, serves more as a general

31 He does, however, not discuss the theological implications of this view!

frame than as a detailed method for particular actions or conclusions. His example of an investigation of heat shows that many additional steps are involved, which are not yet formalized. What he regards, moreover, as being the "further steps of induction" is only enumerated by him, but again not put into a formalized structure. Possibly Bacon does, as Krohn supposes (1987, 150-6), not intend at all to give a really detailed, formalized procedure. His program of science as proceeding from phenomena to laws is explicated only in its very first steps.

What I mentioned as a possible instrumentalistic interpretation of Bacon's "habits of nature" must not be applied, of course, to laws and forms. The forms to be examined are the forms of nature itself, and likewise the laws (II, 15 and 52).[32] It is this unquestioned realistic understanding on which Bacon's firm and often repeated conviction is based that the "twin-objects: human knowledge and human power" are closely and inseparably connected with one another (end of section "Distributio operis", p.144).

V. Considerations touching Laws

In the numerous reports of experimental research presented by Robert Boyle over nearly three decades, a concept of "laws of nature" seems not to play a significant role. The only laws he mentions in this context are what he calls the "known laws of the Hydrostaticks" (1663, 36; 1666a, 743 and 744, for example). His terminology, however, is not very precise: He uses also the terms "principles and theorems of Hydrostaticks" (1666a, 742), or "principles and laws of the Hydrostaticks" (1666a, 745); and in all cases the same things seem to be meant. In other branches of Boyle's physical investigation the term "law" does not play any role; this holds for the refraction of light, the free fall of

32 Hansen [1978: 487] wrongly claims that laws of nature, in Bacon's view, could be broken by art and magic. Such a claim holds only for the habits of nature, but not for the laws.

bodies, the motion of planets, and, in particular, even for the proposition, found by Boyle and Towneley, that the volume and the pressure of air stand in an inverse proportional relation.

In the context of more general considerations, however, a concept of laws of nature becomes apparent, at least in outline. In considering what he regards as a proper notion of nature, the role of God within nature, but also proper investigation and procedure in science, Boyle repeatedly mentions the "catholic laws of motion" (1686, 164 and 222) or the "laws and rules of local motion" (1686, 179). He mentions also, more generally, "laws of nature" (1663, 37 and 1686, 219), "catholic laws of the universe" (1686, 220) or even "general laws whereby things corporeal are guided" (ibid.). His terminology seems to remain essentially unchanged over a long period of time.[33]

It is this concept of laws which has already been treated several times (see above, p.318). What remains, however, is to determine the place Boyle attributes to laws of nature within the whole project of inquiry. The essential aim of his "experimental philosophy" is to find out the "true physical causes"[34] of phenomena (1660, 12; 1662b, 194 or 1686, 165). This needs some explication, since one central point of the sharp controversy between Boyle and Hobbes – as Shapin and Schaffer have readily shown (1985) – is Hobbes' accusing Boyle of having failed to investigate the causes for what he claimed to be experimental results achieved with the air-pump. When Boyle replies that he is first of all interested in phenomena and facts (1660, 12, for example), this must not be understood to mean that he gave up the search for causes. His main point lies in emphasizing that this search needs a more solid foundation than mere speculation; he finds it in facts and phenomena, be they obtained by observation or experiment. In contrast to Hobbes, he insists on

33 Boyle had composed many of his papers long before he published them, see the bibliography in Shapin and Schaffer [1985].

34 Boyle emphasizes the role of God being the primary cause of all phenomena. Nevertheless, natural philosophy is concerned with physical causes.

acknowledging experimental results as reliable facts even when their causes still lie in the dark (1662a, 194).[35]

Considering "physical causes", however, Boyle makes clear that he does not find only one cause of a phenomenon. As an illustrative example, he discusses the phenomenon that all matter, except gold, floats on mercury.[36] In order to explain this phenomenon in a more "naturalistic" way than by a sympathy between gold and mercury, one may observe that gold is the only substance heavier than mercury. This given, "the known laws of the Hydrostaticks" make it necessary that gold sinks in mercury while all other bodies float on it. "But", so Boyle adds,

> though the cause of this effect be thus plausibly assigned, by deducing it from so known and obvious an affection of bodies, as gravity, which every man is apt to think he sufficiently understands; yet will not this put a satisfactory period to a severe inquirer's curiosity, who will, perchance, be apt to alledge, that though the effects of gravity indeed be very obvious, yet the cause and nature of it are as obscure, as those of almost any phenomenon it can be brought to explicate. (1663, 36-37).

Thus the investigators must not be content with this explanation but rather set on to find out the cause of gravity. Boyle is hopeful that

> if we further inquire into the causes of those causes, and desist not from ascending in the scale of causes, till we arrived at the top of it, we shall perhaps find the more catholic and primary causes of things to be either certain, primitive, general, and fixt laws of nature (or rules of action and passion among the parcels of the universal matter,) or else the shape, size, motion, and other primary affections of the smallest parts of matter, and of their first coalitions or clusters ... (ibid., 37).

He adds that of these primary causes a satisfactory account can only be given by acknowledging the wisdom and power of God.

---

35 This aspect is of the greatest importance for his whole approach. It involves, as Shapin and Schaffer [1985] have thoroughly analyzed, new ways of making results commonly accepted as facts.

36 As this example shows, the terms "fact" and "phenomenon" are not only used for singular observations or experimental results, but also for generalized empirical rules.

It is worth taking a closer look at this "scale of causes", which reaches from the most obvious "affections of bodies" to their "primary causes". On every particular step of this scale a cause is given by a certain affection or property of matter together with a law which governs the action and interaction of matter with respect to this affection. In Boyle's example, the cause is given, on a rather elementary level, by the different gravities of matter, together with the laws of hydrostatics. On the highest, most general step, it is given by the most general, primary affections of matter, together with the general laws of nature which govern the action and interaction of matter. Within the framework of Boyle's corpuscular philosophy, the "parcels of matter" are clearly to be understood as corpuscles, and the respecting laws as laws of motion and collision. Thus an intimate connexion of causes and laws becomes clear: causal explanations cannot be constructed without laws.

The scale of causes is, therefore, accompanied by a corresponding scale of laws, according to their different degrees of generality. While the laws of hydrostatics concern only one particular branch of physical experience, the fundamental "laws of nature" – understood as laws of motion and of impact – are thought to govern the whole range of physical phenomena. Indeed Boyle extends this scale into the realm of everyday phenomenological rules. He introduces a distinction between, on the one hand, "the laws of nature, more properly so called" and, on the other, "the custom of nature" or, in other words, between "the fundamental and general constitutions among bodily things" and the "municipal laws ... that belong to this or that particular sort of bodies" (1686, 219). The very terminology recalls immediately Bacon's distinction between "habits" ("consuetudines") of nature and "fundamental laws". Boyle, in contrast to Bacon, seems to be ready to use the term "laws" even with regard to the "customs of nature".

As an example of such a "municipal law", Boyle mentions the common observation that water falls downwards, if not hindered. When it ascends, however, as it does in sucking pumps, "that motion, being contrary to that which is wonted, is made in virtue of a more catholic law of nature" (ibid., 220), namely by what he calls the "laws of the equilibrium of liquors" (ibid., 219). This example renders explicit a difference in power of the various types of laws, as it is already suggested by the very terminology: The properly called laws may dominate or overwhelm the municipal ones.

Regarding the question of how the knowledge of laws can be achieved, Boyle makes abundantly clear that he sees no way of any *a priori* derivation of laws. Criticizing Descartes' derivation of the global conservation of the quantity of motion, he emphasizes that such principles stand in the free will and power of God, not accessible to *a priori* reasoning (1684, 140 and 1688, 396-7; cf. 1666b, 15). He considers that our world might be in a state completely different from the actual one, if only God had arranged one or two fundamental laws in a way slightly different from how he actually did (1684, 139-140). I shall come back to the theological points in the last section.

Just as Bacon, Boyle regards empirical facts as the only firm foundation for the inquiry of causes and laws. He seems to assume at least the possibility of starting with facts and successively coming to more general propositions, rules, or laws (e.g. 1675, 161). In contrast to Bacon, however, who proposes a sort of formalized method at least for the first steps of such a procedure, Boyle does not explicitly discuss how it may work in detail. One may find some hints only in what he gives as actual examples of investigation. In his inquiry concerning the "spring" of the air, he makes clear that knowledge of the causes of that elasticity can be achieved only by framing hypotheses and deriving from them some expected experimental outcomes, which expectations may then be tested experimentally (1660, 12-13). Given the strong connection of causes and laws, such a hypothetico-deductive procedure may likewise give the appropriate way to laws, at least to those fundamental ones which govern the interaction of the microscopic particles of matter. What Boyle calls "municipal laws", however, is typically not to be achieved by a hypothetico-deductive way but rather by simple inductive generalisation, as the example of the water's tendency downward readily shows. These two procedures – hypothetico-deductive reasoning and inductive generalisation – is what I find in Boyle as elements of a larger procedure, possibly leading to the knowledge of laws.

A common characteristic of these procedures is that both of them are inherently uncertain, in that the results are always subject to possible revision by further experience. Absolute necessity cannot be achieved. This might imply that perhaps all knowledge of causes and laws is in principle uncertain. Such an epistemic uncertainty would by distinctly different from the uncertainty inherent in the laws themselves, granted that such laws are always subject to possible change by Divine intervention as happens, for examples, in miracles (1675, 161). It becomes not really clear of whether Boyle thinks of a possible certain knowledge of fundamental principles or not. In contrast to Descartes, he does not give a way by which such certainty might be achieved. What is remarkable here is, in particular, that he does not even consider the possibility, as do others, of essentially introducing mathematical reasoning into natural philosophy, and thereby importing, at least partly, mathematical certainty into this realm.

If, however, the methods of achieving knowlege of both "municipal" and fundamental laws are equally uncertain, then it becomes difficult to find any essential difference between these two types of laws at all. One might think of a principal distinction between qualitative vs. quantitative formulation, but this appears not to make an essential difference for Boyle (see, e.g. his (1671)). As another possible criterion it may be considered that laws in the proper sense perhaps concern only primary affections of things, whereas "municipal" laws may involve secondary qualities. This seems again not to be Boyle's view, since the laws of hydrostatics – being laws in the proper sense – deal with gravity, which Boyle does not count as a primary quality. What remains, then, as a criterion is the differing degree of generality. Given an only gradual difference, however, it appears quite arbitrary to stipulate a point at which laws are divided from non-laws. There seem to be two main possibilities of overcoming this arbitrariness: One may extend the notion of law to cover the whole scale, including empirical rules, or one may simply discard it. Boyle, by introducing his "municipal laws", seems to prefer the first way.

Although he thus preserves the view of God being the governor and legislator of nature, he appears at the same time to weaken the theological commitments of the concept of law. It is not at all clear in which way the "municipal laws", since they are not invariably valid, should be understood as given directly by God. There are laws now which may easily be considered without thinking of a divine legislator.

Simultaneously the possibility of an instrumentalistic view (to use our modern term) is now extended from empirical rules to laws. If mere empirical rules, being the result of an uncertain inductive process, are regarded as laws, then it is no longer clear that such laws are really laws of nature. Boyle himself understands them as our first attempts of bringing order into the vast realm of experience. They are subject to possible revision or even rejection by further experience. It is not clear how Boyle is to exclude the possibility of extending such a view to more fundamental laws. To be sure, this is not at all what he would regard as an appropriate view (see, e.g. (1686) 179 or 222).[37] But again he does not give any criterion which could determine the point of investigation at which such an instrumentalistic view could be with certainty excluded.

Thus Boyle's conception of laws of nature remains unclear and unexplicated in some essential points. This may be due mainly to his considerations not leaving a quite general context. They play almost no role in the course of his actual em-

37 McGuire [1972] does not make completely clear which opinion he ascribes to Boyle regarding the question of *who* imposes the laws to nature: God, by his will and providence (cf. [1972, 537]), or rather the human mind who, in the course of inquiry, projects laws as "abstract conceptions" upon nature (525 or 536). This point is, of course, decisive in thinking of realistic or instrumentalistic interpretations of laws of nature. Contrary to what McGuire considers (537), Boyle's position seems to me, in this respect, to be significantly different from Berkeley's.

pirical investigation, which perhaps would have forced some further explication. Nevertheless, they make particularly clear some problems which may be inherent to all attempts of combining the view of God imposing laws on nature with an empiristic approach to science.

VI. Mathematical Analysis and Empirical Inquiry

In Newton's published works the concept of laws of nature has a preeminent place. As to his terminology, one may roughly distinguish three different uses of the term "law". First, and most prominently, there are his three famous "Axioms, or Laws of Motion", given in the first part of the *Principia Mathematica* (PM, for short)[38]. Newton regards them as the most general propositions about nature. They are most closely connected with the "Definitions" given at the beginning of the *Principia*, in that it is only together with the Definitions that the Laws form a conceptual framework in which all further inquiry takes place.

There is a second, very different use of the term "law" in the *Principia*. Newton mentions, for example, various "laws of centripetal force" (PM I, Prop. 7-13, for example), "laws of attraction" (I, Prop. 67 and 68), "laws of resistance" of a fluid against the motion of a projectile (II, sect.1, scholium, 359), or "laws of condensation" of an elastic fluid (II, Prop. 22, scholium, 428). In all these cases he means mathematically formulated relations between certain quantities: force or attraction as a function of distance, resistance as a function of velocity, or density as a function of the external pressure. Such relations serve as hypotheses for further calculations of the respective physical problem. Newton often assumes many different laws, and calculates what are the respective consequences of such hypotheses. "Laws" are

38 In referring to this work, I give the number of the respective part, the specification of the proposition and, when applicable, of parts of it. If necessary, I give additionally the page-number of the edition of 1972. All English translations are taken of [1934].

not, in these cases, to be understood as laws of nature, for they are not at all supposed to represent any relations actually found in nature; rather they serve as a tool for mathematical practice. The question of which of these laws – if any – are really found in nature, is not a mathematical, but rather a "physical question" (PM II, Prop. 7, scholium, 431). Only once this question is solved can the laws thus signified be regarded as being laws of nature.

This leads to a third use of the term "law". Newton mentions laws of the oscillating pendulum (PM II, Prop. 47, 521), laws of refraction (*Opticks* bk.1, pt.2, Prop.3, exp.7, 129)[39], laws of planetary motions (PM III, Prop.4, scholium, 569; or II, Prop.53, scholium, 546), and, most prominently, the law of gravitation (PM III, Prop.7, Cor.1, 577; or III, scholium generale, 759). These laws are formulated in mathematical terms, like the ones mentioned above. They do not, however, involve the hypothetical character of those ones, but are definitely regarded as being laws of nature. They are, on the other hand, not as general as the three Laws of Motion. Newton himself emphasizes the difference between those particular laws of nature and the three general ones by capitalizing only the latter ones ("Leges Motus") throughout the *Principia*.

In what follows I shall deal with these particular laws of nature, and with the three Laws of Motion. Since the concept of law seems not to be essentially different in Newton's writings on optics and on rational mechanics, I shall use his publications on both these fields. I cannot discuss in detail the question of how Newton's published statements fit to his private scientific practice. Likewise, I shall not deal in its own right with the development of the terminology in the course of Newton's life-

39 In referring to the *Opticks*, I give Newton's classification in books, parts and propositions. The page numbers refer to the commonly available edition [1952] which is based on the fourth edition of 1730. Newton's successive alterations and additions in the various editions, however, are given only in the edition by Horsley in vol. IV of [1779-85].

time. Thus I concentrate on the later Newton as present in the *Opticks* and the second and third edition of the *Principia*.

Which is the place of laws within Newton's project of inquiry? In his famous Query 31 of the *Opticks*, inserted in 1706, Newton describes as the proper course of natural philosophy the way "from Effects to their Causes, and from particular Causes to more general ones, till the Argument end in the most general". These made out, one has, with synthetic methods, to get back again to experience (404-405). This scheme itself of analytical and synthetical procedures is not really new in the traditional philosophical school.[40] What is new, of course, is what Newton counts as causes. Here we find, at first sight, strong similarities to the concepts of both Descartes and Boyle. Besides the very first and fundamental cause – God himself – Newton's search for causes is principally directed toward mechanical processes, regarded as taking place among the microscopic particles of matter.

As the mentioned passage shows, Newton considers, similarly to what one finds in Boyle, the idea of a scale of causes, extending from rather particular ones to increasingly general ones. Here is the point, then, at which his new and original concept of moving forces comes essentially into play. Forces are regarded as being the causes of change of motion. Thus, in order to explain any observed change of motion, the first step of the inquiry has to be the determination of the force causing this change. This program is not restricted to the analysis of local motion of macroscopic bodies; it is – since nature is regarded as being "very conformable to herself and very simple" – intended also for other parts of natural philosophy such as chemical action, optics, electricity, magnetism (*Opticks*, Qu.31, 397). All change – this is a main point of the mechanistic approach – is in principle reducible to the change of motion among the constituent microscopic parcels of matter. Thus in all the mentioned fields the first step of the search for causes consists in reduc-

40 Cf. Crombie (1953). The scheme is also found in Francis Bacon's approach; see *Novum Organum* [1620: II, 10].

ing the observable effects to the action of forces, in particular of central forces (ibid., 401).

Already in this first step, laws are essentially involved. In order to assign the cause for any change of motion, it is not sufficient to assign the type of the acting force. One has, moreover, to give the law according to which it acts. The three Laws of Motion express the effect of *any* acting force of given strength and direction. In order to give an account of particular processes, one has to know, in addition, how strength and direction depend on other relevant parameters. This is what laws of force explain. In the case of central forces, which depend in various ways on the distance only, the main task is to find out the respective distance laws.

The most obvious example of such an inquiry is, of course, Newton's analysis of planetary motion. Starting from what he calls "phenomena", he inquires into the acting forces and comes to the result "that gravity does really exist, and act according to the laws which we have explained" (PM III, scholium generale, 764). This claim is typical in that it involves, first, the assertion of the existence[41] of the gravitational force in all matter and, second, the claim that this force acts according to the inverse-square law. In the course of his successive argumentation for this result, as given in Propositions one to seven of the third book, he argues for both parts conjointly.

Newton makes, however, abundantly clear that with this first step the explanatory aim of science is in no way achieved. Central forces, acting at a distance, are not acceptable as definite explanations, although they may well serve as intermediate ones (PM, Auctoris Praefatio, 16; scholium generale, 764; *Opticks* Qu. 31, 401-2; (1756) 298 and 302-3; or (1962) 304). There is not yet a complete agreement among Newton scholars as to what Newton is ready to accept as causes of central forces. What is certain is that in his early years, and probably in his late years too, he is seriously thinking about some ethereal mechanism

41 I shall deal below with the ontological status of that existence.

as the cause of gravity and of other central forces.[42] At least in those models of ether which do not themselves involve central forces – and thus end in an explanatory regress – the mechanisms to which central forces are reduced involve only forces exerted by immediate contact, i.e. pressure and collision. I shall not go into the details of the question of whether in an intermediate period Newton gives up such considerations and instead accepts only God's power itself as the cause of gravity.[43] What is more important for my purpose, is to observe of what a physical cause of gravity – if considered at all – is thought to consist. In all such attempts made by Newton, the explanation again involves the assumption of both some acting entities, typically particles of ether, as well as of some laws governing the action and interaction of these entities. These are typically laws of collision. Thus Newton's assumption of various levels of causes entails, likewise to what one finds in Boyle, a corresponding hierarchy of laws.[44]

At the top of this scale there are the three Laws of Motion, immediately followed by the laws of collision.[45] In the *Principia*, of course, the laws of collision do not play a prominent role. This is because the work is concentrated on a strict mathematical treatment of the first step – the search for central forces, postponing the question of their causes (PM, Auctoris Praefatio, 16). As soon as this question comes into play, however, laws of collision become prominent.

---

42 See, e.g., Newton's second paper on light and colors of 1675 [1757: 177-90], or his famous letter to Boyle of 1678/9 [1744: 70-4], or Queries 17-24 of his *Opticks*, added in 1717.

43 For discussion and literature, see Corson [1974] and Cohen [1980: ch. 3.8].

44 Newton emphasizes that *any* cause of gravity – may it act mechanically or not – will act "constantly according to certain laws"; 3rd Letter to Bentley, [1757: 303]. For further discussion of the hierarchy of causes and laws, see Cohen [1980: ch. 1.4].

45 In an early manuscript, entitled "The Lawes of Motion", Newton deals with general laws of exchange of motion and, conjointly, with "Laws of Reflection", i.e. collision [1962: 157-64]. He seems to regard these two types as being closely connected.

As to the question of the physical reality of central forces, it is not easy to get a clear view of Newton's ideas. There are, on the one hand, many passages (see above) in which he seems to regard central forces as only mathematical tools, i.e. as provisional instruments of the cognitive process, necessary to enable a strict mathematical progress while determining causes. It is not clear how such an instrumentalistic interpretation could then be avoided with respect to the law governing the action of force. On the other hand, we find in Newton's writings several passages in which he seems to attribute a sort of physical existence to forces, thus implying a realistic interpretation of both forces and their respective laws. The passage quoted above, in which he emphasizes the "real existence" of gravity, may serve as an example. The problem how his approach should be interpreted is intensely discussed already during his lifetime; and his own answers are often not really illuminating.[46] As a result of the debate, many British natural philosophers prefer a realistic interpretation, thus ascribing to central forces a much greater importance than Newton, most probably, ever does. The goal of explaining these forces and their laws by other, more fundamental ones, rapidly loses the importance it has for Newton. A realistic interpretation of the respective force laws can then be maintained unproblematically.

There is another important point in Newton's involving laws in causal explanation. The gravitational law works as a general rule, stating what happens when certain conditions are present. It does not, however, state anything about which conditions are

46 An example is given by Newton's additions to the "Regula III" of the *Principia* [PM III, Regula III, 555]. The rule itself entails that gravity should be ascribed a status similar to other properties of matter. This view is elaborated and emphasized by Cotes in the preface to the second edition. As a reaction, however, Newton makes clear in the third edition that he ascribes to gravity a significantly lesser ontological status than to the "innate" force of inertia. The remark he adds to this end, however, is extremely short; and the argument not really clear. For a discussion, see Koyré [1965: chs. IIIc and VII], Cohen [1980: ch. 3.8] or Steinle [1992: 102-4].

actually found. Newton makes much hay of this dichotomy between laws and initial conditions. He claims that the planetary motions can be readily explained in detail, given the gravitational law plus certain initial conditions, such as the coincidence of their rotation planes, their distances, the masses of planets and of the sun, and so on. The law alone does not give any hint as to why exactly these initial conditions are realized, and not any other, different ones (*Opticks*, Qu. 31, 402; PM III, scholium generale, 759-60; or (1757) 284-7). In this form the dichotomy appears to him to be sharp enough to serve as a strong argument against atheistic tendencies. The actual arrangement – so he emphasizes – becomes understandable only by assuming the action of a wise and powerful Creator. Bentley – as a fruit of his correspondence with Newton – elaborates this argument in remarkable detail. (1693b, 363-6).

It is this emphasis on the dichotomy of laws and initial conditions which makes one think of a criterion enabling a clear separation of those phenomena which should be accepted as inexplicable initial conditions from those of which a further mechanical explanation may be expected. The question is significant since Newton deals with phenomena (such as the equatorial bulge of the earth, for example) which serve as initial conditions in explanations of others (the precession of the equinoxes), but which are themselves explained mechanically (by the daily rotation of the earth). In the case of, for example, the coincidence of the rotation planes of the planets, it is not clear at which point further attempts at mechanical explanations must stop. Descartes had attempted a mechanical explanation of this coincidence, whereas Newton takes it as an initial condition, not further explainable mechanically. Although this problem becomes visible in the context of Newton's considerations, he does not face it in its full sharpness. At least I don't find an explicit dealing with it, or even a solution, in his writings.

It is remarkable that this problem becomes prominent in the context of theological discussions. That it occurs in Newton, but not in Descartes, has indeed to do with their different theological background. Although both of them take a voluntaristic approach, it is very differently fleshed out, particularly with respect to epistemological questions. I shall discuss some aspects in the last section.

Among the five authors discussed in this paper, Newton is the one who makes most extensive use of the concept of laws even in presenting his actual research. The question of how knowledge of laws is thought to be achieved can be analyzed here in more detail. Newton emphasizes in both the *Principia* and the *Opticks* that laws should be inferred by induction from phenomena (PM III, scholium generale, 764; *Opticks*, Qu. 31, 404). This quite general view, being one of the main points of the Baconian tradition, is of course very popular within the Royal Society, in which Newton, from the beginning of the 18th century, plays an important role. In pursuing our question in more detail, the three Laws of Motion have to be considered separately. The first and second of the Laws of Motion form essential constituents of the conceptual framework in which Newton ensuingly argues. What is meant by "impressed force", introduced in the "Definitions", becomes fully understandable only by including these two laws.[47] They cannot, in a direct sense, be derived from experience, since they form parts of the very conceptual tools by which experience is interpreted. Indeed Newton, in considering the relation between these Laws and experience, is rather tentative, stating that the Laws are "confirmed by multiple experience" (PM, scholium to the Laws, 64 and 69). This is remarkably different from what he claims with respect to other propositions and laws, which he regards to be "deduced from the phenomena, and afterwards rendered general by induction" (PM III, scholium generale, 764)[48]. Perhaps Newton would not put this claim with respect to the three Laws of Motion. It remains unclear, however, what he wishes to be regarded as the proper way to these fundamental Laws.[49] As we know from his

47 In Def. 8, Newton implicitly anticipates some aspects of the second Law.

48 The laws considered by Newton in this context are not capitalized!

49 The third Law is even a more complicated case since it at first sight seems not to be as indispensable for the very understanding of the fundamental

manuscripts, his own way to the conceptual framework in mechanics involves at least as many derivations from metaphysical and theological premises, as from any experiential elements. Newton is, however, not willing to present publicly such a procedure which, in some parts, reminds one rather of Cartesian than of empiristic ways of reasoning.

I shall, therefore, not deal in detail with the three Laws of motion, but rather concentrate on the type of more particular laws which he actually claims to be derived from experience. The best-known example is his introduction of the law of gravitation, received among ensuing generations as a paradigmatic case of exact inference from phenomena to laws. Newton's argumentation, given in the third book of the *Principia*, is founded on the results of his geometrical analysis given in the first book. In order to apply these results to the planetary system, he introduces, on the one hand, his four "Rules of Reasoning in Philosophy" and, on the other, what he calls "Phenomena". The "Rules" work in this context as strict inference rules analogous to those of logic; and the "Phenomena" are regarded as providing certain and unquestioned empirical material. Making use in this way of mathematical analysis, inference rules, and phenomena, he claims the strict validity of his above-mentioned result concerning gravity and its law (PM III, Prop. 6 and 7). Cotes, in his preface, elaborates this claim in considerable detail (Editoris Praefatio in Editionem Secundam, 20-7). Newton's result is presented as a clear and certain inference from phenomena to laws.

The main point I am interested in is how Newton manages the transition from experience to the law. If one accepts, for a moment, the mathematical analysis and the inference rules as being valid, then the decisive point lies in his introduction of

---

concepts as the first and second Laws are. Newton has it only in a rather late state of the way to the *Principia*; and in the *Principia* he mentions various experiments exclusively designed to confirming the generality of this Law [PM, scholium to the Laws, 68-9 and 72].

"Phenomena". This very term[50], of course, is to point to the reliability of the empirical base, since phenomena, as understood in the Baconian and Boylean tradition, need not be questioned. What Newton actually presents, however, is quite different. He gives complicated mathematical rules[51], concerning the distances and periodic times of planets and moons, including what we now call Kepler's third law. His derivation of the gravitational law, as presented in the Principia, does not start from experience, but rather from already established generalisations and rules.

As to the empirical foundation of these rules, Newton mentions two points. First, he gives some tables in which the data calculated from these rules are compared with those actually observed. He does, however, not discuss – as Kepler does – possible error in the observational data, or the question of whether this and not any other rule fits best to the data. As a second point, he remarks in two cases that the respective rules are widely accepted by astronomers (PM III, Phaen. 4, 562 and Phaen. 5, 563). Both mentioned procedures are accepted in his time as confirming such rules which Newton introduces as "phenomena". They can be applied, of course, only when some empirical rules have already have formulated. But, if one takes literally Newton's claim to have derived laws from experience, then the problem would seem to arise the other way round: In which way should one, given only the data of astronomical observation, make out at all such empirical rules which Newton takes as given "Phenomena" in the *Principia*? This point, important at the very first stage of empirical investigation, is not addressed by Newton in the *Principia*. He leaves open the question of how knowledge of empirical rules should be achieved.

50 I shall not deal with the problem of why he calls them "Hypotheses" in the first edition of the *Principia*; for analysis and discussion see, e.g., Cohen [1966], or McDonald [1972].

51 At another place, he uses the term "rule" or even "law" for what he here calls "phenomenon"; see PM II, Prop. 52, scholium, 543-4; resp. PM III, Prop. 4, scholium, 569.

There are other parts of Newton's work, however, in which he is directly confronted with this problem. In some fields of his optical investigations he is engaged in gaining and evaluating data of which he has neither an empirical rule nor even an explanation. In dealing, for example, with what we now call "Newton's rings", he tries to formulate empirical rules on various occasions and in different ways. In analyzing, for a series of these rings, the mutual dependency of the angle of incidence of light and the thickness of the air between the two glasses (*Opticks* bk.2, pt.1, Obs.7, 203-205), he gives a table of the measured data. From this table he surmises a rule which he introduces in rather tentative words: "And from these Measures I seem to gather this Rule: ...". The formulation may indicate that he is not really sure about the validity of this rule, which involves, moreover, an arbitrary numerical factor. There are other cases, however, in which his formulation is more decided: When he measures the diameters of a series of rings "at the most lucid parts", he does not give a table of data, but he only tells that

> squaring them, I found their squares to be in the arithmetical Progression of the odd numbers 1, 3, 5, 7, 9, 11. ... I measured also the Diameters of the dark or faint Rings ... and found their Squares to be in the arithmetical Progression of the even numbers 2, 4, 6, 8, 10, 12. And it being very nice and difficult to take these measures exactly; I repeated them divers times at divers parts of the glasses, that by their Agreement I might be confirmed in them. (Bk.2, pt.1, Obs. 5, 200).

The different distinctiveness of the formulations in these two cases is all the more remarkable as I do not see in what respect either the accuracy of the measurement, or the numerical fit between the rule and the data should be more exact in the one than in the other case. The reason for Newton's being much more decided in the second case than in the first one is, however, easy to see: the given rule is equivalent to a series of linearly increasing thicknesses of the interjacent air. This makes it plausible that the diameters should follow exactly this and not another rule. Without this background, it would be not clear why Newton has the idea to compare the very squares of the diameters and not the diameters themselves.

These two examples illustrate how Newton establishes empirical rules. He starts with measured data, achieved by observation or even by experiment. Like Boyle, he seems to find no principal difference between these two. Remaining close to these data he proceeds speculatively in rather small steps. Since he aims, from the very beginning, at a mathematical formulation, direct inductive generalisation, as Boyle had considered, is not essential in this context.

It is remarkable that the very first step of the path from experience to laws involves a speculative element. Newton is well aware that empirical rules are uncertain. As the comparison of the two mentioned examples shows, however, the degree of certainty of such a rule may be enhanced by considerations which do not concern experience. Newton regards, in the second example, the rule as being more reliable because he finds some other considerations which make it appear plausible. The embeddedness in a wider context seems to be of at least as important as the empirical adequacy.

By combining the procedure for determining empirical rules with that for getting from rules to laws, one may have a complete method for proceeding from immediate experience to laws. This seems to be what Newton, at least in principle, presents as his procedure in his mechanical as well as in his optical studies. Due to the very different situations he meets in these two fields, the emphasis of Newton's studies differs as well. I shall come back below to the essential role mathematical analysis plays in this context.

It is remarkable that Newton neither in the first nor in the second example of empirical rules uses the term "law". He seems unwilling to regard those rules – may they be in some way plausible or not – as laws. He would probably not accept Boyle's term "municipal laws" for empirical rules. The question arises, therefore, at which stage of inquiry a switch from rules to laws takes place, and by which criteria laws are distinguished from empirical rules. Newton himself does not explicitly give such a criterion. One has, therefore, to look at what he actually does in this respect. His terminology, however, is not sharp.

Although empirical rules are never called laws, he nevertheless sometimes uses the term "rule" for what at other places he calls "law", as he does, e.g. in the case of the sine law of refraction (*Opticks* bk.1, pt.2, exp.7, 129; or Qu.29, 371).

Concerning possible implicit criteria for separating rules from laws, it quickly becomes clear that neither the question of mathematical formulation, nor the more or less accurate fit between expected and observed data, are decisive points for Newton. He expresses rules as well as laws in mathematical terms. Moreover, a certain fit to observational data is required with respect to laws as well as to rules; but Newton nowhere discusses what we would call the quality of the fit. As another important aspect, the width of the realm of phenomena governed by the rule or law may be considered. There is, however, a large variation in this respect among laws, as one may see by comparing the scope of the law of universal gravitation on the one hand, and of the law of the oscillating pendulum on the other. The proposed criterion would not make really clear, moreover, why the latter one should be called a law, whereas the proportion of the diameters of "Newton's rings" should not. A third, more promising aspect is related to that which, concerning empirical rules, I mentioned as the embeddedness in a more general context. Newton would accept as laws those propositions which he has successfully situated in a wider context, by deriving them from other, already established laws or principles. Vague as this criterion is, it could make plausible why Newton speaks of "laws of planetary motions, detected by Kepler" (PM III, Prop.4, scholium, 569) and of the "laws of refraction" (*Opticks* bk.1, pt.2, Prop.3, exp.7, 129), since in both cases he has a derivation from more general principles: on the one hand, the theory of gravitation, on the other, his corpuscular hypothesis of light. It should, however, be understood as a necessary rather than a sufficient condition, since Newton does not use the term "law" for Boyle's proposition of the compression of air, although he gives a derivation of it from a hypothesis no less firm than the mentioned one of light (PM II, Prop.23). The vagueness of this criterion lies, of course, in that it does not specify sharply what counts as a general law or principle, and to what degree hypothetical principles are allowed. Of course, one may think in this context of other factors as well, as, for example, established uses of terminologies. I fear, however, that it will be difficult to find one thorough and sharp criterion in Newton.

This has consequences in respect to possible instrumentalistic views in Newton's concept of laws. One central point of the concept is its unquestioned realistic interpretation. The laws are imposed by God on nature itself. There are two points, however, at which the problem of a possible instrumentalistic interpretation becomes obvious – and even inevitable – in Newton's concept. First, since Newton regards tentative empirical rules, but not laws, as objects of possible instrumentalistic interpretations, the line which separates instrumentalistic from realistic views coincides with the line separating rules from laws. Since the latter line is defined only very vaguely, this entails a similar vagueness with respect to the former. The second point concerns the more fundamental question, mentioned above, of the ontological status of central forces. If these forces are interpreted as mere instruments of the cognitive procedure, without being ascribed to physical reality, it is not easy to see in which way the laws concerning these forces are to be interpreted in a realistic way. This problem is not discussed by Newton; due to his unclear position regarding the status of forces it remains even more unclear than the former one.

VIII. Laws in the New Sciences

In the preceding sections, I have discussed the ideas of laws of nature of five 17th century authors. The limited space available forced a selection, which has been guided by different factors. All authors discussed are prominently involved in the formation-process of the new sciences, in that they design different programs of inquiry, all of which become influential in the 17th century. Nearly all of them deal with propositions which later are called "laws". This does, however, not hold for

Bacon: due to the high influence of his programmatic sketch, even his aphoristic ideas of laws of nature leave such important traces that they cannot be neglected.

I had to leave aside many authors of interest, among which the most prominent ones are perhaps Kepler and Huygens. Kepler, although he sometimes uses the term "law", seems not to be consistent in his terminology.[52] The different ways in which he uses the term make it probable, moreover, that his ideas of laws do not essentially influence that which, at the end of the century, emerges as a rather common concept. In Huygens, the term "law" occurs quite frequently. As an analysis of his concept shows, it has important similarities to that of Newton. I decided to concentrate on Newton since his concept is both more elaborated and, most probably, more influential.

Newton's concept took considerable more space than that of the other authors for several reasons. It is the most elaborated one, and Newton makes, in contrast to Galileo and Boyle, repeated use of it in the course of actual inquiry. His concept is of extraordinary influence due to the paradigmatic character which both the *Principia* and the *Opticks* quickly achieve. Of the five authors discussed, Newton is the only one to combine, in considering laws, a strictly mathematical approach with the claim of founding laws on systematic empirical inquiry, which combination has become so characteristic for all later concepts. Finally, his concept shows, more than the other ones, important characteristics, as well as problems, which are still involved in modern concepts.

The five authors form an interesting constellation. Descartes discusses Galileo, but not Bacon. Boyle, on the other hand, relies heavily on Bacon, but seems to adopt only very few of Cartesian ideas, although he mentions Descartes at some places. The two very different approaches of Cartesian and empiristic science might be said to meet in Newton. He studies very early and most intensely Descartes' *Principia Philosophiae*, by which

52 For a compilation of relevant passages, see Zilsel [1942: 265-7] and Ruby [1986: 351][this volume, p. 303]

study the formation of his conceptual framework is deeply influenced. On the other hand, he has an extensive exchange on natural philosophy with Boyle, who is then one of the most prominent promoters and representatives of an empiristic approach, institutionalized in the Royal Society. One of the very few other figures of the 17th century of whom I could think as being comparable in this respect is, of course, Huygens.

What I shall do now is to highlight some common aspects, and some essential differences, among the early concepts discussed above. I started my paper by mentioning the common view of a "process by which causes gave way to laws in science" (Drake 1981, xxv). I hope to have made clear that this view has to be seriously qualified. Even in the most elaborated concept of laws in the 17th century – that of Newton – laws serve as a means for determining causes. Laws are established long before causes begin to disappear from their prominent place in scientific inquiry. It is in the course of a fundamental revision of the prevailing conceptions of causes – and not through their simple supercession – that the concept of laws gains its importance. What becomes prominent now, is the idea of physical causes. In order to give an account of the physical cause of a phenomenon, one must stipulate the acting entity together with the law governing its action. These two are always clearly separated. In the context of a mechanistic approach – of whatever variety – the fundamental acting entities are regarded as being parcels of matter. The corresponding fundamental laws are laws of motion, since all change – even qualitative change – is regarded as being reducible to change of motion of matter. Laws of impact are, in this context, of the greatest importance, since action at a distance is regarded as being impossible, or at least highly problematic.

In both Boyle and Newton the concept of causes is more complicated, in that it involves a scale of causes. This is accompanied by an analogous scale of laws, extending from particular laws governing a rather small realm of phenomena, to the fundamental laws, referring to the microsocpic processes underlying all observable phenomena. Such a concept involves the idea of the fundamental laws being more powerful than the particular ones.

What Galileo calls the "inexorability of nature" in obeying the laws imposed on it, entails an aspect of regularity. Whenever certain conditions are present, a certain effect is expected according to a governing law. This aspect is more or less explicitly found again in all other mentioned accounts of laws. A distinctive example is found in Newton who, in the course of the discussions following his first optical paper to the Royal Society, repeatedly emphasizes that the effects described by him do not occur irregularly, but "according to a certain law" (1672a, 100; or 1672b, 125). As this passage illustrates, moreover, the expected effect is thought to be uniquely determined. Regularity is combined with determination. In nature, governed by God's laws, there is no place for chance.

The idea of laws of nature is strongly connected with theological considerations. As has plausibly been argued, the central factor in this respect is a strong voluntaristic approach (Oakley 1961a, 1961b, McGuire 1972, Milton 1981). Galileo's references to the Old Testament may illustrate this point. Laws of nature are regarded not as constituting the fabric of nature, but as being imposed on it. This view is not new. Oakley shows, in the case of Boyle, some strong commitments to an Ockhamian tradition (1961b, 448-9). Summarizing his studies of this tradition in natural theology, Oakley states that "the idea of laws of nature imposed by God upon the world was undoubtedly common coinage in the sixteenth and seventeenth centuries, even before Descartes, Boyle, and Newton made it a commonplace of scientific thinking, ..." (1961b, 446).

But I find among these various authors more essential differences than their common voluntaristic background might suggest. Descartes is the one who deals most explicitly with the difficulty which arises from acknowledging, on the one hand, the unlimited power of God, and, on the other, the point that there are certain truths which we cannot but conceive as being necessary ones. Besides mathematical and logical truths, he includes also the three laws of nature.[53] Descartes emphasizes that even these truths lie within the free will and power of God (1637, 41-3; letters to Mersenne of 15 April 1630, AT I, 145, and to Mesland of 2 May 1644, AT IV, 118). In order to explain the character of necessity included in their being conceived, he claims that God, as he establishes these truths, gives us simultaneously inborn concepts which lead us to understand these very truths as being necessary ones.[54] Only by this assumption is he able to justify his procedure of deriving necessary truths, including laws of nature, directly and with certainty from metaphysical and theological insights. He calls only those propositions of which he has a clear derivation from such principles, "laws of nature". Laws have what he calls metaphysical certainty in contrast to only moral certainty.

Boyle and Newton emphasize, like Descartes, that the laws of nature are imposed by the free will and providence of God. He could have created different ones, he could even change the actual ones if he liked (Boyle 1675, 161-2 and 1688, 396-7; Newton *Opticks*, Qu.31, 404). In contrast to Descartes, however, they do not include in this voluntaristic approach a principle which would warrant the reliability of derivations of laws of nature from metaphysical principles. Such a possibility is, therefore, excluded,[55] at least in what is publicly presented. Thus other sources of knowledge have to be envisaged. As propagated by Bacon, it is experience which is now presented as the foundation of knowledge.[56] While Boyle only assumes the existence of a path leading from experience to laws, without explicating it in detail, Newton gives at least a sketch of all of its stages and, what is more, an elaborated example.

---

53 Cf. the discussion in Loeck [1986: 121-3].

54 There is still a debate on the role of necessary truth in Descartes' reasoning on mathematics; see Frankfurt [1977], Funkenstein [1980], or Curley [1984]. This discussion seems to me to concern likewise laws of nature.

55 This is strongly emphasized by Cotes in his preface to the second edition of Newton's *Principia*, see PM, 33.

56 Oakley [1961b: 442] finds such an approach already in Ockham.

The claim that experience is the only base of knowledge of laws has far reaching repercussions on the concept itself. One has inevitably to envisage rules which can only be regarded as provisional means for the investigator or for practical use. It is this aspect which makes the concept of laws susceptible to possible instrumentalistic views. Thus the Baconian program forms, in a much higher degree than has previously been recognized, an essential and lasting element in the further development.

It is not accidental that Oakley, when highlighting the influence of the voluntaristic background, does not mention Bacon. Indeed Bacon does not easily fit into this tradition. His conceiving laws in intimate connection with forms makes one think of laws being inherent to matter rather than being imposed on it. Bacon regards, moreover, laws of nature as being eternal and immutable. These views cannot easily be combined with a plain voluntaristic approach. Bacon does not discuss these theological problems: indeed, in contrast to the other authors, such considerations do not have a really prominent place in his reasoning.

This difference makes one think of the factors which may be important here. Although such questions are not a main topic of this paper, I would like to briefly mention another medieval intellectual tradition which seems to be involved here, but which has not been sufficiently recognized in previous studies of the 17th century. As Schramm (1962 and, in more detail 1981) points out, Francis's namesake Roger Bacon, in the 13th century, introduces a distinction between "laws of universal nature" and "particular nature".[57] The problem in the course of which he develops this concept concerns the so-called "clepsydra". The main point is that, in using this bottle-like instrument, water can be prevented from falling downwards, although the bottom of the clepsydra is perforated with small holes. Roger Bacon explains this phenomenon, which had already puzzled

57 Ruby (1986) shows that Bacon uses the term law quite frequently in the context of his optical considerations. The specific point, however, seems to be visible mainly in the context discussed by Schramm.

Arabic philosophers, by introducing the above-mentioned distinction. According to the "particular nature", so he reasons, water would fall downwards. There is, however, a "law of universal nature", by which this tendency is overruled, and the water held above (quoted of Schramm 1981, 203). Thus "laws of universal nature" are more powerful than the "particular nature". It is this very aspect which is now again found in Francis Bacon. Roger's terminology of "particular natures" and "laws of universal nature" is reflected in Francis's use of "particular and special habits of nature" and "fundamental and universal laws". In both cases the term law is not applied to particular nature, but only to the more fundamental aspects.

In Boyle, the resemblance becomes even more striking. Not only does he use the same terminology as Francis and Roger Bacon, but he indeed discusses the same problem in the context of which Roger Bacon elaborates his ideas (see above, p.337). Of course, Boyle does not discuss the clepsydra itself, but the modernized version of the problem, as it becomes obvious in sucking pumps. In what he presents as a solution, he emphasizes similar aspects as outlined by the medieval philosopher.

I cannot discuss here the question of a possible continous line of tradition leading from Roger to Francis Bacon and Boyle. The latter conceptual and terminological resemblances make one assume a historical link. This assumption is supported by the observation of other similarities between the approaches of Roger and Francis Bacon, as have been pointed out by Thorndike (1923-58 II, 681-2), Crombie (1953, 300-3) and Rossi (1957, 27-8 and 68). Further investigation may lead to an improved understanding of that development.

The concept of laws of nature as presented by Newton and exemplified by his derivation of the gravitational law, is deeply impressive for his contemporaries, at least in Britain, and becomes paradigmatic for generations. This may, on the one hand, have to do with his influential position within the Royal Society. His method for deriving laws from experience involves, however, an aspect of greatest importance which makes it appear as being reliable. Newton presents his results as being based on a consequent mathematical treatment. His "Rules of reasoning in natural philosophy" are treated, moreover as inference rules, applicable in the course of empirical investigation as strictly as mathematical inference rules and thus leading to results which are to be regarded as certain inferences from phenomena. The program of "subjecting the phenomena of nature to mathematical laws", mentioned in the very first sentence of the *Principia* (PM, Auctoris Praefatio, 15), forms one of the most essential aspects of his approach. In his optical studies, he repeatedly emphasizes that he regards as one of his main achievements to have made accessible some new fields to mathematical analysis. His attempt to combine Galileo's and Descartes' claim for mathematical strictness and certainty with the empiristic claim for basing all inquiry on systematic investigation of phenomena, is reflected in his concept of laws. He presents laws as being, on the one hand, formulated with mathematical rigour, but, on the other, derived from empirical phenomena. This combination is found neither in Galileo and Descartes, nor in Bacon or Boyle.

This is not to say that Newton is the only one to have such an approach. There are, in this respect, some similarities to what one finds in Kepler. As Zilsel emphasizes, moreover, at the end of the 17th century the talk of "laws of nature" is quite common, at least in British circles. The terminology is used, for example, by Wallis, Wren, Hooke, Halley, Cotes, and Bentley, but also outside Britain by Spinoza and, most prominently, by Huygens[58]. Although the concept of laws is not discussed ex-

58 Many references are given by Zilsel [1942: 270-3]. See also Halley's letter to Newton of 29 June 1686 (in Newton [1959-77: II, 442]), or his review of Newton's *Principia* [1687: 405-6], or Cotes' preface to the second edition of the *Principia* [PM, 20, 24, 33, among others], or Bentley's sermons in which he mentions "the Laws of Vegetation and Life" [1693a, 331; resp. 1693b, 368]. For Huygens, see, e.g., his *Horologium Oscillatorium* [1673: 127], or his studies in optics and impact of bodies, published only posthumously in 1703 *[Dioptrique, Oeuvres* XIII, 4-12, 741, 772, 834]; *[De motu corporum ex percussione, Oeuvres* XVI, 33, 139, 150]. Hooke's naming, in 1678, his newly found proposition on the force of elastic bodies a "law" seems to be the first case in which "a physical law, referred to

plicitly, in some of these usages similar characteristics become visible as in Newton's concept; this holds particularly in the case of Huygens. My concentrating on Newton is not to imply that I see him as being the inauguator of this wide use; perhaps he gives it only a particularly clear and influential voice.

The pretension of combining systematic empirical inquiry and strict mathematical formulation is from then on inextricably involved in the concept of laws. The concept outlined by Newton serves as starting point for further developments in the 18th and 19th century. In the course of these developments the concept of law undergoes profound changes. Some essential aspects are altered even disappear, such as the theological commitments; others arise and become important, like statistical considerations. The driving forces of this development are manifold and complicated, all the more so as the field of investigation dramatically widens and spreads in many different directions. Some of the problems which are discussed repeatedly, indeed constantly, concerning laws are, however, already involved in Newton's approach, at least implicitly. I shall conclude my paper by making four points.

First, Newton – and likewise Boyle – claim that laws have to be derived from experience. Investigation has to start with empirical rules. Thus the problem arises of separating mere empirical rules, even of everyday experience, from what properly can be called laws. Descartes' criterion for laws – clear derivability from certain metaphysical principles – is no longer accepted. But it remains quite unclear what Newton gives as an alternative, if, indeed, he gives any.

Closely connected with this problem is the question of a realistic or instrumentalistic interpretation of laws. Newton regards laws always as real laws of nature, whereas empirical rules are allowed to be interpreted in an instrumentalistic way. Again he does not present criteria which would allow one to separate

---

in modern textbooks under the name of its discoverer, is called a law by the discoverer himself" (Zilsel [1942: 273]). The next case is Newton with his gravitational law.

clearly realistic from instrumentalistic options. Both problems are prominent in the discussion on laws until today.

A third point is exemplified by Newton's introducing laws of central forces, all while leaving the ontological status of the forces themselves in the dark. In doing so, he renders unavoidable the possibility of interpreting the laws describing such forces in an instrumentalistic way. This very problem – in combination with the related difficulties concerning theoretical entities – is alive in contemporary debates.

My last point deals with the status of fundamental laws. In what Newton and Boyle introduce as a hierarchy of laws, the more particular laws are thought to be derivable by means of the more fundamental ones. This entails a difference regarding the way in which the respective laws are derived from experience. One may compare in this respect, for example, Kepler's laws with the law of action of an ethereal substance, explaining gravity. The more fundamental a law is, the more indirect its derivation. As Nancy Cartwright discusses in detail (1983), the relation between fundamental laws and experience is significantly different from that between more particular laws and experience. One may well ask what consequences this difference has for our knowledge of laws. Given the empiristic idea, proposed by Newton, of deriving laws from experience, it is not at all clear why laws which are remote from experience should be regarded as certain as – or as even more certain than – those derived from more immediate contact with it. Newton simply does not discuss any further criteria involved here. But the problem seems to be unavoidable. The case is not yet closed.

References

Bacon, F. [1605]: *The twoo bookes of Francis Bacon of the Proficience and Advancement of Learning, Divine and Humane*. London: Henry Tomes. Cf. *Works* III.

Bacon, F. [1609]: *De Sapientia Veterum liber.* London: Baker. Cf. *Works* VI

Bacon, F. [1620]: *Instauratio Magna*. London: Bill. Cf. *Works* I

Bacon, F. [1623]: *De Dignitate et Augmentis Scientiarum libri IX*. Cf. *Works* I

Bacon, F. [1960]: *The New Organon, and Related Writings*. Ed. F.H. Anderson, Indianapolis: Liberal Arts Press

Bacon, F. [Works]: *The Works of Francis Bacon*, coll. and ed. by J. Spedding, R.L. Ellis, and D.D. Heath, London 1857-74; reprinted (vols 1-14) Stuttgart: Frommann-Holzboog 1961-63

Bentley, R. [1693a]: *A Confutation of Atheism from the Origin and Frame of the World. Part II*. London: Mortlock. Cf. Newton, P&L, 313-52.

Bentley, R. [1693b]: *A Confutation of Atheism from the Origin and Frame of the World. The Third and Last Part*. London: Mortlock. Cf. Newton, P&L, 353-94.

Boyle, R. [1660]: *New Experiments Physico-Mechanical, touching the Spring of the Air*. Cf. *Works* I, 1-117

Boyle, R. [1662a]: *A Defence of the Doctrine touching the Spring and Weight of the Air ... against the Objections of Franciscus Linus*. Cf. *Works* I, 118-185

Boyle, R. [1662b]: *An Examen of Mr. T. Hobbes his Dialogus Physicus de Natura Aeris*. Cf. *Works* I, 186-242

Boyle, R. [1663]: *Some Considerations touching the Usefulness of Experimental Natural Philosophy*. Cf. *Works* II, 1-201

Boyle, R. [1666a]: *Hydrostatical Paradoxes, made out by new Experiments* Cf. *Works* II, 738-797

Boyle, R. [1666b]: *The Origin of Forms and Qualities, according to the Corpuscular Philosophy*. Cf. *Works* III, 1-137

Boyle, R. [1671] *Of the Usefulness of Mathematics to Natural Philosophy*; enclosed in the second part of [1663]. Cf. *Works* III, 425-434

Boyle, R. [1675] *Some Considerations about the Reconcileableness of Reason and Religion*. Cf. *Works* IV, 151-191

Boyle, R. [1684]: *Of the High Veneration Man's Intellect owes to God, peculiarly for his Wisdom and Power*. Cf. *Works* V, 130-57

Boyle, R. [1686]: *A Free Inquiry into the Vulgarly Received Notion of Nature*. Cf. *Works* V, 158-254

Boyle, R. [1688]: *A Disquisition about the Final Causes of Natural Things*. Cf. *Works* V, 392-444

Boyle, R. [Works]: *The Works of the Honourable Robert Boyle*. Ed. T. Birch, 6 vol's, London 1772; reprint Hildesheim: Olms 1966

Cartwright, N. [1983]: *How the Laws of Physics Lie*. Oxford: Clarendon Press

Cohen, I.B. [1966]: "Hypotheses in Newton's Philosophy". *Physis 8*, pp. 163-84

Cohen, I.B. [1980]: *The Newtonian Revolution*. Cambridge: Cambridge University Press

Corson, D.W. [1974]: *The Newtonian Aether*. PhD Dissertation Cornell University

Crombie, A.C. [1953]: *Robert Grosseteste and the Origins of Experimental Science*. Oxford: Clarendon Press

Curley, E.M. [1984]: "Descartes on the creation of the eternal truths". *The Philosophical Review 93*. 569-597

Damerow, P., Freudenthal, G., McLaughlin, P. and Renn, J. [1992]: *Exploring the Limits of Preclassical Mechanics*. New York: Springer

Descartes, R. [1637]: *Discours de la Méthode ... plus la Dioptrique, les Météores, et la Géométrie.* Leyde: Jan Maire. Cf. AT VI.

Descartes, R. [1644]: *Principia Philosophiae*. Amsterdam: Elzevir 1644. Cf. AT VIII-1. French edition Paris: Le Gras 1647 (AT IX-2)

Descartes, R. [1664]: *Le Monde de M. Descartes, ou le Traité de la Lumière*. Paris: Le Gras. Cf. AT XI.

Descartes, R. [AT]: *Oeuvres*. Ed. C. Adam et P. Tannery. 11 vols. Nouvelle Présentation, Paris: Vrin 1982-1991.

Descartes, R. [CSM]: *The Philosophical Writings of Descartes*. Ed. and trans. by J. Cottingham, R. Stoothoff, and D. Murdoch. 2 vol's. Cambridge: Cambridge University Press

Drake, S. [1978]: *Galileo at Work*. Chicago: University of Chicago Press

Drake, S. [1980]: *Galileo*. Oxford: Oxford University Press

Drake, S. [1981]: *Cause, Experiment and Science*. Chicago: University of Chicago Press

Drake, S. [1983]: *Telescopes, Tides, and Tactics*. Chicago: University of Chicago Press

Fischer, K. [1983]: *Galileo Galilei*. München: Beck

Frankfurt, H. [1977]: "Descartes on the creation of the eternal truths". *The Philosophical Review 86*, 36-57

Funkenstein, A. [1980]: "Descartes, Eternal Truths and the Divine Omnipotence" in Gaukroger (ed.), 181-95

Galilei, G. [1613]: *Lettera à D. Benedetto Castelli, (21 Dicembre 1613)*. Cf. EN V, 281-8

Galilei, G. [1615]: *Lettera à Madama Cristina di Lorena*. Cf. EN V, 309-48

Galilei, G. [1623]: *Il Saggiatore*. Roma. Cf. EN VI

Galilei, G. [1624]: *Lettera à Sig. Francesco Ingoli*. Cf. EN VI, 509-61

Galilei, G. [1632]: *Dialogo sopra i due massimi sistemi del mondo Tolemaico e Copernicane*. Fiorenza. Cf. EN VII

Galilei, G. [1638]: *Discorsi e dimostrazioni matematiche intorno à due nuove scienze*. Leiden: Elzevir. Cf. EN VIII

Galilei, G. [EN]: *Le Opere di Galilei*. 20 vols; Edizione Nazionale (Ristampa). Firenze: Barbèra 1968

Garber, D. [1992]: *Descartes' metaphysical physics*. Chicago: University of Chicago Press

Gaukroger, S. (ed.) [1980]: *Descartes: Philosophy, Mathematics, and Physics*. Brighton: Harvester Press

Halley, E. [1687]: Review of Newton's *Principia* in *Philosophical Transactions 186*, pp 291-7. Quoted from the reprint in Newton, P&L, 405-11

Hansen, B. [1978]: "Science and Magic" in D. Lindberg (ed.): *Science in the Middle Ages*. Chicago: University of Chicago Press; pp. 483-98

Huygens, C. [1673]: *Horologium Oscillatorium*. Paris: Muguet. Cf. *Oeuvres* XVIII

Huygens, C. [Oeuvres]: *Oeuvres complètes de Christiaan Huygens*. Publieés par la Société Hollandaise des Sciences. 22 vols. La Hague: Nijhoff 1888-1950

Klaaren, E.M. [1985]: *Religious Origins of Modern Science*. Lanham: University Press of America

Koyré, A. [1965]: *Newtonian Studies*. London: Chapmann & Hall

Krohn, W. [1987]: *Francis Bacon*. München: Beck

Larmore, C. [1980]: "Descartes' Empirical Epistemology" in Gaukroger (ed.), pp. 6-22.

Loeck, G. [1986]: *Der cartesische Materialismus: Maschine, Gesetz und Simulation*. Frankfurt/M.: Peter Lang

Maull, N.L. [1980]: "Cartesian Optics and the Geometrization of Nature" in Gaukroger (ed.), pp. 23-40.

McDonald, J.F. [1972]: "Properties and Causes: An Approach to the Problem of Hypotheses in the Scientific Methodology of Sir Isaac Newton". *Annals of Science 28*, pp 217-33

McGuire, J.E. [1972]: "Boyle's Conception of Nature"; *Journal of the History of Ideas 33*, pp. 523-42

Milton, J. [1981]: "The origin and development of the concept of the 'laws of nature"'. *Archives Européennes de Sociologie 22*, pp. 173-95

Needham, J. [1956]: *Science and Civilisation in China, vol.2*. Cambridge: Cambridge University Press

Newton, I. [1672a]: "Mr. Newtons Answer to the foregoing Letter [of Pardies]". *Philosophical Transactions 85*, pp. 5014-8. Cf. P&L, pp. 99-103

Newton, I. [1672b]: "Mr. Isaac Newtons Answer to some Considerations [of Hooke] upon his Doctrine of Light and Colors". *Philosophical Transactions 88*, pp. 5084-103. Cf. P&L, pp. 116-35

Newton, I. [1744]: Newton's letter to Boyle of 28 February 1678/9. Cf. Boyle, *Works* I, pp. 70-3.

Newton, I. [1756]: *Four Letters from Sir Isaac Newton to Doctor Bentley, containing some Arguments in Proof of a Deity*. London: Dodsley. Cf. P&L, pp. 279-312.

Newton, I. [1757]: Newton's second paper on color and light, read at the Royal Society in 1675/6 in vol.3 of T. Birch: *The History of the Royal Society of London*. London: Millar, pp. 247-305. Cf. P&L, pp. 177-235.

Newton, I. [1779-85]: *Isaaci Newtoni Opera quae exstant omnia*. Ed. S. Horsley, 5 vols, London: Nichols. Reprint Stuttgart: Frommann-Holzboog 1964

Newton, I. [1934]: *Sir Isaac Newton's Mathematical Principles of Natural Philosophy and his System of the World*. Transl. A. Motte, rev. F. Cajory, 2 vol's, Berkeley: University of California Press

Newton, I. [1952]: *Opticks*. Based on the fourth edition 1730, New York: Dover

Newton, I. [1959-77]: *The Correspondence of Isaac Newton*. Ed. H.W. Turnbull, J.F. Scott, A.R. Hall and L. Tilling, 7 vol's, Cambridge: Cambridge University Press

Newton, I. [P&L]: *Isaac Newton's Papers and Letters on Natural Philosophy*. Ed. I.B. Cohen, Cambridge (Mass.): Harvard University Press 1958

Newton, I. [PM]: *Isaac Newton's Philosophiae naturalis Principia mathematica*. 3rd ed. with variant readings; ed. A. Koyré and I.B. Cohen, 2 vol's, Cambridge (Mass.): Harvard University Press 1972

Newton, I. [1962]: *Unpublished Scientific Papers of Isaac Newton*. Ed. A.R. Hall and M.B. Hall, Cambridge: Cambridge University Press

Oakley, F. [1961a]: "Medieval theories of natural law: William of Ockham and the significance of the voluntaristic tradition". *Natural Law Forum 6*, pp. 65-83

Oakley, F. [1961b]: "Christian theology and the Newtonian science: the rise of the concept of the laws of nature". *Church History 30*, pp. 433-57

Oakley, F. [1984]: *Omnipotence, Covenant and Order.* Ithaca: Cornell University Press

Rodis-Lewis, G. [1993]: "From Metaphysics to Physics" in Voss (ed.), pp. 242-58

Rossi, P. [1957]: *Francesco Bacone: Dalla Magia alla scienza*. Bari: Laterza. Quoted of the English translation: *Francis Bacon: From Magic to Science.* London: Routledge 1968

Ruby, J.E. [1986]: "The origins of scientific 'law' ". *Journal of the History of Ideas 47*, pp. 341-359

Sabra, A.I. [1967]: *Theories of Light from Descartes to Newton*. Oldourne. Repr. Cambridge: Cambridge University Press 1981

Schimank, H. [1949]: "Der Aspekt der Naturgesetzlichkeit im Wandel der Zeiten" in *Das Problem der Gesetzlichkeit II: Naturwissenschaften*; ed. Joachim Jungius-Gesellschaft der Wissenschaften; Hamburg: Meiner, pp. 139-186

Schramm, M. [1962]: "Aristotelianism: basis and obstacle to scientific progress in the middle ages. Some remarks on A.C. Crombie's 'Augustine to Galileo' ". *History of Science 2*, pp. 91-113

Schramm, M. [1981]: "Roger Bacons Begriff vom Naturgesetz" in Weimar, P. (Hrsg.): *Die Renaissance der Wissenschaften im 12. Jahrhundert*. Zürich: Artemis, pp. 197-209

Schuster, J.A. [1993]: "Whatever should we do with Cartesian Method? Reclaiming Descartes for the History of Science" in Voss (ed.), pp. 195-223

Shapin, S. and Schaffer, S. [1985]: *Leviathan and the Air-Pump: Hobbes, Boyle and the Experimental Life*. Princeton: Princeton University Press

Steinle, F. [1992]: "Was ist Masse? Newtons Begriff der Materiemenge". *Philosophia Naturalis 29*, pp. 94-117

Thorndike, L. [1923-58]: *A history of Magic and Experimental Science*. New York: Columbia University Press, 8 vol's.

Urbach, P. [1987]: *Francis Bacon's Philosophy of Science.* La Salle: Open Court

Voss, S. (ed.) [1993]: *Essays on the Philosophy and Science of René Descartes.* New York: Oxford University Press

Wheeler, H. [1983]: "The invention of modern empiricism: juridical foundations of Francis Bacon's philosophy of science". *Law Library Journal 76*, pp.78-120

Zilsel, E. [1942]: "The genesis of the concept of physical law". *The Philosophical Review 51*, pp. 245-79

Peter Galison

Theory Bound and Unbound: Superstrings and Experiment[1]

1. Introduction: Theory Unbound

In mid-July 1986, Steven Weinberg stood before the International Conference on High Energy Physics, then meeting in Berkeley, and delivered what should have been welcome news: theory matched experiment to a T. In virtually every sector of experiment the collocation of theories known as the Standard Model matched the results of a generation of accelerator work from the Stanford Linear Accelerator Center in Palo Alto, Fermilab outside Chicago and the Brookhaven National Laboratory in New York; all the way through CERN bordering Geneva and the Deutsches Elektronen Synchrotron (DESY) near Hamburg to Novosibirsk in Siberia. One quark (of six) still was missing, but the search had by no means been exhausted. Challenges had come and gone: monojets (isolated pencil-thin bursts of particles) "shouldn't" have existed but they appeared at CERN, then disappeared back into the noise; massive neutrinos with no place in the model had been detected, then dropped; a new ultra-weak force might have been seen, but it looked increasingly dubious. Despite these "peculiarities," or perhaps because of their number and short half-life, theorists were growing increasingly inured to oddities in the data, less willing to plunge into new speculation. And while the strongly-interacting sector of the Standard Model could

1 I would like to thank Howard Georgi, Caroline Jones, Cumrun Vafa, and Steven Weinberg for many helpful discussions. Richard Beyler, Pamela Butler, and Michael Gordin ably assisted in tracking down sources and preparing the manuscript.

not be used to derive quantitatively the complexities of nuclear physics – for example – the theoretical community chalked that circumstance up to the known calculational tribulations of a theory poorly suited to the perturbative expansions used so successfully in quantum electrodynamics. If the goal of theoretical physics was the prediction, retrodiction, and quantitative account of stabilized, recognizably "relevant" phenomena, the Standard Model was a spectacular, indeed unrivalled success. Everyone "should" have been content. After decades of unexplained tracks in cloud chambers, bubble chambers, and nuclear emulsions, after countless thousands of nights staring at the photographic remnants of the subvisible and the data tapes from electronic experiments, theory, at last, had caught up with experiment.

Yet (Weinberg continued) in the years since the Standard Model's consolidation in 1974-75, a deep unease had settled over the theoretical community. But this was not the unease of being unable to account for experimental anomalies, new particles, interactions, or effects. No, what Weinberg reported was dissatisfaction of another sort: in what we might call *the intratheoretical* domain. Why, he and other colleagues had long demanded, were there so many, ("embarrassingly many") free parameters in the theory, such as the strength by which quarks and leptons were bound to the Higgs particle (the entity responsible for separating the electromagnetic from the weak force). Why do the particle masses have the values they do? Why is the symmetry of the theory just SU(3) x SU(2) x U(1) and none other? Why are there three versions of each of the exclusion principle-obeying particles – an electron, for example, is accompanied by exactly two heavier versions, the muon and an even heavier partner, the tau? After a decade of furious work on these and other intratheoretical lacunae, the list of proposed solutions had grown ever longer.

One proposed solution to some of these issues was known as technicolor (a scheme to extend, analogically, the dynamics of the newly articulated color force that explained nuclear particles as composites made of quarks). Technicolor modeled a composite Higgs on just this sort of scheme. Another explanatory scheme involved attributing a substructure to the quarks and leptons themselves. Yet another theoretical reassessment had been dubbed supersymmetry; it paired every known whole-spin particle with a postulated half-spin partner, and every known half-spin particle with a postulated whole-spin particle. An electron (half-spin) got a hypothesized whole-spin mate, the *selectron*; the whole spin photon got a half-spin partner, the *photino*. Grand Unified Theories (GUTs) were yet another scheme that would explain much of what the standard model had to postulate. In it the symmetries that had brought the weak and electromagnetic forces together would be extended to embrace the strong (color) force as well. "You know," Weinberg lamented, "just listening to myself reciting that list, I feel a sense of tremendous frustration. We've been working on these ideas [the theoretical "loose ends"] for more than a decade, since the mid-1970s, and we have almost nothing to show for it in terms of hard agreement between predictions and experiments." (Weinberg, [1987a], p. 271). No theoretical mechanism, that is, that would first explain why the parameters of the Standard Model were what they were, why there were precisely three generations, and so on, then go further to make a concrete, confirmed, and novel experimental prediction. (The one exception to this being one parameter predicted by grand unified theories.) Despite this gloomy assessment, Weinberg went on to fly the banner of a theory that offered new hope – even while it lacked any confirmed experimental test: "superstrings."

Indeed, in the months preceding Weinberg's summary talk, and almost exactly a decade after the events of November 1974 that heralded the "gauge revolution" that had ushered in the Standard Model, theoretical physicists had begun speaking of a "string revolution." Launched in August 1984 (Gross, [1989], p. 311), this was not a revolution that took hold by predicting a raft of new entities; it was not remotely like the heady days of the mid-1970s in which theorists hung on each latest experimental development at the Stanford Linear Accelerator Center, the Brookhaven National Laboratory or the Deutsches Elektronensynchrotron to interpret data and suggest new regions in which to search. Perhaps just because the events of 1984 and what followed were so *unlike* the recent experience of gauge physics, advocates and detractors alike began to invoke history to ground their claims.

According to some, the introduction of strings was unlike Dirac's introduction of his relativistic equation, others contended that string theory was similar to Einstein's discovery of general relativity, and yet others likened string theory to the fruitless quest Einstein had undertaken for a Unified Field Theory – and these were but a few of the analogies and counter-analogies drawn on each side. Never before had the history of physics been invoked so liberally in defense of current research. While to some, theory had come unbound from its necessary and historical moorings in the details of experiment, to others theory had finally arrived at its telos. At long last, advocates asserted, theorists could derive the foundation of physics largely on the basis of mathematical self-consistency alone.

My concern here is to explore how it was possible for Weinberg, at the end of his 1986 Berkeley address, to embrace string theory as a fundamental account of all known physics, a "final theory," *and yet* in the same breath, aver that "the phenomenological success of superstring theory is not part of its justification so far." (Weinberg [1987a], p. 280). Baldly put, the superstring theorists had a final theory of nature that had little traffic with experiment beyond the existence of gravity and the handedness of certain particles. It is this combination of ambition and detachment that would have been so unimaginable in almost any other time in the nineteenth or twentieth century, and that commands our attention. For what the late twentieth-century is witnessing in the string controversy is a profound and contested shift in the position of theory in physics.

For decades, perhaps centuries, theorists and experimentalists have viewed theory as irreducibly constrained by quite immediate features of experimental results – it was always possible to make vastly more theories than were consistent with laboratory results, and news from the benchtop or accelerator floor halted, guided, and confined theoretical work. That constraint structure has now shifted. String theorists are not, in the first instance, looking to the daily produce of laboratories for the elimination, suggestion, or constraint of theories. Instead, string theory has turned to mathematics, not simply to the long-established domains of Lie algebras, Fourier transforms, and the linear algebra of quantum mechanics and gauge theory, but to a vast array of mathematical techniques and disciplines some of which are being worked out simultaneously by physicists and mathematicians. In this process the disciplinary and social location of theoretical physics has been, often painfully, displaced.

Closely tied to this shift in disciplinary location is a philosophical displacement in the character of realism as defended by practitioners. In the past a standard defense of the realistic character of physical law invoked the sense that experiment matched theory in ways and to a degree that lay outside the theorist's control. Now, with string theory, that effect behind realism, the sense that a selection takes place outside the domain of theory, has been shifted away from the specifics of masses and lifetimes. Instead, the broad features of physical theory serve to provide certain constraints on any theory, but even more importantly is the role of the much more abstract notion of consistency. Weinberg: "the real reason so far for being interested in superstring theories is that they are mathematically consistent[,] finite ... relativistic quantum theories, of which there are pretty few, and these are the only ones we know that contain gravity." (Weinberg, [1987a], p. 280).

Weinberg may have been a forceful spokesperson for the superstring revolution, but he was certainly not alone. In theoretical physics groups across the United States, Europe, and the fragmenting Soviet Union, the race for string theory shot forward after 1984. Behind the enthusiasm lay a view of nature as a unified whole with underlying principles that we can grasp largely through mathematics. Opposing this view lay another image of nature: nature as always surprising, nature marked by pockets of order hard won from the intense study of experimental work. On the pro-string view experiment, while needed, could wait; the order of the world would reveal itself through that single theory that will pass through the narrow gates of consistency. On the antistring view, experiment was necessary not down the road but right now; without experiment, theory would wander aimlessly.

To understand this divide is to glimpse the changing role of theory in the physics of the late twentieth century.

2. Superstrings: The Final Theory

Behind the idea of string theory is this conviction: to unify gravitational, nuclear, and electromagnetic physics, it is necessary to move beyond a world of point particles and quantum field theory. Instead of structureless point particles as basic, string theory has one-dimensional objects (strings) held under enormous tension – some 10^{39} tons – with vibratory states that correspond to the different elementary particles. These tiny strings are to be on the order of 10^{-33} cm long (some 10^{20} smaller than a proton). Just as point particles arc out one-dimensional "worldlines" in space-time, so these strings carve out two-dimensional "worldsheets." A closed string, for example, would trace the surface of a distorted cylinder in spacetime; the coming together in spacetime of two such objects would construct a surface something like that of a pair of pants. These and related geometrical objects open onto vast areas of modern mathematics – Riemann surfaces, algebraic topology, and even number theory.

The string debates revolve around an agreement and a profound disagreement. By consensus, theorists agree that theory must operate under a series of constraints, and that those constraints must confine theoretical work to a very few, if not a single representation of some sector of the world. The dispute is over the source and appropriateness of those constraints. To particle theorists like Howard Georgi, Sheldon Glashow and others, the constraints must come, in large measure, from the nitty-gritty of experiment. To string theorists like David Gross, Edward Witten, John Schwarz and Michael Green, such experimental constraints might be desirable, even possible in the long run, but in the here and now, theory must look to its own set of constraints, those provided by the most basic features of the world combined with powerful mathematical and aesthetic constraints. These, string advocates insist, can or at least might isolate a single "theory of everything."

String theory had its origin far from unified theories. For during the 1950s and 1960s literally millions of bubble chamber photographs had flooded theorists with new particles, resonances, and lifetimes. Among the many theoretical responses to the deluge was a series of empirical formulae and conjectures that led in the late 1960s and early 1970s to a picture of the hadrons (strongly interacting particles) predicated on quarks bound together by one-dimensional strings, elastic bands on the order of 10^{-13}cm. Vibrations of these strings became the different resonance of, for example, the pion. Philosophically, the early string program issued from the attempt to discuss the strong interactions in the light of the general properties of scattering processes – rather than the detailed mechanisms modeled on quantum electrodynamics sought by quantum field theorists. These early string results, designed to account for even-half-spin particles (bosons) were then expanded in 1971 to include a string account of odd-half-spin (fermion) particles like the proton.[2] This program came to a screeching halt in 1973-74, with the extraordinary success of gauge physics, both in theory (in particular the discovery that under certain conditions quarks inside nucleons acted free of one another) and experimental success (the discovery of neutral currents and shortly thereafter the J/psi and other charmed particles).[3] In addition to the success of the competition, string theories of nuclear particles had their own, internal failings: inter alia, they predicted the existence of a new massless spin-2 particle that was nowhere to be seen in the hadronic world.

---

2 String models of hadrons are reviewed in Scherk [1975]; many of the original string papers are reprinted in Schwarz [1985].

3 On the discovery of neutral currents, see Galison, 1987, chapters 4-6.

Unwilling to part with the remarkable mathematical-physical structure of string theories, Schwarz and the young French physicist, Joël Scherk, brought their creation back to life later in 1974 by changing the scale of the theory, and by reinterpreting the misbegotten spin-2 particle as the quantum of gravitational force, the graviton. Quite independently, T. Yoneya came to the same conclusion. (Yoneya [1974]). This quite radically reformulated the meaning of the formalism – overnight strings went from lying at the interior of hadrons and stretching across a proton diameter (10^{-13}cm) to being objects at the Planck length: a scale one gets by combining the fundamental constants of relativity, quantum mechanics, and gravity, $L_p = (hG/c^2)^{1/2} = 10^{-33}$cm. By moving their physics up in energy by a factor of twenty powers of ten, Schwarz and Scherk advanced on two goals: 1) they hoped to save their hadron theory by finding work for the otherwise unemployed spin-2 massless particle, and 2) they hoped to save quantum gravity by employing string theory to make a finite theory of quantum gravity. Strings were no longer a form of intrahadronic glue, they were now the defining fabric of spacetime itself. (Scherk and Schwarz [1974]). Few listened. For many of the next ten years, the development of string theory was slow, involved only a handful of people, and had a highly restricted audience, even among theoretical particle physicists. First, a few authors began exploring the possibility that string theories might be made supersymmetric, that is, that the theories might contain a symmetry that postulated a fermion for every boson. Such theoretical objects were known as superstrings. By itself, the constraint of supersymmetry does not tightly limit the construction of string theories, but there were other considerations as well. (See figure 1.)

In 1983, at the Fourth Conference on Grand Unification, Edward Witten [1983] spoke persuasively about the promise of the new theories, and highlighted a further set of constraints: the mixed gauge and gravitational superstring anomalies had to be eliminated. Quantum field theorists had, of course, long known about other kinds of anomalies: terms generated when the classical version of a theory has a certain symmetry, but where that

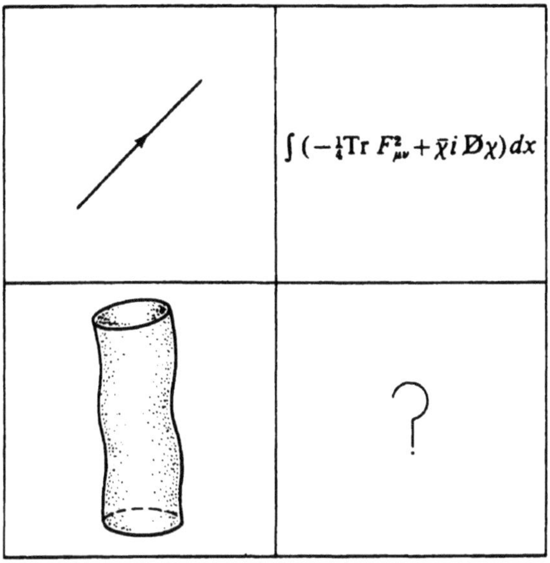

Figure 1. Superstring's Magic Square. In the upper left hand corner, Green, Schwarz, and Witten have placed the worldline (in spacetime) of a point particle. Made supersymmetric (every exclusion-principle-obeying particle has a non-exclusion-principle- obeying partner), the equation for a point particle sits in the upper right. In the bottom left is the world sheet of a closed string – a deformed cylinder in spacetime. Needed, Green, Schwarz, and Witten insisted in 1987, was a full theory, grounded on a cleanly stated physical principle that would combine string theory, supersymmetry and Einstein's general theory of relativity. These physical principles would then ground the mathematical successes of string theory as it was known now. Source: Green, Schwarz, and Witten [1987], p. 27.

symmetry was violated by quantum corrections. The danger of anomalies lies in the symmetries they destroy. Gauge theories, for example, have certain symmetries: in the simplest case, if one takes every quantum field and modifies each wave function by an arbitrary phase at each point in spacetime, the theory remains unchanged. More generally, the quantum fields can be altered in more complex ways, for example by mathematical operators that form a group, but where the order of application matters (simple phase changes, by contrast, can be made in arbitrary order).[4] Such non-commuting symmetries can be vitally important; gauge symmetries assure the renormalizability of the theory, and without renormalizability, a theory is hardly a theory at all. (In a nonrenormalizabile theory, quantum calculations lead to infinite quantities, and the theory loses its predictive capacity.) In the case of the standard model, anomalies break the gauge symmetry of the theory, spoil its renormalizability, and so render it inconsistent.

Cancelling the gauge anomalies in the Standard Model became an absolute necessity to salvage the consistency of the theory. It turned out that for the anomalies to cancel, there had to be an equal number of types of leptons (such as electrons, neutrinos, or muons) and types of quarks (such as the up quark, the down quark, the strange quark, and the bottom quark). In this way, the constraint of renormalizability led historically to a powerful prediction: given the six known leptons, there had to be sixth quark.

Similarly, superstring theory can be constrained by the condition that the dreaded anomalies must be cancelled. For example, for the theory to be consistent, it must be possible to relabel the points on the world sheet of a string without changing the physics. This symmetry, known as conformal invariance, has an anomaly – that is, even if the theory itself has confor-

---

4 Symmetry groups can be classified exhaustively. A simple phase symmetry is dubbed the unitary group U(1) – it is this symmetry that characterizes electromagnetism. Rotations in two dimensions are known as the orthogonal group, O(2), and higher-dimensional extensions of the rotation group follow the nomenclature O(3), O(4), ... O(n). Another infinite series of order-dependent (non-commuting) groups find a two-dimensional representation in the complex two-by-two Pauli spin matrices of nonrelativistic quantum mechanics. Known as the special unitary group, these are designated SU(2), SU(3), ... SU(n). Combined symmetries such as the SU(2)xU(1) symmetry of the electroweak interaction play a central role in the last 25 years of unified field theories. In addition to these infinite series of groups, there are particular symmetry groups which include the exceptional group, E_6 that play an important role in grand unified theories joining the strong and electroweak forces, and E_8, which has particular importance in the string theories.

mal symmetry, quantum mechanical corrections threaten to destroy it. In the nonsupersymmetric string theory, cancellation of this conformal anomaly meant that the background space (in which the strings live) has to have precisely 26 dimensions; in the supersymmetric case, a similar cancellation of the conformal anomaly forced the background spacetime to have 10 dimensions (nine spatial, one temporal). Contemplating these "constraints on possible string theories," Michael Green remarked, "It is in this sense that superstring theory is immensely elegant." (Green [1986], p. 55).

Consider the aesthetic virtue of the conformal symmetry constraint a bit further by contrasting strings with point particles. In the quantum field theory of point particles, as quintessentially exhibited by quantum electrodynamics, the theory is typically organized into a perturbation expansion in powers of the electric charge, e. For any given process – such as the scattering of a positron from an electron, certain terms in the perturbation expansion are relevant; corresponding to each such term is a Feynman diagram which (in this case) has a positron worldline and an electron line coming in and one of each coming out. Between the arrival and departure of the electron and the positron, anything that can happen (is allowed by conservation laws to happen) will happen. The electron and proton can annihilate and form a virtual photon, the virtual photon can then disintegrate into a positron and an electron. Or the same thing might happen except the virtual photon might spend part of its ephemeral life as another positron and electron along the way. As the number of particles involved increases, or as the order of perturbation theory increases, the number and complexity of these diagrams rapidly increases. Nowhere is the challenge of such diagrammatic analysis so well illustrated as in the work of T. Kinoshita, the acknowledged master of higher-order corrections to the muon and electron magnetic moments. In these calculations, which lead to the ten-significant figure predictions that have been matched to experiment, twentieth-century science, finds one of its most astonishing results. But the cost in complexity is astonishing. In 1982,

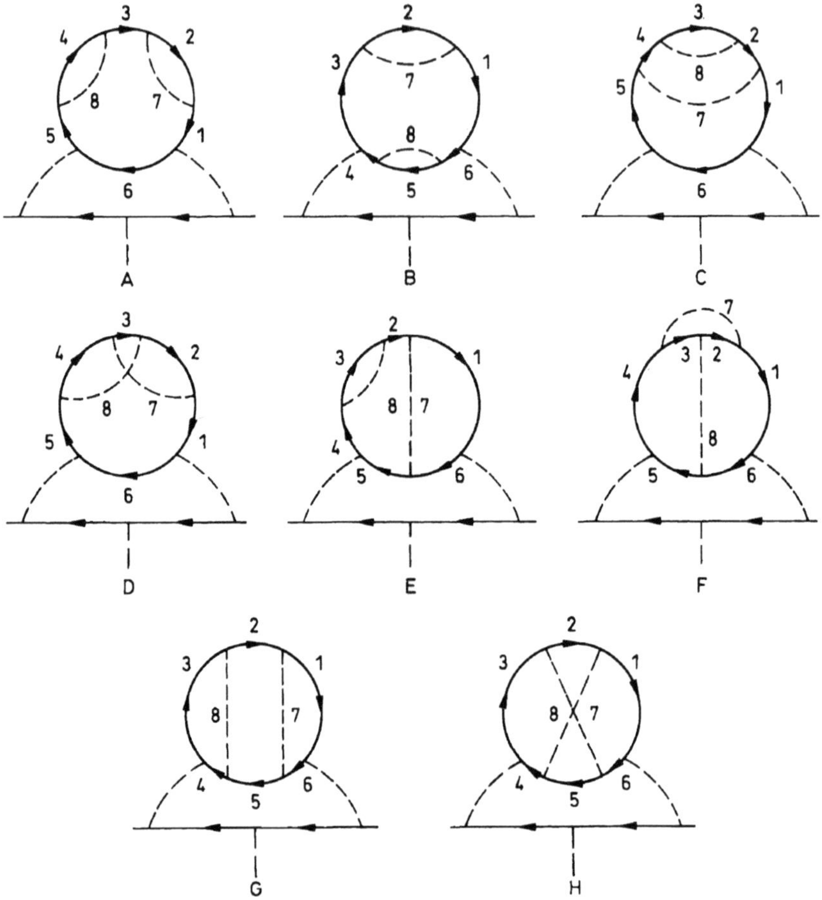

Figure 2. Kinoshita's Feynman diagrams. Kinoshita and Lindquist [1983], p. 868.

Kinoshita published eighth-order corrections to the muon magnetic moment – and had to face 891 diagrams, each distinct from the other. (See figure 2).

In string theories, the world line of a point particle is, as noted above, replaced by a worldsheet. For closed strings, the worldsheet forms a cylinder, and two strings in interaction form something like figure 3a. Conformal symmetry, the mapping of a complex point z into f(z), can take any topologically equivalent shape into the shape of figure 3a. Only surfaces with (for ex-

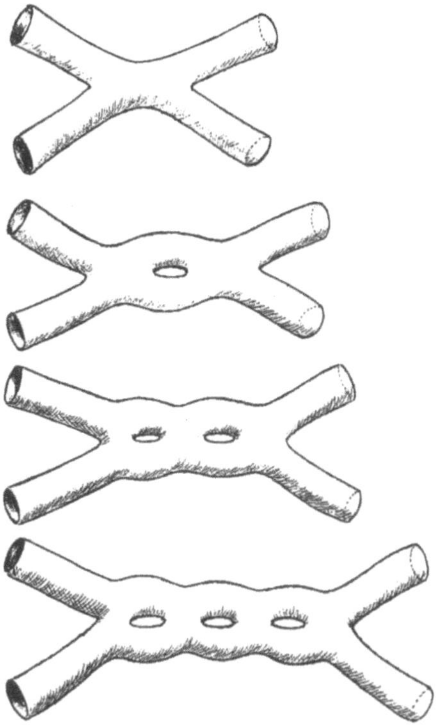

Figures 3a-3d. Feynman diagrams for two closed strings scattering into two closed strings

ample) a hole in the middle of the shape, such as figure 3b, cannot be so altered. As it turns out, for important classes of string theories the perturbation expansion of the string theory corresponds precisely to the series of such doughnuts with ever more holes, figures 3a-3d. Kinoshita's 891 diagrams become one. So it was that mathematics – as here with the topological and algebraic properties of two-dimensional Riemann surfaces – radically reshaped the constraints, simplicity, and as Green put it, the "elegance" of physics.

Witten in particular wanted to know whether the combined symmetries of gauge theory and gravitational interaction would lead to anomalies.

Soon, Witten, working with a younger collaborator, Luis Alvarez-Gaumé, could demonstrate that the dangerous version of the anomalies, the "hexagon anomaly" canceled in one version of the theory. (Alvarez-Gaumé and Witten [1983]). Though this was a calculational tour-de-force, became widely known as one of the "miracles" of string theory, and provided a new theoretical technology for the string theorists, the particular model Witten and Alvarez-Gaumé explored was not phenomenologically compatible with the particles of the standard model.

Anomalies played one final role in setting the constraint structure of any possible string theory. It was a long-established feature of "low-energy" particle physics (low compared with the energy of string theory) that many particles came either left- or right- handed. This "chirality" can be illustrated by the neutrino, which only occurs in the "left-handed" form. That is, the angular momentum of a neutrino is always left-handed with respect to the neutrino's linear momentum. If we were to look at the relation of the angular momentum of a neutrino to the momentum of that neutrino in a mirror, however, we would have a "right-handed" neutrino, an object not to be found in nature. So any theory of particles must be preferentially left- or right-handed in certain respects. Since even space dimensions (e.g. 2-dimensional space, 4-dimensional space, etc.) are always reflection invariant, the number of space dimensions had to be odd. But even in theories with odd spatial dimensions, quantum corrections frequently lead to anomalies in chiral theories, and these anomalies can destroy crucial symmetries. Disastrously for superstrings, some theorists believed in the early 1980s that it was impossible in 10 dimensions to have a chiral superstring theory. (Green, [1986], p. 56-57). Since superstrings had to live in 10-dimensional space, the constraints seemed to leave no exit. Then came the string revolution.

It was in August 1984 that Schwarz and Green, building on the work by Witten and Alvarez-Gaumé, showed how to get a chiral, anomaly-free theory using the specific gauge symmetries, SO(32) and $E_8 \times E_8$. For the former, they could actually produce a full string theory. (Green and Schwarz [1985a, 1985b]). Their efforts were followed shortly by those of the

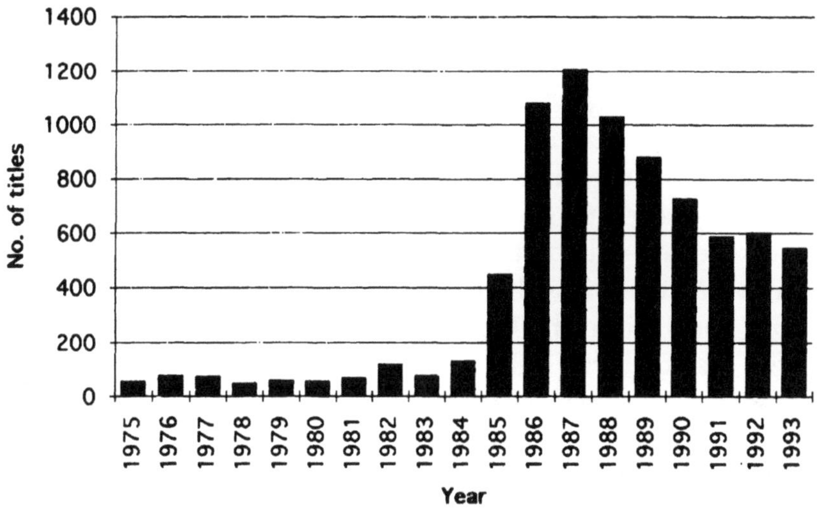

Figure 4. Number of articles on strings and superstrings, 1974-1993.

"Princeton string quartet," David Gross, Jeffrey Harvey, Emil Martinec, and Ryan Rohm who, (in [1985]), constructed a hybrid variant of string theory that went with E_8 x E_8. Theirs was a mixture of two closed string theories, one was the older bosonic string in 26 dimensions and the other was the newer superstring theory in ten dimensions. Dubbed "heterotic" after the strength of mixed biological species, the new string theory promised phenomenological success along with theoretical value. Hopes ran high as the group E_8 clearly contained as a subgroup the symmetry associated with one of the known grand unified theories. The quartet concluded that "Although much work remains to be done there seem to be no insuperable obstacles to deriving all of known physics from the E_8 x E_8 heterotic string." (Gross, Harvey, Martinec, and Rohm [1985], p. 283). Or again, as Schwarz [1985] put it a bit later that same year, "Superstring theory – or, more specifically, the E_8 x E_8 heterotic superstring – could be a 'theory of everything' (p. v)."

Reaction among theorists was extraordinary, as figure 4 indicates.

The steady stream of string and superstring articles from 1974 to 1984, was followed by an explosive growth – up to 1,200 articles in 1985 – followed by a gentle decline to a roughly steady-state number of 500-600 articles per year in the period following the self-proclaimed "revolution."

Now there were only a handful of possible string theories, but for some theorists even this handful was too much. One problem was that there might be a multitude of compactification schemes – theoretically driven compression mechanisms that would compress the 10 or 26 dimensions of the full superstring theory into the four spacetime dimensions that we observe. As Schwarz [1985] wrote, anomaly conditions, perhaps at the two-loop level, might winnow the race. "Additional conditions, such as these might provide, are hoped for since five theories, each with many possible compactifications, is more than we really desire. It is conceivable that the correct theory is the only one possible" (pp. 942-943). And elsewhere, in 1987: "Three theories are known, but it would be extremely nice if the number could be reduced to one. Then we could argue that there is a unique consistent theory that accounts for all of fundamental physics." (Schwarz, [1987b], p. 40).

This search for uniqueness and what Weinberg called "rigidity" is the attribute of a theory that constraints fully secure. Supersymmetry by itself would not do the job; indeed, the wide variety of mixed gravitational and particle theories had so far been available in many variants, none of which were uniquely determined, and none of which were completely consistent. Ordinary Lorentz invariance, taken by itself, provided Weinberg with an example of an incompletely rigid constraint: it is not difficult to combine vectors, axial vectors, and tensors to form a wide variety of relativistically invariant theories, an infinite number of which have strictly nothing to do with the physical world. Superstring theories, Weinberg insisted in October 1985, are different: they are "rigid." There are, he added, "almost no string theories at all." A free field string theory has interactions that only arise from the topology of two-dimensional manifolds – and these are specified uniquely by the simplest of properties: the number of handles that you can put on them. "[S]o there is nothing you can tinker with in these theories; they are either right or wrong as they stand." (Weinberg [1986], pp. 236-237). Superstring theory, he continued a few months later, "has the smell of inevitability about it." In superstrings, according to Weinberg, we find "a theory that cannot be altered without messing it up; that is, if you try to do anything to this theory, add additional terms, change any of the ways that you define things, then you find you lose these symmetries, and if you lose these symmetries the theory makes no sense. The sums over histories, the sums over surfaces break down and give nonsensical results. For this reason, quite apart from the fact, string theory incorporates gravitation, we think that we have more reason for optimism now about approaching the final laws of nature than we have had for some time." (Weinberg, [1987b], p. 105).

In the measure that these constraints – anomaly freedom, GUT compatibility, unitarity, chirality, and finiteness – exclude all but a single account of the world, superstring theorists were inclined to adopt that account, *in virtue of its unique consistency over and above phenomenological success*. As they compiled their two-volume textbook on strings in 1987, Green, Schwarz and Witten [1987] made this quite plain. "Quantum gravity," they argued,

has always been a theorist's puzzle *par excellence*. Experiment offers little guidance except for the bare fact that both quantum mechanics and gravity do play a role in natural law. The characteristic mass scale in quantum gravity is the Planck mass [10^{19} GeV]. This is so far out of experimental reach that barring an unforeseeable stroke of good luck (like the discovery of a stable Planck particle left over from the big bang) we can hardly hope for direct experimental tests of a theory of quantum gravity. The real hope for testing quantum gravity has always been that in the course of learning how to make a consistent theory of quantum gravity one might learn how gravity must be unified with other forces. (p. 14).

Mathematics has a new role here. For the claim made for this mathematized physics was not simply that mathematics rendered physics more rigorous, as Laurent Schwarz's work on distributions had rigorously defined the Dirac delta function. Nor was the claim by string theorists that they had used mathematics to rid the discipline of an objectional concept the way Lagrange or Hamilton had cut "force" from Newton's mechanics. Reflecting in 1986 on the two years since the revolution, John Schwarz called this period "a new era in theoretical physics, quite unlike anything before ... the beginning of an era of unification – not only of forces and particles, but of concepts and disciplines." Mathematics and physics had converged in their interests, and only the "loss of contact with our experimental colleagues and the real world" had to be avoided. (Schwarz, [1987], p. 201). As string theorists exploited this new admixture of algebraic topology, quantum field theory, and general relativity, they could "test" theories, until a single, inalterable and fundamental account of physical reality remained.

This kind of rigidity, Weinberg contended, was not *just* a matter of aesthetic or philosophical satisfaction. Having read the critical articles and comments made about superstrings by other prominent theorists, Weinberg [1987a] responded in 1987:

> I don't really understand these attacks. At present superstring theory provides our only hope of understanding physics at the Planck scale, the scale where gravity is important. Furthermore it is beautiful, it has a kind of rigidity that you look for in a kind of physical theory that will in the end turn out to have something to do with the real world (p. 282).

"[S]omething to do with the real world," for Weinberg, meant the very real possibility that "fundamental physics" the theory in terms of which all the other entities of science were ultimately built, might soon come to an end. (Galison [1983]).

Precisely the separation of theoretical concepts from experimental analogues has, for some physicists, broken the sense of "interpretation," that held concepts to correspond to intuitions grounded in experience. Application, however, was not what Weinberg had in mind; it was not by deriving specific results from the lower energy theory that the concepts of string theory would find their place. In one interview, the physicist P. C. W. Davies asked Weinberg how superstring theory would describe the electron and the neutrino. Weinberg responded: "I find your question an awkward one. It's like asking 'how in general relativity do you work out the shape of a suspension bridge?" (Davies and Brown [1988], p. 216). It is not in girders that the curvature tensor of spacetime finds its sense. And it was not in reinterpreting experimentally accessible objects like the electron that Weinberg found the meaning of superstrings.

But beyond the question of applicability – the inability of a high-level theory to reproduce the results of a lower level phenomena – there is the question of how to interpret the concepts used. Could we give a physical interpretation to the variables beyond the four that correspond to ordinary spacetime? Weinberg's answer:

> I think not. The final theory is going to be what it is because it's mathematically consistent. Then the physical interpretation will come only when you solve the theory and see what it predicts for physics at accessible energies. This is physics in a realm which is not directly accessible to experiment, and the guiding principle can't be physical intuition because we don't have any intuition for dealing with that scale. The theory has to be conditioned by mathematical consistency. (Davies and Brown [1988], p. 221).

On Weinberg's view, the removal of the theory from proximity to experiment was therefore not simply a statement about how we ought proceed, strategically, in building theory. The energy chasm between the two branches of physics removed the intuition of experimental entities from our understanding of the basic entities: superstrings.

I end this brief excursion into string theory with some remarks by Princeton's David Gross, who chose an exploration on foot as the basis for his extended metaphor about the relation of theory to experiment. It was a relation far different from the close cooperation that had marked the mid-1970s. "One of the important tasks of theorists is to accompany our experimental friends down the road of discovery; walk hand in hand with them, trying to figure out new phenomena and suggesting new things that they might explore." (Zichichi [1990], p. 237). Burnt into Gross's memory, indeed into the collective memory both of pro- and antistring theorists, was the example of the J/psi and other "charm" particle discoveries of November 1974 and its aftermath. For during those frenetic days experimentalists tossed new particles into the ring and theorists worked furiously to explain them; theorists postulated new particles, new effects, and new theories – experimentalists responded with tests that could be prosecuted within months, sometimes in days. Now, in the late 1980s, this close relation had been broken – not very many new experimental results were coming out of the accelerators, and the discoveries that were being reported had a wickedly short life: the neutrino oscillations (indicating that the neutrino might have a mass) came and went, proton decays were reported and retracted, monojets spurted momentarily from CERN, then vanished, the fifth force grabbed attention for a while and then faded. Under these circumstances some theorists – Gross included – were less and less inclined to theorize furiously after each new sighting. These were no longer the days of new charmed particles and the myriad of discoveries that followed the "November Revolution" launched by SLAC and Brookhaven/MIT in 1974. Looking back in 1989, Gross continued his metaphorical trek:

> It used to be that as we were climbing the mountain of nature the experimentalists would lead the way. We lazy theorists would lag behind. Every once in a while they would kick down an experimental stone which would bounce off our heads. Eventually we would get the idea and we would follow the path that was broken by the experimentalists. Once we joined our friends we would explain to them what the view was and how they got there. That was the old and easy way (at least for theorists) to climb the mountain. We all long for the return of those days. But now we theorists might have to take the lead. This is a much more lonely enterprise. (p. 328).

Without knowing the location of the summit, or how far it was, the theorists could promise little by way of reassurance to themselves or to the experimentalists. In the meantime, experimentalists were not only left behind, they were left out altogether.

3. Antistrings: Keeping Physics Between Chemistry and Metaphysics

Not surprisingly, many experimentalists were shocked, not so much by the idea of theorists leading the way up an uncertain trek up an uncharted mountain, but because they did not see how they could even gain a toe-hold in the foothills. Carlo Rubbia, who only a few years before had taken home a Nobel Prize for his team's discovery of the intermediate vector bosons, the W and the Z, sadly reflected:

> I am afraid I am one of the few experimentalists here. In fact, I can see we are really getting fewer and fewer. I feel like an endangered species in the middle of this theoretical orgy. I am truly amazed. The theories are inventing particle after particle and now for every particle we have there is a particle we do not have, and of course we are supposed to find them. It is like living in a house where half the walls are missing and the floor only half finished ... (Zichichi [1990], p. 232).

After the bruising W and Z search, and a contentious struggle with the top quark, Rubbia had little appetite for an unknown zoo of particles as numerous as the known. Even one or two particles were terribly hard to find – Rubbia's UA1 collaboration had employed some 150 physicists for years at a cost of hundreds of millions of dollars to find the W and Z. Now the theorists were ordering a supersymmetric partner of every known entity: the selectron, partner to the electron, and so on all the way down the line.

Not only was this half-missing world overwhelming in its mandate for experimental discovery, the very motivations cited by the theorists had moved ever further from the accelerator floor. Gosta Ekspong, a senior European experimentalist who often worked at CERN, looked down on the podium itself on which was inscribed the Dirac equation, $(\delta + m)\,\psi = 0$. Reflecting on these spartan symbols he saw a line of poetry, "beautiful," "full of life and beauty," and above all open to "true predictions," including the existence of the positron, and to variances with experiment, in the subtleties of spectroscopy. Every theorist, he reflected, must aspire to the heights of Dirac's achievement, and in particular to infer truth from the presence of beauty. But cautioning his theorist colleagues, he argued that this poetic elision from beauty to truth often will not hold, and so hesitated before the purported aesthetic satisfactions of the superstring theorists:

I would like to address the question of truth and beauty; truth being experiment, beauty being theory ... The problem is that the latest theories are so remote from experiment that they cannot even be tested. Therefore they don't play the same role as Dirac's equation ... I hope that this search for beauty does not drive theorists from experiments, because experiment has to be done at low energies, from one accelerator to the next and so on. Big jumps do not seem to be possible. (Zichichi [1990], p. 232).

For Ekspong, the path from superstring theory to experiment was too long, and if compactification multiplied its branches, the path was also too poorly defined to carry beauty the great distance to truth.

In the 1970s and 1980s, "theory" and "experiment" were categories (better: subcultures) with intermediate categories of phenomenologists straddling the fence. Ahmed Ali was one of these, having worked at both DESY (*Deutsches Elektronensynchrotron*) and CERN, and he invoked the idiom of the experimentalist, when he declaimed: "The present superstring theories are like letters of intent written by a lobby of theoretical physicists. They are very good in intent; but often what is said in the letter of intent and what is measured in the experiment are two very different things. The figure of merit of a theory is its predictive power which could be tested in an experiment in a laboratory." (Zichichi [1990], p. 232).

In a sense, the discomfiture of experimentalists, and those working hand in glove with them, could be expected. New techniques in theory had left experimentalists ill at ease with gauge theories in their early stages, though by 1974, experimentalists had found the gauge theories suggestive of a wide range of predictions, tests, and new directions for empirical work. Now the case of superstrings seemed much worse; there was no clear avenue for the accelerator laboratory to follow, and the theories themselves offered precious little to hold on to in the way of physically "intuitable" entities. Less expected, perhaps, at least to an outsider, was the vehement reaction against string theory from *within* the theoretical high-energy physics community.

I now turn to the most considered part of the opposition, an opposition grounded not on a hostility to wide-ranging claims about unification, nor on an opposition to theoretical particle physics. Quite the contrary, Georgi and Glashow were as central to the gauge revolution of the 1970s as anyone. No, the dispute centered on something much deeper, on a vision of what physics should be.

Before 1984, the *annum mirabilis* of strings, Howard Georgi opened the 1983 Fourth Workshop on Grand Unification with a transparency of a recent advertisement he had spotted:

HELP WANTED
Young Particle Theorist
to work on
Lattice Gauge theories
Supergravity
and
Kaluza-Klein Theories

Here, Georgi asserted, was a telling sign of the times, a position caught between "chemistry" (particle physicist nomenclature for a calculational activity in which fundamental principles were no longer at stake) and "metaphysics and mathematics" (activities disconnected from any contact with experiment) (p. 3). Superstrings had not yet emerged but the problem had. These highly mathematized theories did not fix experimental consequences – by contrast, the SU(5) GUT had fixed the parameter relating the strength of electromagnetism to the weak nuclear force, $sin\theta_W$.

The next years polarized the situation further. In 1986, Paul Ginsparg (who had himself contributed a computer analysis to the Alvarez-Gaumé and Witten no-anomaly demonstration superstrings in 1983), collaborated with his Harvard colleague Sheldon Glashow to bemoan the loss in superstrings of a productive tension between experiment and theory:

In lieu of the traditional confrontation between theory and experiment, superstring theorists pursue an inner harmony where elegance, uniqueness and beauty define truth. The theory depends for its existence upon magical coincidences, miraculous cancellations and relations among seemingly unrelated (and possibly undiscovered) fields of mathematics. Are these properties reasons to accept the reality of superstrings? Do mathematics and aesthetics supplant and transcend mere experiment? Will the mundane phenomenological problems that we know as physics simply come out in the wash in some distant tomorrow? Is further experimental endeavor not only difficult and expensive but unnecessary and irrelevant? (Ginsparg [1986], p. 7).

This was an altogether different view than that which Glashow had taken in the euphoric moments after he and Georgi had produced the first grand unified theories. GUTs, at least the SU(5) and SO (10) versions, did have very high energies – only a few orders of magnitude less than the Planck scale. And like the string theories, GUTs too forecast a "desert" in which no new experimental results could be found. But differentiating GUTs and superstrings were several crucial items. GUTs forecast a crucial parameter in the electroweak theory – $sin\theta_W$; at least SU(5) and SO(10) predicted a decay of the proton that should be measurable in deep mine experiments; and by construction the new grand unified theories reproduced all of the known phenomenology of both the electroweak and quantum chromodynamic theories. Strings could do neither, that is they could not make new predictions (such as $sin\theta_W$) and they could not reproduce the known phenomenology of SU(2)xU(1) and SU(3). Finally, GUTs, at least in their original formulation, simply continued the gauge program as it had already been formulated. That is, grand unified theories were "conservative" in the sense that they operated under essentially the same theoretical constraints (those of renormalizable gauge theories) as the rules of the game for SU(2)x U(1) and color SU(3).

In fact, by 1989, for Georgi the proliferation of GUTs, especially their assimilation into even larger unification schemes of the superstring, was the source of a mighty ambivalence:

I feel about the present state of GUTs as I imagine that Richard Nixon's parents might have felt had they been around during the final days of the Nixon administration. I am very proud that the grand unification idea has become so important. [But] I cannot help being very disturbed by the things which GUTs are doing now.

GUTs, he insisted, had been motivated by the desire to complete the unification of forces by accounting for the weak mixing angle and explaining the quantization of charge.

They [GUTs] were certainly not an attempt to emulate Einstein and produce an elegant geometrical unification of all interactions including gravity, despite the parallels which have been drawn in the semipopular press. Einstein's attempts at unification were rearguard actions which ignored the real physics of quantum mechanical interactions between particles in the name of philosophical and mathematical elegance. (p. 446).

Imitating Einstein in the late 20th century was, Georgi judged, a losing proposition.

Georgi argued that it was the nature of physics itself at stake, not an incidental question of style, but the defining quality of the discipline. "Theorists," Georgi [1989] insisted, "are, after all, parasites. Without our experimental friends to do the real work, we might as well be mathematicians or philosophers. When the science is healthy, theoretical and experimental particle physics track along together, each reinforcing the other. These are the exciting times." (p. 452). His argument was that when experimentalists get ahead, as the bubble chamber experimentalists did in the 1960s, the discipline becomes eclectic, overrun with results without order or explanation. At other times theory gets ahead and the path is strewn with the irrelevant speculations of rationalization out of touch with reality.

Georgi's assessment of the state of theoretical physics is reflected in the practice of his physics, and in particular from the understanding he has advocated of effective field theories. Effective field theories (EFTs) have their roots in work by Weinberg and others quite some time ago. But over the last twenty years they have evolved into more than a specific tool, a way of thinking about physics that characterizes theoretical work by a generation of physicists whether they follow a phenomenological or a string theoretical path through the world. The idea is this: the mass of a particle and its wavelength are interconnected through the deBroglie relationship: the heavier the mass, the shorter the wavelength. And just as a surfer is concerned about waves on the scale of meters (not on the scale of millimeters), a physicist working at an energy scale corresponding to a wavelength of 10^{-13} cm need not be concerned with physics that takes place at the energy of 10^{-19} cm. This observation is formalized within the language of quantum field theory: in a given account of a quantum process – such as the scattering of particle A off of particle B – there will always be terms that involve the exchange of much more massive particles. For example, an electron scattering from another electron does so (in largest measure) through the exchange of a photon. But it is possible, quantum dynamically, for the electrons to exchange the much heavier particle Z^0 postulated in the Weinberg-Salam-Glashow theory, and experimentally defended in work culminating at CERN in 1983. Effective field theorists say yes, it is true that quantum electrodynamics left out the Z^0 in its classic formulation, but these additional Z^0 terms Z^0 will be suppressed by a factor of 1/M, where M is the mass of the Z^0. Should we discover new particles using bigger accelerators (RIP, SSC), these new particles would give rise to even smaller corrections because 1/M would be that much smaller.

Just because new particles of higher mass have a decreasing effect on the calculations made at lower energies, effective field theories can be thought of as a theoretical means of vouchsafing existing knowledge, a reason to take seriously quantum electrodynamics in its domain, come what may. It is this feature of epistemic insulation that has drawn the attention of Cao and Schweber [1993].

But for most physicists the retrospective justification of theories such as QED is not the main function of effective field theories; instead, their interest is in using the method to think about future theories. Suppose one wants to design an accelerator. In the case of the superconducting supercollider, most theoretical physicists were primarily interested in exploring the way in which the weak and electrodynamic forces separated. The unbuilt accelerator might have produced the Higgs particle, a leading contender for the symmetry breaking mechanism. Effective field theories told them that the physics of much higher energies would not play any immediate role in those dynamics, even if ultimately the Higgs particle could itself be explained by reference to theories grounded at yet higher energies.

> If we adopt the effective field theory point of view, we must try to work our way down to short distances from what we know at longer distances, working whenever possible in the effective theory which is *appropriate* to the scale we are studying. We should not try to guess the ultimate theory at infinitely small distances. Even if we could do it, it would probably be about as useful as explaining biology in the language of particle physics. (Georgi, [1989], p. 456).

This is an antireductionist argument. It addresses (as I understand it) the possibility that a Planck-scale string theory might be found that satisfies the theoretical constraints imposed on it. Georgi's fear is that even with such a theory, the compactification (that takes the many- dimensional theory into our accessible four-dimensional one) might remain unsolved, and therefore the gap between the Planck scale and the scale at which our accelerator-based particles live might remain so large that we would be uninformed about the accessible physics. We might have some reason not to rule out the high-energy theory, but we could not ever feel confident that the gap could be closed.

Now suppose, optimistically, that a nonperturbative string theory is found, that this theory dynamically determines the compactification, and that the resulting low-energy theory does match known phenomenology. Under such assumptions, Georgi [1989] was still not persuaded that other theories were ruled out, perhaps an infinite class of such theories might also produce what we know:

> [I]f we allow ourselves to be beguiled by the siren call of the 'ultimate' unification at distances so small that our experimental friends cannot help us, then we are in trouble, because we will lose that crucial process of pruning of irrelevant ideas which distinguishes physics from so many other less interesting human activities. (p. 457).

This is an underdetermination argument, one that I take to be distinct from the antireductionist one.

To develop his view that superstring theory is not necessarily unique, Georgi has often brought in different metaphors, one about the man who lost his keys one dark night and was seen hunting furiously beneath a street lamp. Asked whether he thought he lost them in just that circle of light, the despondent fellow said, no, but that was the only place he was ever to have a chance at finding them. Outside the circle of light cast, so to speak, by the energy regime around symmetry breaking of the weak and electromagnetic forces (around 1 TeV), Georgi saw physics as blundering in the dark – why search in one place, under one set of constraints, and not another? Especially in the energy region of the Planck mass (where we have no idea how physics will alter), why should anyone even suspect that string theory, developed for use in 1960s strong interaction physics, is relevant? For Georgi, the depth of our ignorance at the Planck scale is practically infinite. We know that at these energies ordinary relativistic quantum field theory breaks down; everyone agrees that is so. Under these circumstances, we have no grip on how to generalize the theory; according to Georgi, we do not even know how to formulate the problem, let alone the solution. It is his view that the uniqueness of superstring theory is an artifact of history; had the particular alternative of hadronic string theories not been available as the legacy of strong interaction theories, Georgi contended, superstring theory might well not exist in the 1980s. We see this theory as unique, but its uniqueness issues from our (perfectly understandable) delight in having *any* consistent extension of quantum field theory, not in the inevitability of this particular route beyond current knowledge. But (Georgi maintained) that was an historical accident deriving from a failed strong interaction theory resuscitated to fill a desperate need: string theory was unique within the frame of history, not *sub specie aeternitatis*.

To Weinberg, Georgi's history was flawed. Hadronic string theory ought not be seen merely as an abortive account of 1960s strong-interaction theory; it should rightly be thought of as an attempt to satisfy the constraints of analyticity, unitarity, Regge asymptotic behavior, and other symmetries demanded of any scattering process depicted by the S-matrix formalism. As such, string theory was (and remains) more abstract than a theory of matter; it is a characterization of features that must hold good for any theory. Perhaps, Weinberg suggested, superstring theory is best thought of as the unique way of satisfying the fundamental conditions imposed by scattering processes that include both gravity and familiar forces of particle interaction. On his view, these S-matrix constraints were still the relevant criteria for any scattering theory, including any theory embracing gravity. (Weinberg, [1986], p. 237).

While further exchange is not on the record, the question haunting the field remains: in the Planck limit, where we know "ordinary" quantum field theory breaks down, why choose to abandon just the constraint of theorizing with point particles? loosen this particular assumption and maintain intact the rest of the quantum field theoretical structure? Knowing that ordinary quantum field theories break down, how, in the absence of experiment, is one, among all the assumptions of field theory, to choose rightly?

Precisely this absence of connection to the "everyday accelerator physics" (if you will pardon this oxymoronic construction) set strings apart, and now in a more extreme form than ever before. Ginsparg and Glashow [1986] continued in their May 1986 manifesto:

> For the first time since the Dark Ages, we can see how our noble search may end, with faith replacing science once again. Superstring sentiments eerily recall "arguments from design" for the existence of a supreme being. Was it only in jest that a leading string theorist suggested that "superstrings may prove as successful as God, Who has after all lasted for millennia and is still invoked in some quarters as a Theory of Nature?" (p. 7).

Julian Schwinger, himself one of the creators of quantum electrodynamics (and Glashow's mentor in the early 1960s), was likewise appalled by the hubris of the theory, and the concomitant abandonment of experimentation. "[P]hysics has been, until recently, an experimental subject, the theories also ought to have more contact with experiment. Or, to put it another way, theories should be more modest. They should try to make contact with what we know and to extend it a modest amount. Let us see if we can't avoid theories that attempt to be theories of everything and thereby announce that we have in our hands the theory of the whole Universe." (Schwinger, [1990], p. 236).

For Witten, the accusation that string theory ignored experiment was premature and historically unreasonable. After all, he argued in September of 1986, relativity and relativistic astrophysics offered a striking three-fold instance of a theory confirmed by experience in ways utterly unavailable at the time Einstein sent his field equations to *Annalen der Physik*. Neutron stars, proposed in the 1930s, were well within the known physics of the day. Weak in their emission of light, small in cross section, it seemed hopeless to expect their discovery. Who could have divined the future invention of radioastronomy. "[O]ne can be very close to the truth and still be caught unawares by a sudden and almost accidental discovery. Wheeler around 1960 guessed that a spinning (and slowing down) neutron star was powering the Crab nebula, but no one realized that this meant that the neutron star could be seen as a pulsar." (Witten, [1987], p. 77). Gravitational lenses formed a second example of a powerful confirmatory phenomenon proposed by Einstein in the 1930s but in no way expected to be a significant astrophysical tool in the 1970s and beyond. Finally, gravitational waves, known in the 1910s but seemingly unobservable for decades, now loom on the horizon as nearly detectable. "[T]echnology develops faster and farther than we think." Experiment and observation, Witten [1987] is arguing, proceed down a twisted path, and we should not prejudge what may be observed down the road on the basis of what appears discoverable now: "So," he concluded, "we should be optimists." (p. 77).

At the *Superworld I* conference, which took place as Witten prepared these remarks for publication, his collaborator on the anomaly paper, Luis Alvarez-Gaumé, presented Witten's gravitational example of a "hidden experiment." Glashow was unmoved: "I am familiar with that example from your prophet. He was talking about 1930. Let us remember that in 1932 the positron and the neutron were discovered so there were a few other things happening aside from waiting around for gravitational lenses. That is not so obviously true today." (Zichichi [1990], p. 239).

Now it might be thought that Georgi and Glashow, authors of the very project of a grand unified theory that sat a million times above the presently observable mass scale, were hardly in a position to criticize strings on the grounds of scale alone. From their perspective, however, the two cases were hardly analogous. What marked off the superstring from GUTs was precisely the absence of experimentation; "Only a continued influx of experimental ideas and data can allow the paths from top [that is Planck-scale, superstring theories] and bottom [that is experimentally grounded theorizing] to meet." (Ginsparg and Glashow [1986], p. 9). By 1988, Glashow [1988] judged that

> the historical connection between experimental physics and theory has been lost. Until the string people can interpret perceived properties of the real world, they simply are not doing physics. Should they be paid by universities and be permitted to pervert impressionable students? Will young Ph.D.'s whose expertise is limited to superstring theory be employable if and when the string snaps? Are string thoughts more appropriate to departments of mathematics or even to schools of divinity than to physics departments?" (p. 25).

Reiterating his discomfiture with the string theory in 1986, Glashow added an historical argument.

> The downwards path [from rational explanation of theory to phenomenology] hadn't worked very well. However elegant it would have been had planets moved in circles, they simply don't. The revolution of special relativity emerged from observation and not from the power of positive thinking. Without the careful experimental work that led Maxwell to his equations, no philosopher could have discovered the real nature of space and time. Quantum mechanics, too, crept out of an experimental morass which was irreconcilable with classical theory. It was not an abstract mathematical creation. (Glashow [1990], p. 250).

By 1990, to his dismay, Glashow reported that Princeton piety had spread to Harvard. (Zichichi [1990], p. 236).

Using history rather differently, Weinberg claimed that superstrings in the late 1980s, was like general relativity in the 1920s: Eddington and Pauli must have found Einstein's theory compelling on similar grounds, even though, like string theory, general relativity had the barest of confirmatory evidence.

That, at present, physicists were a "long way" from being able to provide the details of the nonperturbative theory or a quantitative account of phenomenology is understandable. After all, he noted, we are still "a long way from being able to understand turbulence." (Weinberg, [1987a], p. 282). Historical examples flew back and forth across the divide. To the string theorists, patience was needed; Einstein's theory of general relativity had scant support in its original formulation, the precession of the perihelion of mercury and not much else. Arrayed against the Einstein analogy were the myriad of dead ends that theory had pursued without their experimental colleagues. As Gross acknowledged in [1989], "The arguments against success are easy – history teaches us that without direct experimental clues and tests theorists tend to go wrong." But in favor lay the extraordinarily successful theory of "low energy physics," (relative to the Planck scale) alongside gravity, which offered a window into Planck mass physics. "Finally," he added, "we can be lucky." (p. 327).

4. Conclusion

For all too long, we physicists, philosophers, and historians have found it convenient to speak about the history of the relation of experiment, theory, and mathematics as if these categories of inquiry passed from premodern chaos to a permanent and quasi-stable "modern" science sometime in the 17th-century. But why should we assume such a static relationship? Why expect that there is a fixed model when so much else has changed in the practice of physics. Already in the mid-19th century, theoretical physics began to separate from experimental physics. By the time of H. A. Lorentz, Wilhelm Wien, Ludwig Boltzmann, and Einstein, recognizable sites of distinctly theoretical work had emerged in a form distinct from the much older tradition of mathematical physics and celestial mechanics. Quantum mechanics undoubtedly further separated experiment and theory as institutional settings for theory arose, perhaps none better known than Niels Bohr's in Copenhagen. Enrico Fermi, who established himself on paper and in the laboratory, is remembered precisely because his crossover was one of the last. String theory continues this separation, but has done so in sufficient measure that the difference in degree has been registered by theoretical physicists as a difference in kind: the practice of superstring theory is a practice far closer to mathematics than to experiment.

Sociologically, the separation of superstring from gauge physics done at accelerator energies has been remarkable. Much as graduate students entering their training quickly sort out into experimentalists and theorists, now theoretical students quickly divide into those destined for strings and those heading for point particles. And once in place, few make the transition. Theorists, even some string theorists, watch in wonder at virtuoso string theorists who move with ease in Calabi-Yau spaces, yet have little sense of pions, muons, and other entities whose hard-won understanding characterized so much of twentieth-century physics. At the other extreme from experiment, mathematicians have greeted the string theories with a mixture of wonder and horror at the productive speculations offered by string theorists about topics previously thought thoroughly mathematical. One much-discussed proposal by Arthur Jaffe and Frank Quinn [1993] sought to divide mathematics into "theoretical" and rigorous subfields on the model of theory and experiment in physics:

> [P]hysicists are not in fact isolated. They have found a new "experimental community": mathematicians. It is now mathematicians who provide them with reliable new information about the structures they study. Often it is to mathematicians that they address their speculations to stimulate new "experimental" work. And the great successes are new insights into mathematics, not into physics. What emerges is not a new particle, but a description of representations of the "monster" sporadic group using vertex operators in Kac-Moody algebras. What is produced is not a new physical theory, but a new view of polynomial invariants of knots and links in 3-manifolds using Feynman path integrals or representations of quantum groups ... (p. 3).

Jaffe and Quinn touch on the myriad of partly epistemic and partly sociological tensions that sometimes strain the border region between mathematics and physics. Physicists can, at times, disdain the mathematical work as rigorous, but ultimately a matter of "cleaning up" after the real work of understanding is done. Mathematicians, conversely occasionally view the physicists' speculations as nothing more than empty promises. By modeling the relation of speculation and proof on the theory/experiment division of labor, Jaffe and Quinn hope to make room for both. But however this cultural boundary area is negotiated, experiment – the activity of manipulating gases, magnets, and data tapes – has receded from view.

This shift in theory is the second shoe to fall; as I have argued elsewhere, high-energy experimentation too is barely recognizable to someone trained in an early modality of work. Boyle's efforts at Gresham House differ not merely in scale from CERN and Fermilab, but in kind. When a collaboration of two, three or five hundred physicists undertakes a five- to ten-year $500 million inquiry in particle physics, it is not a 17th-century air pump or 19th-century Hertzian coil writ large: the founding epistemic principles of experiment have altered. There are no gentlemanly "neutral" witnesses to experiment even in the 19th century, let alone in the late twentieth; there is no possibility of any direct "experimental repetition" when the device costs more than the net assets of most major universities; there is not even a very well-defined notion of an experimental author when no single individual can conceivably be fundamentally involved with all sectors of experimental construction, data taking, data analysis or interpretation. Even when the results are in, comparing data in a colliding beam experiment often must pass through triggers, filters, and phenomenological analysis before it faces the basic equations of quantum chromodynamics. Monte Carlo simulations, which stand at the threshold between experiment and theory, again challenge any fixed notion of what counts as an experiment. What would Maxwell have called a young particle physicist, who qualified professionally by doing a simulation of a drift chamber deep within a projected colliding beam detector? An experimentalist? A theorist? These categorical boundaries have shifted.

Though the large-scale experiment has been the focus of my concern for many years (Galison e.g. [1985], [1987], [1992]), I have here deliberately turned away from these matters and towards the practice of theory to explore the opposite end of twentieth-century physics: theory pushed maximally far from what Glashow called "the experimental morass."[5] As string theory (and string theorists) occupied an increasingly significant part of contemporary particle physics, this distance from the laboratory has raised deep questions about the very nature of physics as a discipline. To superstrings advocates, the new theories represent the best hope for a final theory of nature, the most important physics development since quantum mechanics. As these advocates see the world, hearkening back to the back-and-forth between experiment and theory is a nostalgic recollection not unlike the colliding beam experimentalist who wishes he was blowing glass cloud chambers at the Cavendish. To superstring detractors, string theory posed a threat to the very existence of an experimentally-based inquiry they call physics.

This is a debate about the nature of physical knowledge. It is a struggle within the community of particle theorists to understand the nature of the constraints that define the character of physical law, and make good its purchase on nature. The string vision takes quantum field theory and extends it in what advocates consider to be the minimal possible way: generalizing the point particle to the one-dimensional string. By doing so, while maintaining other constraints on quantum field theory, it became possible in the mid-1980s for physicists to explore a unified theory of the electromagnetic, weak, and strong forces along with gravity, and to do so without the disabling infinities that hamstrung all previous attempts at a quantum theory of

5 For the argument that theory is not a categorical opposite from practice, but rather that theoretical practices and constraints should be considered on a par with experimental practices and constraints, see Galison, 1987, esp. chapter 1 and 5; Galison, 1988a,b; and Warwick (1992).

gravity. The combination of theoretical virtues – anomaly freedom, chirality, unitarity, finitude – with at least the possibility of finding grand unified phenomenology as a low-energy limit – elevated string theory to a "theory of everything." For the superstring theorists, mathematical constraints (from algebraic topology to a host of related disciplines) served the role of delimitation the way experiment had at an earlier time. Mathematical self-consistency promised uniqueness and in that singularity the hope of access to a final, realistic picture of the foundation of nature.

This substitution of mathematical constraint for experimental constraint is precisely what stunned much of the particle physics world outside the cadre of superstring theorists. Absent the world of experiment, theorists dubious of the superstring program argued that the mathematically-generated intratheoretical constraints would be dangerously arbitrary. True, the string theorists conceived of their substitution of strings for point particles as the minimal modification of quantum field theory. But without experiment to adjudicate among starting assumptions, unbelievers did not see this as such a modest move.

Alongside the contending visions of nature came different strategies for the pursuit of physics. Top-down theory or bottom-up physics? Perhaps both will be productive. Faced with both sides' frustration at making new and experimentally confirmable results, Glashow sardonically remarked at one point that the choice was much crueler, either build castles in the air or chase ambulances on the street.

One among the many interesting features of the debate is the near-universal invocation of history by all participants. The ghost of Einstein hovers over the scene, but appears in many guises. To Green, Witten, Gross, and Weinberg, Einstein's apparition is that of the youthful physicist, struggling to develop general relativity. His success, with only the barest of observational data – really no more than the precession of the perihelion of mercury – offers hope for string theorists as they labor far from the laboratory. The promise of many future confirmations in yet undreamt-of accelerators or astrophysics girds against the charge of being outside of experience altogether. To Georgi and Glashow, the spectre of Einstein is not the creative 36-year old of 1915, but the struggling 75-year old Einstein, disconnected from experiment, cut off from the power of quantum mechanics, quantum field theory, and accelerator-based particle physics that animated so much of the physics of the early 1950s. Wounded by this disconnection, Georgi and Glashow's ghostly Einstein wanders without direction in the futile search for a unified field theory that would grasp all physics and bring the unruly discipline to an end.

As physics comes to the millenium, the path of physics divides. Which does the theorist follow? Theorists drove Einstein's ghost both ways.

References

Achinstein, Peter, and Hannaway, Owen (*eds.*) [1985]: *Observation, Experiment, and Hypothesis in Modern Physical Science*. Studies from the Johns Hopkins Center for the History and Philosophy of Science. Cambridge, Mass.: MIT Press.

Alvarez-Gaumé, Luis, and Witten, Edward [1983]: 'Gravitational Anomalies'. *Nuclear Physics*, *B234*, pp. 269-330.

Cao, Tian Yu, and Schweber, Silvan S. [1993]: 'The Conceptual Foundations and the Philosophical Aspects of Renormalization Theory'. *Synthese 97*, pp. 33-108.

Davies, P. C. W., and Brown, Julian [1988]: *Superstrings: A Theory of Everything?* Cambridge: Cambridge University Press.

Galison, Peter [1983]: 'Re-reading the Past from the End of Physics; Maxwell's Equations in Retrospect'. In L. Graham, W. Lepenies, and P. Weingart (*eds.*), *Functions and Uses of Disciplinary Histories*, pp. 35-51. Dordrecht: D. Reidel.

Galison, Peter [1985]: 'Bubble Chambers and the Experimental Workplace'. In Peter Achinstein and Owen Hannaway (*eds.*), *Observation, Experiment and Hypothesis in Modern Physical Science*, pp. 309-373. Cambridge, Mass.: MIT Press.

Galison, Peter [1987]: *How Experiments End*. Chicago: University of Chicago Press.

Galison, Peter [1988a]: 'History, Philosophy, and the Central Metaphor'. *Science in Context 2*, pp. 197-212.

Galison, Peter [1988b]: 'Multiple Constraints, Simultaneous Solutions'. *PSA 2*, pp. 157-163.
Galison, Peter, and Hevly, Bruce. (*eds.*) [1992]: *Big Science: The Growth of Large-Scale Research*. Stanford: Stanford University Press.
Georgi, Howard [1983]: 'Opening Remarks'. In H. A. Weldon, P. Langacker, and P. J. Steinhardt (*eds.*), *Fourth Workshop on Grand Unification: University of Pennsylvania, Philadelphia, April 21-23, 1983*, pp. 3-12. Progress in Physics, vol. 9. Boston: Birkhäuser.
Georgi, Howard M. [1989]: 'Effective Field Theories'. In Paul Davies (*ed.*), *The New Physics*, pp. 446-467. Cambridge: Cambridge University Press.
Ginsparg, Paul, and Glashow, Sheldon [1986]: 'Desperately Seeking Superstrings'? *Physics Today 39*, no. 5, pp. 7-9.
Glashow, Sheldon Lee [1980]: 'The New Frontier'. In Paul H. Frampton, Sheldon L. Glashow, and Asim Yildiz (*eds.*) *First Workshop on Grand Unification: New England Center, University of New Hampshire, 10-12 April 1980*, pp. 3-8. Lie Groups: History, Frontiers and Application, vol. 11. Brookline, Mass.: Math Sci Press.
Glashow, Sheldon [1988]: 'Tangled in Superstring: Some Thoughts on the Predicament Physics Is In'. *The Sciences 28*, no. 2, pp. 22-25.
Glashow, Sheldon [1990]: 'The End of the Superworld'. In Antonino Zichichi (*ed.*), *The Superworld I*, pp. 247-254. Proceedings of the 24th Course, International School of Subnuclear Physics, Erice, Italy, August 1986. New York: Plenum Press.
Gooding, David, Pinch, Trevor, and Schaffer, Simon (*eds.*) [1989]. *The Uses of Experiment: Studies in the Natural Sciences*. Cambridge: Cambridge University Press.
Green, Michael B. [1986]: 'Superstrings'. *Scientific American 255*, pp. 48-60.
Green, Michael B., and Schwarz, John H. [1985a]: 'The Hexagon Gauge Anomaly in Type 1 Superstring Theory'. *Nuclear Physics, B255*, pp. 93-114.
Green, Michael B., and Schwarz, John H. [1985b]: 'Infinity Cancellations in SO(32) Superstring Theory'. *Physics Letters, 151B*, pp. 21-25.
Green, Michael B., Schwarz, John H., and Witten, Edward [1987]: *Superstring Theory*, vol. 1, *Introduction*. Cambridge Monographs on Mathematical Physics. Cambridge: Cambridge University Press.
Gross, David J., Harvey, Jeffrey A., Martinec, Emil and Rohm, Ryan [1985]: 'Heterotic String Theory'. *Nuclear Physics, B256*, pp. 253-284.
Gross, David J. [1989]: 'Superstrings and Unification'. In R. Kotthaus and J. H. Kühn (*eds.*), *Proceedings of the XXIV International Conference on High Energy Physics: Munich, Fed. Rep. of Germany, August 4-10, 1988*, pp. 310-333. Berlin: Springer-Verlag
Hacking, Ian [1983]: *Representing and Intervening: Introductory Topics in the Philosophy of Natural Science*. Cambridge: Cambridge University Press.

Holton, Gerald [1973]: 'Mach, Einstein, and the Search for Reality'. In *Thematic Origins of Scientific Thought: Kepler to Einstein*, pp. 69-90. Cambridge, Mass.: Harvard University Press.

Jaffe, Arthur, and Quinn, Frank [1993]: " 'Theoretical Mathematics': Toward a Cultural Synthesis of Mathematics and Theoretical Physics". *Bulletin (New Series) of the American Mathematical Society 29*, pp. 1-13.

Kinoshita, T. and Lindquist, W.B. [1983]: 'Eighth-order Magnetic Moment of the Electron. I. Second-order Vertex containing Second-, Fourth-, and Sixth-order Vacuum Polarization Subdiagrams'. *Physical Review D27*, pp. 867-868.

Scherk, J., and Schwarz, John H. [1974]: 'Dual Models for Non- Hadrons'. *Nuclear Physics, B81*, pp. 118-144.

Scherk, J. [1975]: 'An Introduction to the Theory of Dual Models and Strings'. *Reviews of Modern Physics, 47*, pp. 123- 164.

Schwarz, John J. (*ed.*) [1985]: *Superstrings: The First 15 Years of Superstring Theory*. Singapore: World Scientific, 1985.

Schwarz, John J. [1987a]: 'The Future of String Theory '. In Lars Brink et al (*eds.*), *Unification of Fundamental Interactions: Proceedings of Nobel Symposium 67, Marstrand, Sweden, June 2-7, 1986*, pp. 197-201. Stockholm: Physica Scripta.

Schwarz, John J. [1987b]: 'Superstrings'. Physics Today 40, no. 11, November, pp. 33-40.

Schwinger, Julian S. [1990]: In Antonini Zichichi, (*ed.*), *The Superworld I*, p. 236. Proceedings of the 24th Course, International School of Subnuclear Physics, Erice, Italy, August 1986. New York: Plenum Press.

Shapin, Steven and Schaffer, Simon [1985]: *Leviathan and the Air-Pump: Hobbes, Boyle, and the Experimental Life*. Princeton: Princeton University Press.

Warwick, Andrew [1992]: Cambridge Mathematics and Cavendish Physics: Cunningham, Campbell, and Einstein's Relativity 1905-1911 Part I: The Uses of Theory. *Studies in the History and Philosophy of Science 23*, pp. 625-656.

Weinberg, Steven [1986]: 'Particles, Fields, and Now Strings'. In Jorrit de Boer, Erik Dal, and Ole Ulfbeck (*eds.*), *The Lesson of Quantum Theory: Niels Bohr Centenary Symposium, October 3-7, 1985*, pp. 227-239. Amsterdam: North-Holland.

Weinberg, Steven [1987a]: 'Summary and Outlook'. In Stewart C. Loken (*ed.*), *Proceedings of the XXIII International Conference on High Energy Physics, 16-23 July 1986, Berkeley, California*, pp. 271-282. Singapore: World Scientific.

Weinberg, Steven [1987b]: 'Towards the Final Laws of Physics'. In Richard P. Feynman and Steven Weinberg, *Elementary Particles and the Laws of Physics: The 1986 Dirac Memorial Lectures*, pp. 61-110. Cambridge: Cambridge University Press.

Weinberg, Steven [1992]: *Dreams of a Final Theory*. New York: Pantheon Books.

Witten, Edward [1983]: 'D=10 Superstring Theory'. In H. A. Weldon, P. Langacker, and P.J. Steinhardt (*eds.*), *Fourth Workshop on Grand Unification: University of Pennsylvania, Philadelphia, April 21-23, 1983*, pp. 395-408. Progress in Physics, vol. 9. Boston: Birkhäuser.

Witten, Edward [1987]: 'Some Remarks about String Field Theory'. In Lars Brink et al.(*eds.*), *Unification of Fundamental Interactions: Proceedings of Nobel Symposium 67, Marstrand, Sweden, June 2-7, 1986*, pp. 70-77. Stockholm: Physica Scripta.

Yoneya, Tomiaki [1974]: 'Connection of Dual Models to Electrodynamics and Gravidynamics'. *Progress of Theoretical Physics 51*, pp. 1907-1920.

Zichichi, Antonino (moderator) [1990]: 'A Discussion of the Superworld'. In Antonino Zichichi (*ed.*), *The Superworld I*, pp. 231-246. Proceedings of the 24th Course, International School of Subnuclear Physics, Erice, Italy, August 1986. New York: Plenum Press.

List of Contributors

Francisco Ayala, Department of Ecology and Evolutionary Biology, University of California, Irvine CA 92717, USA.
John Bigelow, Department of Philosophy, Monash University, Melbourne, Clayton, Victoria 3168, Australia.
Paul Davies, Department of Physics and Mathematical Physics, The University of Adelaide, GPO Box 498,
Adelaide, South Australia 5001.
Allan Franklin, Condensed Matter Laboratory, Department of Physics, Campus Box 390, Boulder, Colorado 80309-0390, USA.
Peter Galison, Department of History of Science, Science Center 235, Harvard University, Cambridge MA 02318, USA.
Ronald Giere, Minnesota Center for Philosophy of Science, 309 Ford Hall, 224 Church Street S.E., Minneapolis, Minnesota 55455, USA.
Hermann Haken, Institut für theoretische Physik und Synergetik, Universität Stuttgart, Pfaffenwaldring 57/4, 70550 Stuttgart, Germany.
Karl Lambert, Department of Philosophy, University of California, Irvine CA 92717, USA.
Martin Leckey, Department of Philosophy, Monash University, Melbourne, Clayton, Victoria 3168, Australia.
Margaret Morrison, Department of Philosophy, Trinity College, University of Toronto, Toronto M5S1H8, Canada.
Jane E. Ruby, Wheaton College, Norton MA 02766, USA.
Erhard Scheibe, Philosophisches Seminar, Ruprecht-Karls-Universität, Schulgasse 6, 69117 Heidelberg, Germany.

Brian Skyrms, Department of Philosophy, University of California, Irvine CA 92717, USA.

Friedrich Steinle, Institut für Wissenschaftsgeschichte der Georg August Universität, Humboldtallee 11, 37073 Göttingen, Germany.

Norman Swartz, Department of Philosophy, Simon Fraser University, Burnaby, British Columbia V5A1S6, Canada.

Friedel Weinert, Department of Interdisciplinary Human Studies, University of Bradford, West Yorkshire BD7 1DP, UK.

Subject Index

Names Index

Bei Fragen zur Produktsicherheit wenden Sie sich bitte an:
If you have any questions regarding product safety,
please contact:

Walter de Gruyter GmbH
Genthiner Straße 13
10785 Berlin
productsafety@degruyterbrill.com